Selected Titles in This Series

150 **B. Ya. Levin,** Lectures on entire functions, 1996
149 **Takashi Sakai,** Riemannian geometry, 1996
148 **Vladimir I. Piterbarg,** Asymptotic methods in the theory of Gaussian processes and fields, 1996
147 **S. G. Gindikin and L. R. Volevich,** Mixed problem for partial differential equations with quasihomogeneous principal part, 1996
146 **L. Ya. Adrianova,** Introduction to linear systems of differential equations, 1995
145 **A. N. Andrianov and V. G. Zhuravlev,** Modular forms and Hecke operators, 1995
144 **O. V. Troshkin,** Nontraditional methods in mathematical hydrodynamics, 1995
143 **V. A. Malyshev and R. A. Minlos,** Linear infinite-particle operators, 1995
142 **N. V. Krylov,** Introduction to the theory of diffusion processes, 1995
141 **A. A. Davydov,** Qualitative theory of control systems, 1994
140 **Aizik I. Volpert, Vitaly A. Volpert, and Vladimir A. Volpert,** Traveling wave solutions of parabolic systems, 1994
139 **I. V. Skrypnik,** Methods for analysis of nonlinear elliptic boundary value problems, 1994
138 **Yu. P. Razmyslov,** Identities of algebras and their representations, 1994
137 **F. I. Karpelevich and A. Ya. Kreinin,** Heavy traffic limits for multiphase queues, 1994
136 **Masayoshi Miyanishi,** Algebraic geometry, 1994
135 **Masaru Takeuchi,** Modern spherical functions, 1994
134 **V. V. Prasolov,** Problems and theorems in linear algebra, 1994
133 **P. I. Naumkin and I. A. Shishmarev,** Nonlinear nonlocal equations in the theory of waves, 1994
132 **Hajime Urakawa,** Calculus of variations and harmonic maps, 1993
131 **V. V. Sharko,** Functions on manifolds: Algebraic and topological aspects, 1993
130 **V. V. Vershinin,** Cobordisms and spectral sequences, 1993
129 **Mitsuo Morimoto,** An introduction to Sato's hyperfunctions, 1993
128 **V. P. Orevkov,** Complexity of proofs and their transformations in axiomatic theories, 1993
127 **F. L. Zak,** Tangents and secants of algebraic varieties, 1993
126 **M. L. Agranovskiĭ,** Invariant function spaces on homogeneous manifolds of Lie groups and applications, 1993
125 **Masayoshi Nagata,** Theory of commutative fields, 1993
124 **Masahisa Adachi,** Embeddings and immersions, 1993
123 **M. A. Akivis and B. A. Rosenfeld,** Élie Cartan (1869–1951), 1993
122 **Zhang Guan-Hou,** Theory of entire and meromorphic functions: Deficient and asymptotic values and singular directions, 1993
121 **I. B. Fesenko and S. V. Vostokov,** Local fields and their extensions: A constructive approach, 1993
120 **Takeyuki Hida and Masuyuki Hitsuda,** Gaussian processes, 1993
119 **M. V. Karasev and V. P. Maslov,** Nonlinear Poisson brackets. Geometry and quantization, 1993
118 **Kenkichi Iwasawa,** Algebraic functions, 1993
117 **Boris Zilber,** Uncountably categorical theories, 1993
116 **G. M. Fel′dman,** Arithmetic of probability distributions, and characterization problems on abelian groups, 1993
115 **Nikolai V. Ivanov,** Subgroups of Teichmüller modular groups, 1992
114 **Seizô Itô,** Diffusion equations, 1992

(See the AMS catalog for earlier titles)

Lectures on Entire Functions

Boris Yakovlevich
LEVIN
1906–1993

Translations of
MATHEMATICAL
MONOGRAPHS

Volume 150

Lectures on Entire Functions

B. Ya. Levin

In collaboration with

Yu. Lyubarskii
M. Sodin
V. Tkachenko

American Mathematical Society
Providence, Rhode Island

Б. Я. ЛЕВИН

ЦЕЛЫЕ ФУНКЦИИ

При участии Ю. И. Любарского, М. Л. Содина, В. А. Ткаченко

Translated by Vadim Tkachenko from an original Russian manuscript

2000 *Mathematics Subject Classification.* Primary 30Dxx, 30D20.

For additional information and updates on this book, visit
www.ams.org/bookpages/mmono-150

Library of Congress Cataloging-in-Publication Data

Levin, B. ĨA. (Boris ĨAkovlevich)
Lectures on entire functions / B. Ya. Levin; in collaboration with Yu. Lyubarskii, M. Sodin, V. Tkachenko.
p. cm. — (Translations of mathematical monographs, ISSN 0065-9282 ; v. 150)
Translated from the author's Russian manuscript.
Includes bibliographical references and indexes.
ISBN 0-8218-0282-8 (alk. paper; hard cover)
ISBN 0-8218-0897-4 (alk. paper; soft cover)
1. Functions, Entire. I. Title. II. Series.

QA351.E5L484 1996 96-318
515′.98—dc20 CIP

♾ The paper used in this book is acid-free and falls within the guidelines
established to ensure permanence and durability.
Visit the AMS home page at http://www.ams.org/

10 9 8 7 6 5 4 3 2 10 09 08 07 06 05

Contents

Preface

Boris Yakovlevich Levin was born on December 22, 1906 in Odessa, the beautiful port city on the Black Sea, one of the main trade and cultural centers in the South of the Russian Empire. "I am an Odessit by social origin and nationality[1]," he used to joke. His father was a clerk for a Black Sea steamer company, whose work often took him to various beach ports on long-term missions. B. Ya. (as he was called by his colleagues and friends, rather than more formal Boris Yakovlevich) spent his youth moving with his family from one port town to another. He kept his devotion to the sea all his life, was an excellent swimmer and longed for the Black Sea, while living far from it in Kharkov.

Being of nonproletarian social origin, B. Ya. had no right to higher education in post-revolutionary Russia after graduating from secondary school. For some time he worked as an insurance agent and newspaper dispatcher, and as a welder during the construction of oil pipe-lines in the North Caucasus. This gave him the right to enlist as a univerisity student, and in 1928 he started his first year at the Department of Physics and Engineering of Rostov University, Russia. He and his friend decided, before concentrating on physics, to widen and improve their knowledge of mathematics. It was the choice of destiny: once entering mathematics they never parted with it. Both became famous experts: Boris Levin in analysis and Nikolai Efimov in geometry. To a great extent they were influenced by Dmitrii Mordukhai-Boltovskoi, an interesting and original mathematician with wide interests who worked at the Rostov University at that time.

While a second year student B. Ya. obtained his first mathematical result when he solved a problem proposed by Mordukhai-Boltovskoi. He investigated the functional equation

$$\Phi\left(\frac{\alpha x+\beta}{\gamma x+\delta}\right) = R(x)\Phi(x),$$

where $R(x)$ is a given rational function. This equation generalizes the functional equation $\Gamma(x+1) = x\Gamma(x)$ of the Euler Γ-function. B. Ya. proved that, apart from some exceptions, all solutions to this equation are hypertranscendental, just as $\Gamma(x)$. All exceptional cases were explicitly described by him. This theorem generalizes the famous Hölder theorem.

In 1932 B. Ya. graduated from the university, and for the next three years worked on his dissertation and taught mathematics at a technical institute in Rostov. His close personal and scientific friendship with Naum Akhiezer (Kharkov) and

[1] "Social origin" (or "class origin") and "nationality" were obligatory questions in all application forms in the Soviet Union. Unlike in other countries, the notions "nationality" and "citizenship" had different meanings there.

Mark Krein (Odessa) started at that time.[2] In 1936 B. Ya. submitted his Candidate of Science dissertation "On the growth of an entire function along a ray, and the distribution of its zeros with respect to their arguments" to Kharkov University, but was awarded with the highest degree of Doctor in Mathematics, which was an extremely rare event. In this dissertation B. Ya. founded the general theory of entire functions of completely regular growth, whose creation he shared with Albert Pfluger.

In 1935 B. Ya. moved to Odessa and began teaching mathematics at the Odessa Institute of Marine Engineering. In due time he got a Chair of Mathematics of this institute. Parallel to his teaching B. Ya. spent a lot of time and effort in advising his colleagues who worked on hydrodynamical problems of ships and mechanics of construction. In his later years he would say that teaching and communicating with engineers in a serious technical university is an important experience for a mathematician.

Starting from the middle 1930s a new school of functional analysis has been forming around Mark Krein in Odessa, and B. Ya., as he later used to say, experienced its strenghtening influence. He became interested in almost periodic functions, quasianalytic classes and related problems of completeness and approximation, algebraic problems of the theory of entire functions, and Sturm-Liouville operators. These remained the main fields of interest during his life.

In Odessa, the first students of B. Ya. have started their own research. Moshe Livshits and Vladimir Potapov, who became well-known specialists in functional analysis, were in equal measure students of Mark Krein and Boris Levin. Today, the family tree of B. Ya.'s mathematical children, grandchildren, and great-grandchildren contains more than a hundred mathematicians.

During World War II, B. Ya. worked with his institute in Samarkand (Uzbekistan). His attempts to join active military service failed, since Full Professors were exempt from the draft. After the war B. Ya. returned to Odessa. At that time a destruction of mathematics at Odessa University began. Mark Krein and his colleagues were not permitted to return to work at the university, and very soon an anti-semitic campaign waged against Mark Krein and B. Ya. forced the latter to leave Odessa. On invitation of Naum Akhiezer in 1949, B. Ya. moved to Kharkov. During several decades after the end of World War II, some other mathematicians moved from Odessa to Kharkov: Izrail Glazman, Mikhail Dolberg, Moshe Livshits, Vladimir Potapov. However, B. Ya. has kept close ties with Odessa, Mark Krein, and the mathematicians of Krein's circle for the whole life.

Despite all difficulties, the period from late forties to late sixties was the time of blossoming of the Kharkov mathematical school. At that time Naum Akhiezer, Boris Levin, Vladimir Marchenko, Aleksandr Povzner, and Aleksei Pogorelov worked in Kharkov, and their impact determined the image of Kharkov mathematics for many years.

From 1949, B. Ya. worked at Kharkov University. In addition to undergraduate courses of calculus, theory of functions of a complex variable and functional analysis, he taught advanced courses on entire functions, quasianalytic classes, almost periodic functions, harmonic analysis and approximation theory, and Banach algebras. The lectures were distinguished by their originality, depth and elegance.

[2]Reminiscences of Mark Krein, written by B. Ya., were published in the Ukrainian Mathematical Journal, 46, no. 3, 1994.

B. Ya. used to include his own, yet-unpublished results as well as new original proofs of known theorems. He attracted a very wide audience of students of various levels and also research mathematicians. This book emerged from notes of one of such courses and it is a great pity that notes of other courses are not available.

In 1956 he published his monograph "Distribution of zeros of entire functions", which greatly influenced several generations of analysts. It was translated into German and English and revised in 1980. Even now the book is the main source on the subject.

During the same year B. Ya. started his Thursday seminar at Kharkov University. For about 40 years it has been a school for Kharkov mathematicians working in analysis and has been a center of active mathematical research. The major part of seminar talks concerned complex analysis and its applications. Nevertheless, there was no restriction on the subject: there were talks on Banach spaces, spectral theory of operators, differential and integral equations, and probability theory. A meeting of the seminar usually lasted more than two hours, with a short break. In most cases detailed proofs were presented. Its active participants included Vladimir Azarin, Aleksandr Eremenko, Sergei Favorov, Aleksandr Fryntov, Anatolii Grishin, Vladimir P. Gurarii, Illich Hachatryan, Mikhail Kadets, Victor Katsnelson, Vladimir Logvinenko, Yurii Lyubich, Vladimir Matsaev, Iossif Ostrovskii, Igor' Ovcharenko, Victor Petrenko, Lev Ronkin, and many others. B. Ya. has always been proud and delighted with achievements of the participants of his seminar.

In 1969, without interrupting his teaching at the university, B. Ya. organized and headed the Department of the Theory of Functions at the Institute for Low Temperature Physics and Engineering of the Academy of Sciences of Ukraine, where he gathered a group of his former students and young colleagues. He worked there to the last days of his life. A well-known western mathematician working in complex analysis once said: "It is a typical Soviet habit to make secrets from everything; evidently, "Low Temperature Physics" is just a code for function theory."

The name of the founder and first director of the institute Boris Verkin must be mentioned here. A specialist in experimental physics, he held mathematics in high esteem and gave a lot of support to its progress. "Mathematicians ennoble the institute," Verkin used to say. Due to his initiative, Naum Akhiezer, Izrail Glazman, Vladimir Marchenko, Anatolii Myshkis, Aleksei Pogorelov joined the institute in the early sixties and very soon the Mathematical Division of the institute became one of the leading mathematical centers in the former Soviet Union, with the wonderful creative athmospere.

It is not our intention to give here a detailed description of mathematical activities of B. Ya. We only mention that he knew how to find unexpectedly simple ways leading to a solution of a problem which from the beginning seemed to be extremely complicated. After his talks and works one would be puzzled why other mathematicians who attacked the same problem did not have the same insight? The participants of his seminar remember that sometimes after somebody's "hard" talk B. Ya. proposed his simple and elegant solution. At the same time B. Ya. mastered the fine analytic techniques, which he successfully used if required.

The main part of results obtained by B. Ya. are related to the theory of entire functions. Being interested in the central problems of this theory, he found new and important connections with other domains of analysis. His results helped to transfer applications of the theory of entire functions to functional analysis and the spectral theory of differential operators to a deeper level. Often B. Ya. expressed the viewpoint that the theory of entire functions remains of importance due to its numerous applications.

Boris Levin lived a long life, full of mathematical quest and discoveries. He experienced many difficult periods, but despite all strokes of fate remained faithful to his highest moral principles which he defended openly and selflessly. He did not have, and did not try to seek favours from officialdom. Until mid-80s he was not allowed to travel abroad and had very scarce possibilities to contact foreign colleagues. In Kharkov he lived in a small and wet ground-floor appartment. Nevertheless, very often B. Ya. would invite his colleagues and students to his home. Several hours would be devoted to mathematics. Then supper time would arrive, and his wife Liya, a woman of great charm, kindness, and benevolence, joined the guests. After traditional strong tea which B. Ya. always made himself, there was the time for discussing politics and politicians, for storytelling and poetry, in which B. Ya. was the expert and connoisseur.

Outstanding mathematician, brilliant lecturer and storyteller, witty companion, B. Ya. radiated some kind of energy that attracted to him even people who were very far from mathematics. He was a person of the highest quality to the many people who knew him.

For many years B. Ya. has planned a book based on his lecture course at the Moscow University in 1969. It was intended for a reader interested in applications of the theory of entire functions.

During the last two years of his life, we worked with B. Ya. on preparation of this book. The material of the lectures was shaped, extended, and augmented with a bibliography. Initially, B. Ya. planned to include some important applications of the theory of entire functions to the spectral theory of operators, discovered in various directions by Louis de Branges and Vladimir Matsaev, but this task was never completed.

B. Ya. hoped to see his book published both in Russian and English and worked on the manuscript until the last days of his life. He died on August 24, 1993. His daughter Natalya and wife Liya had preceded him, passing away in 1980 and 1992. He is survived by his son Mikhail and two grandchildren.

David Drasin carefully read the entire manuscript and made many corrections and suggestions related both to mathematics and English. Gundorph Kristiansen called our attention to some misprints. We are very grateful to them for that.

Yurii Lyubarskii, Mikhail Sodin, Vadim Tkachenko

Introduction

This monograph originated from a lecture course which I gave at the Moscow University in 1969. The lectures were printed by the University in an edition of 500 copies.

The title of the monograph remains the same, but it is not a second edition, since the main part of lectures has been written anew, some new material has been added, and some old topics have been extended resulting in a substantial increase of the entire volume.

The theory of entire functions has a multitude of applications in calculus and functional analysis. I made it my goal to present the main facts of the theory of entire functions from that point of view and tried, as much as possible, to develop a connection between applications and the general theory. I hope that such an exposition will help in mastering the methods of the theory of entire functions. Many sections of this monograph contain problems with applications related to these topics. Their solution is not necessary to comprehend subsequent parts, but may be of some use. No special knowledge is required to read this book, except a conventional university course on the theory of functions of a complex variable.

Yu. Lyubarskii, M. Sodin, and V. Tkachenko helped me very much in writing the monograph; it would not have been written without their support.

I. Ostrovskii read the whole manuscript and made several remarks, which were taken into account in the final text. A. Eremenko made several useful remarks to Part I. I would like to express my deep gratitude to these colleagues.

This monograph was written with a partial support by a grant from the American Mathematical Society, which is highly appreciated.

B. Levin

Part I. Entire Functions of Finite Order

A function $f(z)$ analytic in the whole complex plane, i.e., a function represented by a power series of the form

$$f(z) = \sum_{n=0}^{\infty} c_n z^n \,, \qquad \lim_{n\to\infty} \sqrt[n]{|c_n|} = 0 \,,$$

is called an entire function. This is the simplest class of analytic functions containing all polynomials. Polynomials are classified according to their degree, i.e., according to their growth as $|z| \to \infty$. An entire function can grow in various ways along different directions. For a general characterization of the growth, the function

$$M_f(r) = \max_{|z|=r} |f(z)|$$

is introduced. It follows from the Maximum Principle that this function increases monotonically.

The more roots a polynomial has, the faster it grows. This property is extended to entire functions, but it is much more complex. The relationship between the growth of an entire function and the distribution of its roots is the main subject matter of the theory of entire functions.

Here we would like to point out, without giving precise definitions, that there is a large cycle of theorems which state that if an entire function f "grows slowly enough" and its roots are "arranged very densely", then $f(z) \equiv 0$. These are uniqueness theorems similar to the simplest uniqueness theorem for polynomials (a polynomial of degree n having more than n roots is identically equal to zero). The solution of many completeness problems for various functional systems, in particular eigenfunctions of boundary value problems, reduces to such theorems.

Another cycle of questions involves the study of relationship between the growth (or decrease) of a function along different directions, and its global growth characterized by the function $M_f(r)$. A polynomial grows in all directions uniformly. The asymptotic behavior of an entire function as $z \to \infty$ is much more complicated. The main facts pertaining to this problem can be stated in the following way: an entire function having a "small" global growth cannot "decrease too fast in some direction", but must "grow on a large enough part of the complex plane". The simplest fact of this type, directly implied by the Liouville theorem, can be formulated in the following way: if $f(z) = O(1)$, $z \to \infty$, and $f(z_n) \to 0$ for some sequence $z_n \to \infty$, then the function $f(z)$ is identically equal to zero. More refined estimates are usually based on various versions of Phragmén-Lindelöf theorems. Theorems of this type are used in functional analysis (in particular, in the theory of nonselfadjoint operators and in the theory of Banach algebras), in harmonic analysis, and in some problems of mathematical physics.

Finally, some problems of expanding functions of a real or complex variable into special functional series (problems of bases) reduce to certain questions of the theory of interpolation by entire functions.

Thus, the theory of entire functions provides us with a powerful tool to solve many problems of classical and functional analysis.

This is the approach which will be used to present the theory of entire functions in this monograph.

LECTURE 1

Growth of Entire Functions

1.1. The growth scale for entire functions

We shall start by considering an important question: how fast can the function $M_f(r)$ grow?

THEOREM 1. *If, for a nonnegative λ, the equation*

$$\liminf_{r\to\infty} \frac{M_f(r)}{r^\lambda} = 0$$

holds, then $f(z)$ is a polynomial whose degree does not exceed λ.

PROOF. We shall use the Cauchy inequalities

$$|c_n| \le \frac{M_f(r)}{r^n}.$$

For $n > \lambda$ we obtain

$$|c_n| \le \liminf_{r\to\infty} \frac{M_f(r)}{r^n} = 0.$$

Thus, in order to classify entire functions according to their growth, we must construct a scale of monotonic functions that grow faster than any polynomial.

Can the function $M_f(r)$ grow arbitrarily fast?

PROBLEM 1. Let $\varphi(r)$ be a function growing as $r \to \infty$. Construct an entire function $f(z)$ to satisfy the inequality $M_f(r) > 1 + \varphi(r)$.

On the other hand, there exist entire functions with a slow rate of growth.

PROBLEM 2. Let $\psi(r)$ be an arbitrary function increasing unrestrictedly as $r \to \infty$. Construct an entire function $g(z)$ which is not a polynomial and satisfies the inequality $M_g(r) < 1 + r^{\psi(r)}$.

HINT. Look for a function in the form of a power series with positive coefficients.

1.2. Order and type of entire functions

Let us introduce the following notation. If an inequality $h(r) < \varphi(r)$ holds for sufficiently large values of r, we shall call it an asymptotic inequality and write $h(r) \overset{\text{as}}{<} \varphi(r)$. If the same inequality holds for some sequence of values $r_n \to \infty$, then we shall write $h(r) \overset{\text{n}}{<} \varphi(r)$.

An entire function $f(z)$ is called a *function of finite order* if $M_f(r) \overset{\text{as}}{<} \exp(r^k)$ for some $k > 0$. The *order* (or the *order of growth*) of an entire function f is the greatest lower bound of those values of k for which this asymptotic inequality is fulfilled. We shall usually denote the order of an entire function f by $\rho = \rho_f$. It follows from the definition of the order that

$$e^{r^{\rho-\varepsilon}} \overset{\mathrm{n}}{<} M_f(r) \overset{\mathrm{as}}{<} e^{r^{\rho+\varepsilon}} .$$

By taking the logarithm twice we obtain

$$\rho - \varepsilon \overset{\mathrm{n}}{<} \frac{\log\log M_f(r)}{\log r} \overset{\mathrm{as}}{<} \rho + \varepsilon ,$$

whence

$$\rho = \limsup_{r\to\infty} \frac{\log\log M_f(r)}{\log r} .$$

Do entire functions of any nonnegative order exist? This question will be answered at the end of the lecture.

Note that among the functions of the same order there are functions growing in different ways. For instance, take $e^{r/\log r}$, e^r and $e^{r\log r}$. These functions are not entire, but it is not difficult to find entire functions for which $M(r)$ grows in the same way. Such functions are distinguished by using another characteristic, namely the type.

Let ρ be the order of an entire function f. The function is said to have a *finite type* if for some $A > 0$ the inequality

$$M_f(r) \overset{\mathrm{as}}{<} e^{Ar^\rho}$$

is fulfilled.

The greatest lower bound for those values of A for which the latter asymptotic inequality is fulfilled is called the *type* $\sigma = \sigma_f$ *(with respect to the order* ρ*) of the function* f. It follows from the definition of the type that

$$e^{(\sigma-\varepsilon)r^\rho} \overset{\mathrm{n}}{<} M_f(r) \overset{\mathrm{as}}{<} e^{(\sigma+\varepsilon)r^\rho} .$$

Having taken the logarithm and divided by r^ρ, we obtain

$$\sigma - \varepsilon \overset{\mathrm{n}}{<} \frac{\log M_f(r)}{r^\rho} \overset{\mathrm{as}}{<} \sigma + \varepsilon,$$

and therefore

$$\sigma_f = \limsup_{r\to\infty} \frac{\log M_f(r)}{r^\rho} .$$

PROBLEM 3. Prove the inequalities $\rho_{fg} \le \max(\rho_f, \rho_g)$, $\rho_{f+g} \le \max(\rho_f, \rho_g)$, $\sigma_{fg} \le \sigma_f + \sigma_g$, and $\sigma_{f+g} \le \max(\sigma_f, \sigma_g)$.

If, for a given $\rho > 0$, the type of a function is infinite, then the function is of *maximal type*; for $0 < \sigma_f < \infty$ the type is *normal* or *mean*; for $\sigma_f = 0$ the type is *minimal*. In the last case, for any $\varepsilon > 0$ the asymptotic inequality

$$M_f(r) \overset{\mathrm{as}}{<} e^{\varepsilon r^\rho}$$

is fulfilled.

Entire functions of order $\rho = 1$ and normal type σ are called *entire functions of exponential type* σ.

EXAMPLES. Verify that $\sin Az$ is of order $\rho = 1$ and type $\sigma = |A|$, which means that it is an entire function of exponential type $|A|$; $\sin\sqrt{z}/\sqrt{z}$ is of order $1/2$ and type 1; $\exp\{a_0 z^n + \ldots + a_n\}$, $a_0 \neq 0$, is of order n and type $|a_0|$.

1.3. The relation between the growth of an entire function and the decrease of the coefficients of its power series expansion

Let

$$f(z) = \sum_{n=0}^{\infty} c_n z^n \tag{1}$$

be an entire function.

LEMMA 1. *If the asymptotic inequality*

$$M_f(r) \overset{\text{as}}{<} e^{Ar^K} \tag{2}$$

is fulfilled, then

$$|c_n| \overset{\text{as}}{<} \left(\frac{eAK}{n}\right)^{n/K}. \tag{3}$$

PROOF. By the Cauchy inequality, it follows from (2) that

$$|c_n| \leq \frac{M(r)}{r^n} < e^{Ar^K - n\log r}, \quad r \geq r_0. \tag{4}$$

Minimizing the exponent with respect to r, we obtain $KAr^{K-1} - n/r = 0$ and $r_n^K = n/(AK)$. For sufficiently large n we have $r_n \geq r_0$. After substituting r_n in (4) we obtain (3).

LEMMA 2. *If the asymptotic inequality* (3) *is fulfilled, then*

$$M_f(r) \overset{\text{as}}{<} e^{(A+\varepsilon)r^K}, \quad \forall\varepsilon > 0. \tag{5}$$

PROOF. First, note that if an entire function f satisfies inequality (5), then so does the function $f + Q$, where Q is a polynomial. Therefore, we can assume that $c_0 = 0$ and (3) holds for all $n \geq 1$. Thus, we have

$$\begin{aligned} |f(z)| &\leq \sum_{n=1}^{\infty} |c_n| r^n \leq \sum_{n=1}^{\infty} \left(\frac{eAK}{n}\right)^{n/K} r^n \\ &= \sum_{n=1}^{\infty} \left(\frac{eAr^K}{n/K}\right)^{n/K}, \quad |z| = r. \end{aligned}$$

Set $m = [n/K]$. Then, for sufficiently large r, we have

$$\left(\frac{eAr^K}{n/K}\right)^{n/K} \leq \left(\frac{eAr^K}{m}\right)^{m+1}.$$

Hence

$$|f(z)| \leq \sum_{m=1}^{\infty} \left(\frac{eAr^K}{m}\right)^{m+1}.$$

By using the Stirling formula

$$m! \sim \left(\frac{m}{e}\right)^m \sqrt{2\pi m}, \quad m \to \infty$$

and the inequality

$$\sqrt{2\pi m} < C\left(\frac{A+\varepsilon}{A}\right)^{m+1}, \quad m \geq 1,$$

we obtain

$$|f(z)| \leq C_1 \sum_{m=1}^{\infty} \frac{e^m}{m^m}\left(Ar^K\right)^{m+1} < C_1 \sum_{m=1}^{\infty} \frac{(A+\varepsilon/2)^{m+1} r^{K(m+1)}}{m!}$$
$$= C_1 e^{(A+\varepsilon/2)r^K}\left(A+\frac{\varepsilon}{2}\right) e r^K \overset{\text{as}}{<} C_2 e^{(A+\varepsilon)r^K},$$

where C, C_1, C_2 are constants. Thus,

$$|f(z)| \overset{\text{as}}{<} e^{(A+\varepsilon)r^K}.$$

The lemma is proved.

These lemmas enable us to express the order and the type of an entire function in terms of the rate of decrease of the coefficients of its power expansion. Indeed, the order ρ equals the greatest lower bound of those K for which (5) holds for any $A > 0$, in particular, for $A+\varepsilon = 1$. By Lemmas 1 and 2 we have

$$\left(\frac{e(\rho-\varepsilon)}{n}\right)^{\frac{n}{\rho-\varepsilon}} \overset{\text{n}}{<} |c_n| \overset{\text{as}}{<} \left(\frac{e(\rho+\varepsilon)}{n}\right)^{\frac{n}{\rho+\varepsilon}}$$

for each $\varepsilon > 0$. Having taken the logarithms, we obtain

$$\frac{n}{\rho-\varepsilon}[\log e(\rho-\varepsilon) - \log n] \overset{\text{n}}{<} \log|c_n| \overset{\text{as}}{<} \frac{n}{\rho+\varepsilon}[\log e(\rho-\varepsilon) - \log n],$$

or

$$\frac{n\log n}{\rho+\varepsilon}[1+o(1)] \overset{\text{as}}{<} \log\frac{1}{|c_n|} \overset{\text{n}}{<} \frac{n\log n}{\rho-\varepsilon}[1+o(1)].$$

Thus we have proved

THEOREM 2. *The order of the entire function* (1) *is determined by the formula*

$$\rho = \limsup_{n\to\infty} \frac{n\log n}{\log(1/|c_n|)}. \tag{6}$$

Likewise, the type σ equals the greatest lower bound of those A for which estimate (3) holds with $K = \rho$. From this we deduce

THEOREM 3. *The type of the entire function* (1) *is determined by the formula*

$$\sigma = \frac{1}{\rho e}\limsup_{n\to\infty}\left(n\sqrt[n]{|c_n|^{\rho}}\right). \tag{7}$$

EXAMPLES. Let $0 < \rho < \infty$, $0 < \sigma < \infty$. The entire function

$$f(z) = \sum_{n=1}^{\infty} \left(\frac{e\sigma\rho}{n}\right)^{n/\rho} z^n$$

is of order ρ and of type σ. The function

$$f(z) = \sum_{n=2}^{\infty} \left(\frac{e\sigma\rho}{n \log n}\right)^{n/\rho} z^n$$

is of order ρ and of minimal type, whereas the function

$$f(z) = \sum_{n=2}^{\infty} \left(\frac{e\rho \log n}{n}\right)^{n/\rho} z^n$$

is of order ρ and of maximal type. The entire function

$$f(z) = \sum_{n=2}^{\infty} \left(\frac{1}{\log n}\right)^{n} z^n$$

is of infinite order, and the function

$$f(z) = \sum_{n=0}^{\infty} e^{-n^2} z^n$$

is of zero order.

PROBLEM 4. Using formulas (6) and (7), prove that the order and type of an entire function do not change under differentiation.

PROBLEM 5. If $f(z)$ is an entire function and the numbers $f^{(n)}(0)$ are integers, then either $f(z)$ is a polynomial, or the type σ_f of this function with respect to the order $\rho = 1$ is at least 1.

LECTURE 2

Main Integral Formulas for Functions Analytic in a Disk

To investigate the relation between the growth of an entire function and its zeros we shall need several formulas.

2.1. The Poisson formula and the Schwarz formula

We assume that the reader knows the classical Poisson formula which represents a function harmonic in a disk $\{z: |z| < R\}$ and continuous in its closure:

$$u(z) = \frac{1}{2\pi}\int_0^{2\pi} u(Re^{i\psi})\frac{R^2 - r^2}{R^2 - 2Rr\cos(\psi - \theta) + r^2}\,d\psi\,, \qquad z = re^{i\theta}\,.$$

The same formula may be written in the form

$$(1) \qquad u(z) = \frac{1}{2\pi}\int_0^{2\pi} u(\zeta)\frac{|\zeta|^2 - |z|^2}{|\zeta - z|^2}\,d\psi = \frac{1}{2\pi}\int_0^{2\pi} u(\zeta)\,\mathrm{Re}\,\frac{\zeta + z}{\zeta - z}\,d\psi\,,$$

where $\zeta = Re^{i\psi}$.

To represent a function $f = u + iv$ holomorphic in a disk $\{z: |z| < R\}$, whose real part u is continuous in $\{z: |z| \le R\}$, we shall be using the Schwarz formula:

$$(2) \qquad f(z) = \frac{1}{2\pi}\int_0^{2\pi} u(\zeta)\frac{\zeta + z}{\zeta - z}\,d\psi + iv(0)\,.$$

The latter formula follows from (1). Indeed, by (1) the real parts of the functions on the left and right sides of (2) coincide in $\{z: |z| < R\}$. Hence, the functions differ by a purely imaginary constant, and for $z = 0$ they coincide: $u(0) + iv(0) = f(0)$.

2.2. The Poisson-Jensen formula

If $f(z) \neq 0$ in a disk $\{z: |z| \le R\}$, then $\log f(z)$ is a holomorphic function in the disk, and by formula (2) we have

$$(3) \qquad \log f(z) = \frac{1}{2\pi}\int_0^{2\pi} \log|f(Re^{i\psi})|\frac{Re^{i\psi} + z}{Re^{i\psi} - z}\,d\psi + iC\,.$$

Formula (1), as well as formula (3), implies

$$(4) \qquad \log|f(z)| = \frac{1}{2\pi}\int_0^{2\pi} \log|f(Re^{i\psi})|\frac{R^2 - r^2}{R^2 - 2Rr\cos(\theta - \psi) + r^2}\,d\psi\,.$$

Now let $a_1, a_2, \ldots, a_n$ be the zeros of $f(z)$ in $\{z : |z| < R\}$ arranged according to increasing modulus. We shall make a permanent convention to write down each zero as many times as its multiplicity. Let $f(z) \neq 0$ for $|z| = R$, and let

$$\varphi(z) = f(z) \prod_{m=1}^{n} \frac{R^2 - \bar{a}_m z}{R(z - a_m)} . \tag{5}$$

It is evident that $|\varphi(Re^{i\psi})| = |f(Re^{i\psi})|$ and $\varphi(z) \neq 0$ for $|z| \leq R$. Let us apply formulas (3) and (4) to φ. It follows from (3) that

$$\log \varphi(z) = \frac{1}{2\pi} \int_0^{2\pi} \log |f(Re^{i\psi})| \frac{Re^{i\psi} + z}{Re^{i\psi} - z} \, d\psi + iC$$

or

$$\log f(z) = \frac{1}{2\pi} \int_0^{2\pi} \log |f(Re^{i\psi})| \frac{Re^{i\psi} + z}{Re^{i\psi} - z} \, d\psi + \sum_{|a_m|<R} \log \frac{R(z - a_m)}{R^2 - z\bar{a}_m} + iC . \tag{6}$$

For the function $\log \dfrac{R(z - a_m)}{R^2 - z\bar{a}_m}$ to be single-valued it is necessary to cut the complex plane along the rays $\{z = re^{i \arg a_m}, \ r \geq |a_m|\}$. If some cut meets the point z, we shall slightly deform it counterclockwise. After separating the imaginary parts in (6) we obtain

$$\begin{aligned} \log |f(z)| &= \frac{1}{2\pi} \int_0^{2\pi} \log |f(Re^{i\psi})| \frac{R^2 - r^2}{R^2 - 2Rr\cos(\theta - \psi) + r^2} \, d\psi \\ &\quad + \sum_{|a_m|<R} \log \left| \frac{R(z - a_m)}{R^2 - z\bar{a}_m} \right| , \qquad z = re^{i\theta} . \end{aligned} \tag{7}$$

Formula (7) was derived by R. Nevanlinna who named it after Poisson and Jensen. It forms a foundation of the Nevanlinna theory of distribution of values of meromorphic functions.

2.3. The Jensen formula

Let us assume, first, that $f(0) \neq 0$. If we set $z = 0$ in equation (7) we obtain

$$\log |f(0)| = \frac{1}{2\pi} \int_0^{2\pi} \log |f(Re^{i\psi})| \, d\psi + \sum_{|a_m|<R} \log \frac{|a_m|}{R} .$$

The second term on the right can be written as a Stieltjes integral. Denoting by $n(t)$ the number of points a_m satisfying the inequality $|a_m| \leq t$, we obtain a left continuous, monotonic, integer-valued and piecewise constant function. It is called a *counting function* of zeros. We have

$$\sum_{|a_m|<R} \log \frac{R}{|a_m|} = \int_0^R \log \frac{R}{t} \, dn(t) = n(t) \log \frac{R}{t} \Big|_0^R + \int_0^R \frac{n(t)}{t} \, dt ,$$

and, finally,

$$\int_0^R \frac{n(t)}{t} \, dt = \frac{1}{2\pi} \int_0^{2\pi} \log |f(Re^{i\psi})| \, d\psi - \log |f(0)| . \tag{8}$$

This is the famous Jensen formula.

If $f(0) = 0$ (and, of course, $f \not\equiv 0$) we denote by k the multiplicity of the root at $z = 0$. Then formula (8) takes on the form

$$\int_0^R \frac{n(t) - n(0)}{t}\, dt + n(0)\log R = \frac{1}{2\pi}\int_0^{2\pi} \log|f(Re^{i\psi})|\, d\psi - \log\left|\frac{f^{(k)}(0)}{k!}\right| .$$

To prove the latter formula it is sufficient to apply (8) to the function $f(z)/z^k$.

2.4. The Nevanlinna characteristics

If $f(z)$ is a meromorphic function in the disk $\{z : \ |z| \le R\}$ with neither zeros nor poles on the circumference $\{z : \ |z| = R\}$ and at $z = 0$, then the function $\varphi(z)$ must be chosen different from (5). Namely, let

$$\varphi(z) = f(z) \prod_{|b_m|<R} \frac{R(z - b_m)}{R^2 - \bar{b}_m z} \left\{ \prod_{|a_m|<R} \frac{R(z - a_m)}{R^2 - \bar{a}_m z} \right\}^{-1} ,$$

where $\{b_m\}$ are poles of a function f in the disk $\{z : \ |z| < R\}$. Then formula (6) is replaced by

$$\log f(z) = \frac{1}{2\pi}\int_0^{2\pi} \log|f(Re^{i\psi})| \frac{Re^{i\psi} + z}{Re^{i\psi} - z}\, d\psi + \sum_{|a_m|<R} \log \frac{R(z - a_m)}{R^2 - z\bar{a}_m} - \sum_{|b_m|<R} \log \frac{R(z - b_m)}{R^2 - z\bar{b}_m} + iC .$$

The Poisson-Jensen formula (7) takes the form

$$\begin{aligned} \log|f(z)| &= \frac{1}{2\pi}\int_0^{2\pi} \log|f(Re^{i\psi})| \frac{R^2 - r^2}{R^2 - 2Rr\cos(\theta - \psi) + r^2}\, d\psi \\ &\quad + \sum_{|a_m|<R} \log\left|\frac{R(z - a_m)}{R^2 - z\bar{a}_m}\right| - \sum_{|b_m|<R} \log\left|\frac{R(z - b_m)}{R^2 - z\bar{b}_m}\right| , \end{aligned} \tag{9}$$

and the Jensen formula becomes

$$\int_0^R \frac{n(t,0)}{t}\, dt - \int_0^R \frac{n(t,\infty)}{t}\, dt = \frac{1}{2\pi}\int_0^{2\pi} \log|f(Re^{i\psi})|\, d\psi - \log|f(0)| . \tag{10}$$

Here $n(t, 0)$ is the counting function of zeros, and $n(t, \infty)$ is the counting function of poles of the function f . Following R. Nevanlinna, let us introduce

$$m(R, f) = \frac{1}{2\pi}\int_0^{2\pi} \log^+ |f(Re^{i\psi})|\, d\psi ,$$

where $a^+ = \max(a, 0)$, and

$$\begin{aligned} N(R, f) &= \int_0^R \frac{n(t,\infty) - n(0,\infty)}{t}\, dt + n(0,\infty)\log R , \\ T(R, f) &= m(R, f) + N(R, f) . \end{aligned}$$

The function $T(R, f)$ is called the Nevanlinna characteristic. In this notation, the Jensen formula (10) becomes

$$T(R, f) = T(R, 1/f) + C \,. \tag{11}$$

Here C is a constant, and if $f(0) \neq 0$ and $f(0) \neq \infty$, then $C = \log|f(0)|$.

It is easy to see that

$$T(R, af + b) = T(R, f) + O(1)\,, \quad a \neq 0\,,$$

and hence equation (11) implies

$$T\Big(R, \frac{af+b}{cf+d}\Big) = T(R, f) + O(1)\,, \quad ad - bc \neq 0\,.$$

This relation is called the *First Main Theorem of Nevanlinna.*

PROBLEM 1. Prove the following estimates:

$$\begin{aligned} T\Big(R, \sum_{\nu=1}^{n} f_\nu\Big) &\le \sum_{\nu=1}^{n} T(R, f_\nu) + \log n\,, \\ T\Big(R, \prod_{\nu=1}^{n} f_\nu\Big) &\le \sum_{\nu=1}^{n} T(R, f_\nu)\,. \end{aligned} \tag{12}$$

Here f_ν, $1 \le \nu \le n$, are meromorphic functions.

PROBLEM 2. Prove the following statements (H. Cartan).

1. If $f(z)$ is a meromorphic function, and $f(z) = f_1(z)/f_2(z)$, where $f_1(z)$, $f_2(z)$ are entire functions without common zeros, then

$$T(r, f) = \frac{1}{2\pi}\int_0^{2\pi} \max(\log|f_1(re^{i\psi})|, \log|f_2(re^{i\psi})|)\, d\psi + O(1)\,.$$

2. If $f(z)$ is a meromorphic function, then

$$T(r, f) = \frac{1}{2\pi}\int_0^{2\pi} N\Big(r, \frac{1}{f - e^{i\theta}}\Big)\, d\theta + C_f\,,$$

where the constant C_f does not depend on r.

HINT. Use the formula

$$\frac{1}{2\pi}\int_0^{2\pi} \log|w - e^{i\theta}|\, d\theta = \log^+|w|\,.$$

The function $T(R, f)$ plays an important role in the study of entire and meromorphic functions. If f is an entire function, then

$$T(R, f) = m(R, f) = \frac{1}{2\pi}\int_0^{2\pi} \log^+|f(Re^{i\psi})|\, d\psi \le \log^+ M_f(R)\,.$$

On the other hand, using the Poisson-Jensen formula (7) we have

$$\begin{aligned} \log|f(z)| &\le \frac{1}{2\pi}\int_0^{2\pi} \log|f(Re^{i\psi})| \frac{R^2 - r^2}{|Re^{i\psi} - z|^2}\, d\psi \\ &\le \frac{R+r}{R-r}\, \frac{1}{2\pi}\int_0^{2\pi} \log^+|f(Re^{i\psi})|\, d\psi\,, \quad r = |z|\,, \end{aligned}$$

and

$$\log M_f(r) \le \frac{R+r}{R-r} m(R, f) .$$

For $R = 2r$ we obtain

$$\log M_f(r) \le 3m(2r, f). \tag{13}$$

If, in particular, $m(r, f) \overset{\text{as}}{<} Ar^k$, then

$$\log M_f(r) \overset{\text{as}}{<} 3 \cdot 2^k Ar^k.$$

Hence, in defining the order of an entire function one can use the Nevanlinna characteristic $T(r, f)$ instead of $\log M_f(r)$.

Let f_1, f_2 be entire functions such that the quotient $\varphi = f_1/f_2$ is an entire function. Then the First Main Theorem of Nevanlinna and estimates (13) and (12) imply

$$\begin{aligned}\log M_\varphi(r) &\le 3T(2r, \varphi) \le 3[T(2r, f_1) + T(2r, f_2) + O(1)] \\ &\le 3 \log M_{f_1}(2r) + 3 \log M_{f_2}(2r) + O(1) .\end{aligned}$$

A theorem on the growth of a quotient of entire functions now directly follows from the latter inequality.

THEOREM 1. *If the quotient of two entire functions of order not greater than ρ is an entire function, then its order is also at most ρ. If, in addition, the numerator and denominator are of mean type with respect to ρ, then the quotient is of mean type with respect to ρ.*

We refer the reader to the monographs by Nevanlinna [**102**], Hayman [**51**], Goldberg and Ostrovskii [**43**] where the Nevanlinna theory of meromorphic functions and its applications can be found.

2.5. Some corollaries of the Jensen formula

Let f be an entire function. Then it follows directly from the Jensen formula that

$$\log |f(0)| \le \frac{1}{2\pi} \int_0^{2\pi} \log |f(Re^{i\psi})| \, d\psi . \tag{14}$$

If $|f(0)| = 1$, then for $r > 0$ we have

$$\log M_f(er) \ge \int_0^{er} \frac{n(t)}{t} \, dt \ge \int_r^{er} \frac{n(t)}{t} \, dt \ge n(r),$$

and hence

$$n(r) < \log M_f(er) . \tag{15}$$

The modulus of an entire function may decrease in some directions, and inequality (14) shows that "in the mean" it decreases not faster than it grows. The latter inequality shows that an entire function with an upper bound for $M_f(r)$ cannot have too many zeros. We remark that if $|f(0)| \neq 1$, then (15) must be replaced by $n(r) < \log M_f(er) + \text{const}$.

LECTURE 3

Some Applications of the Jensen Formula

3.1. A theorem on (I)-quasianalyticity

A class C of functions defined on some interval is called (I)-quasianalytic if each function $g \in C$ vanishing almost everywhere on an interval, no matter how small is its length, vanishes almost everywhere on its domain of definition. We shall use the Jensen formula to prove a theorem on (I)-quasianalyticity.

THEOREM 1 (Pólya). *If a 2π-periodic function $f \in L^2[-\pi, \pi]$ is represented by a lacunary Fourier series*

$$f(t) \sim \sum_{k=-\infty}^{\infty} c_{n_k} e^{in_k t}, \qquad \lim_{k\to\pm\infty} \frac{n_k}{k} = +\infty$$

and $f(t) = 0$ almost everywhere on an arbitrarily small interval, then $f(t) = 0$ almost everywhere on $[-\pi, \pi]$.

PROOF. Making a shift of the periodic function $f(t)$, we obtain a function $\varphi(t) = f(t+h)$ equal to zero for $\pi - \delta < |t| \le \pi$ and represented by a series

$$\varphi(t) \sim \sum_{k=-\infty}^{\infty} c_{n_k} e^{in_k h} e^{in_k t} = \sum_{k=-\infty}^{\infty} d_{n_k} e^{in_k t}.$$

By a well-known formula we have

$$d_n = \frac{1}{2\pi} \int_{-\pi}^{\pi} \varphi(t) e^{-int}\, dt.$$

According to the conditions of Theorem 1 "many" coefficients d_n are equal to zero. We shall prove that in this case all coefficients vanish.

Let $\varphi(t) \not\equiv 0$. We define the function

$$\Phi(z) = \frac{1}{2\pi} \int_{-\pi}^{\pi} \varphi(t) e^{-izt}\, dt = \frac{1}{2\pi} \int_{-\pi+\delta}^{\pi-\delta} \varphi(t) e^{-izt}\, dt.$$

It is easy to see that $\Phi(z)$ is an entire function, and that

$$|\Phi(x+iy)| \le \frac{1}{2\pi} \int_{-\pi+\delta}^{\pi-\delta} |\varphi(t)|\, dt\, e^{(\pi-\delta)|y|} \le \frac{1}{\sqrt{2\pi}} \|\varphi\|_{L^2[-\pi,\pi]} e^{(\pi-\delta)|y|}$$

or

(1) $$\log|\Phi(re^{i\theta})| \le (\pi - \delta) r |\sin\theta| + C_\varphi.$$

Let $n(t)$ be the counting function for zeros of the function $\Phi(z)$. We have $\Phi(n) = 0$ for $n \neq n_k$, and hence

$$n(t) \geq 2[t] + 1 - n_1^+(t) - n_1^-(t) ,$$

where $n_1^+(t)$ and $n_1^-(t)$ are the numbers of points n_k located inside intervals $[0, t)$ and $[-t, 0)$ respectively. Let $n_0 < 0 \leq n_1$. Then, for $n_k \leq t < n_{k+1}$, $k > 0$, we have

$$n_1^+(t) = k = o(n_k) = o(t) , \quad t \to \infty .$$

In the same way, $n_1^-(t) = o(t)$, $t \to \infty$. Hence

$$n(t) \overset{\text{as}}{>} (2 - \varepsilon)t$$

for all $\varepsilon > 0$. Therefore,

$$\int_{r_0}^{r} \frac{n(t)}{t}\, dt \overset{\text{as}}{>} (2 - 2\varepsilon)r . \tag{2}$$

Applying the Jensen formula (equation (8), Section 2.3), and estimates (1) and (2), we obtain

$$N(r) \overset{\text{as}}{\leq} \frac{1}{2\pi}\int_0^{2\pi} \left(\pi - \frac{\delta}{2}\right) r|\sin\theta|\, d\theta = \frac{4}{2\pi}\left(\pi - \frac{\delta}{2}\right) r = (2 - \frac{\delta}{\pi})r ,$$

and so

$$2 - 2\varepsilon < 2 - \delta/\pi .$$

This is a contradiction! The function $\Phi(z)$ has too many roots for its insufficiently fast growth. Thus, $\Phi(z) \equiv 0$, all Fourier coefficients d_n are equal to zero, and $f(t) = 0$ almost everywhere. The theorem is proved.

PROBLEM 1. Let

$$f(t) = \sum_{n=-\infty}^{\infty} c_n e^{i\lambda_n t} , \qquad \sum_{n=-\infty}^{\infty} |c_n| < \infty ,$$

where λ_n are real numbers, and let

$$\frac{\lambda_n}{n} \to +\infty , \qquad |n| \to \infty .$$

If $f(t) = 0$ on an interval, no matter how small, then $f(t) \equiv 0$.

HINT. Use the identity

$$\int_0^{\infty} f(t)e^{-tz}\, dt = \sum_{n=-\infty}^{\infty} \frac{c_n}{z - i\lambda_n} , \qquad \operatorname{Re} z > 0 ,$$

and apply the Jensen formula to the meromorphic function on the right-hand side of the latter formula.

The reader can find more sophisticated theorems related to the same field in the monographs by Levin [**82**] (Appendix 2), Levinson [**84**], Koosis [**72**].

3.2. The convergence exponent and the upper density of the sequence of zeros

DEFINITION. Given a sequence $a_1, a_2, \ldots, a_n, \ldots$, $a_n \neq 0$, $\lim_{n\to\infty} a_n = \infty$, the greatest lower bound of λ's such that the series

$$\sum_{n=1}^{\infty} \frac{1}{|a_n|^\lambda}$$

converge is called the *convergence exponent.*

Let $n(r)$ be the counting function of a sequence $\{a_n\}$. We denote by ρ_1 its order; i.e.,

$$\rho_1 = \limsup_{r\to\infty} \frac{\log n(r)}{\log r}\ .$$

The number

$$\overline{\Delta} = \limsup_{r\to\infty} \frac{n(r)}{r^{\rho_1}}$$

is called the upper density of the sequence $\{a_n\}$ with respect to the given order ρ_1. It is evident that

$$r^{\rho_1-\varepsilon} \overset{\text{n}}{<} n(r) \overset{\text{as}}{<} r^{\rho_1+\varepsilon}$$

and

$$(\overline{\Delta}-\varepsilon)r^{\rho_1} \overset{\text{n}}{<} n(r) \overset{\text{as}}{<} (\overline{\Delta}+\varepsilon)r^{\rho_1}$$

for every $\varepsilon > 0$. The number

$$\underline{\Delta} = \liminf_{r\to\infty} \frac{n(r)}{r^{\rho_1}}$$

is called the lower density of the sequence $\{a_n\}$ with respect to the given order ρ_1.

PROBLEM 2. Prove the identities

$$\overline{\Delta} = \limsup_{n\to\infty} \frac{n}{|a_n|^\rho},$$
$$\underline{\Delta} = \liminf_{n\to\infty} \frac{n}{|a_n|^\rho}\ .$$

LEMMA 1. *Let a series*

$$\sum_{n=1}^{\infty} \frac{1}{|a_n|^\lambda}$$

be convergent for some $\lambda > 0$. Then the integral

$$\int_0^\infty \frac{n(t)}{t^{\lambda+1}}\, dt$$

converges, and

$$\lim_{t\to+\infty} \frac{n(t)}{t^\lambda} = 0\ .$$

PROOF. Since

$$\sum_{n=1}^{\infty} \frac{1}{|a_n|^\lambda} = \int_0^\infty \frac{dn(t)}{t^\lambda}, \tag{3}$$

upon integrating by parts we find

$$\int_0^r \frac{dn(t)}{t^\lambda} = \frac{n(r)}{r^\lambda} + \lambda \int_0^r \frac{n(t)}{t^{\lambda+1}}\, dt. \tag{4}$$

The convergence of the series in (3) implies that both summands on the right-hand side of (4) are bounded from above. The second summand does not decrease, and therefore tends to the finite limit, which together with the inequality

$$\frac{n(r)}{r^\lambda} \le \lambda \int_r^\infty \frac{n(t)}{t^{\lambda+1}}\, dt$$

proves Lemma 1.

LEMMA 2. *The convergence exponent of the sequence* $\{a_n\}$ *is equal to the order* ρ_1 *of its counting function.*

PROOF. Let K be the convergence exponent, and let $\lambda > K$. Then the series in (3) converges, and by Lemma 1 we have $\lim_{r\to\infty} \frac{n(r)}{r^\lambda} = 0$. Hence $\rho_1 \le \lambda$ and $\rho_1 \le K$. On the other hand,

$$n(t) \overset{\text{as}}{<} t^{\rho_1+\varepsilon/2}, \quad \varepsilon > 0.$$

Hence for $\lambda = \rho_1 + \varepsilon$ the integral

$$\int_0^\infty \frac{n(t)}{t^{\lambda+1}}\, dt$$

converges and $n(t)/t^\lambda \to 0$, $t \to \infty$. It follows from (3) and (4) that the series in (3) converges, and therefore $K \le \rho_1$, proving Lemma 2.

PROBLEM 3. Prove that the convergence of the series in (3) is equivalent to the convergence of the integral on the right-hand side of (4).

PROBLEM 4. Prove that if the terms of a converging series form a decreasing sequence $a_1 \ge a_2 \ge a_3 \ge \dots$, then $na_n \to 0$ (E. Borel).

THEOREM 2 (Hadamard). *The convergence exponent of zeros of an entire function does not exceed its growth order.*

PROOF. According to a corollary to the Jensen formula, Section 2.5, we have

$$n(r) \le \log M_f(er) + O(1).$$

It follows that

$$\limsup_{r\to\infty} \frac{\log n(r)}{\log r} \le \limsup_{r\to\infty} \frac{\log\log M_f(er)}{\log r} = \limsup_{r\to\infty} \frac{\log\log M_f(r)}{\log r},$$

i.e., $\rho_1 \le \rho$. The theorem is proved.

We shall give another corollary of the Jensen formula, namely, a uniqueness theorem that does not permit an entire function to vanish on a "dense set".

THEOREM 3. *Let $f(z)$ be an entire function of type not greater than σ with respect to an order ρ. If $f(z)$ vanishes on a set Λ and at least one of the inequalities*

$$\overline{\Delta}(\Lambda) > e\rho\sigma \,, \tag{5}$$

$$\underline{\Delta}(\Lambda) > \rho\sigma \tag{6}$$

holds, where $\overline{\Delta}(\Lambda)$ and $\underline{\Delta}(\Lambda)$ are the upper and lower density of the sequence Λ with respect to the order ρ, then $f(z) \equiv 0$.

PROOF. Let us assume, for example, that (5) holds. We denote by $n_\Lambda(r)$ the counting function of the sequence Λ and set $n(r) = n_f(r)$.

For every $\lambda > 1$, we have

$$n_\Lambda(r) \le n(r) \le \frac{1}{\log\lambda}\int_r^{\lambda r}\frac{n(t)}{t}\,dt \le \frac{1}{\log\lambda}N(\lambda r)\,.$$

By the Jensen formula, with $f \not\equiv 0$,

$$N(\lambda r) \le \log M_f(\lambda r) + O(1) \overset{\text{as}}{<} (\sigma+\varepsilon)\lambda^\rho r^\rho\,.$$

Hence

$$\overline{\Delta}(\Lambda) \le \frac{\sigma\lambda^\rho}{\log\lambda}.$$

Minimizing with respect to λ, we obtain $\overline{\Delta}(\Lambda) \le e\rho\sigma$, a contradiction.

If $\underline{\Delta}(\Lambda) > \rho\sigma$, then for some $\varepsilon > 0$ we have $n(r) \ge n_\Lambda(r) \overset{\text{as}}{>} (\rho\sigma + 2\varepsilon)r^\rho$. Hence $N(r) > \dfrac{1}{\rho}(\rho\sigma+\varepsilon)r^\rho$, and using the Jensen formula we obtain

$$\log M_f(r) \overset{\text{as}}{>} \frac{1}{\rho}(\rho\sigma+\varepsilon)r^\rho\,, \quad \sigma > \frac{1}{\rho}(\rho\sigma+\varepsilon)\,,$$

a contradiction again.

3.3. Completeness of a system of exponential functions

DEFINITION. A system $\{x_k\}$ of elements of a linear topological space E is said to be complete if the closure of its linear hull coincides with E. In other words, each element $x \in E$ may be approximated by finite linear combinations of elements of a complete system $\{x_k\}$.

If a system is not complete, then the closure of its linear hull is a proper subspace $L \subset E$. If E is a locally convex space, then by the Hahn-Banach theorem there exists a nonzero linear functional $f \in E^*$ such that $f(x) = 0$ for every element $x \in \{x_k\}$. The existence of such a functional is a necessary and sufficient condition of the noncompleteness.

Let a system $\{e^{i\lambda_k t}\}$ be given with real exponents λ_k.

THEOREM 4. *Let $n(t)$ be the counting function of a sequence $\{\lambda_k\} = \Lambda$. If*

$$\liminf_{t\to\infty}\frac{n(t)}{t} > 2\,, \tag{7}$$

then the system $\{e^{i\lambda_k t}\}$ is complete in the space of continuous functions $C[-\pi,\pi]$.

PROOF. If the completeness fails, then by the F. Riesz theorem on the form of a linear functional on the space of continuous functions there exists a nonconstant function $\sigma(t)$ of bounded variation such that

$$\int_{-\pi}^{\pi} e^{i\lambda_k t}\, d\sigma(t) = 0\,, \quad \lambda_k \in \Lambda\,.$$

The function

$$\Phi(\lambda) = \int_{-\pi}^{\pi} e^{i\lambda t}\, d\sigma(t)$$

is entire, not identically equal to zero and satisfies the inequality

$$|\Phi(s+i\tau)| \le (\text{Var}\ \sigma) e^{\pi|\tau|}\,.$$

Since $\Phi(\lambda_k) = 0$, we obtain, by the Jensen formula, Section 2.3,

$$\int_{r_0}^{r} \frac{n(t)}{t}\, dt \le \frac{\pi r}{2\pi} \int_0^{2\pi} |\sin\theta|\, d\theta + O(1) = 2r + O(1)\,, \quad r \to \infty\,.$$

Since for some $\varepsilon > 0$ we have

$$\frac{n(t)}{t} \overset{\text{as}}{>} 2 + \varepsilon\,,$$

we conclude that

$$(2+\varepsilon)r + O(1) \le 2r + O(1)\,, \quad r \to \infty\,.$$

This is a contradiction. The theorem is proved.

PROBLEM 5. Let λ_0 be an arbitrary real number and let $\lambda_n = n - \delta_n$, $\lambda_{-n} = -n + \delta_n$ be pairwise distinct real numbers with $|n| > \delta_n \ge \delta > 0$, $n = \pm 1, \pm 2, \ldots$. Prove that the system of exponential functions $\{e^{i\lambda_n t}\}$ is complete in the space $C[-\pi,\pi]$.

Some more sofisticated theorems on the completeness of a family $\{e^{i\lambda_n t}\}$ will be proved in the second part of the monograph. In particular, it will be proved that the assertion of Theorem 4 remains in force if the *lower* limit in (7) is changed to the *upper* limit.

3.4. Completeness of a special system of functions in countably normed spaces

Let us consider the space $A(D)$ of all analytic functions in a simply connected domain $D \subset \mathbb{C}$. Let us choose an expanding sequence $G_1, G_2, \ldots$ of compact sets which exhaust D from the inside and are such that every G_m is compactly imbedded in D:

$$G_1 \Subset G_2 \Subset \cdots \Subset G_m \Subset \cdots\,; \quad G_1 \cup G_2 \cup \cdots \cup G_m \cup \cdots = D\,,$$

and let us introduce the system of norms

$$\|f\|_m = \sup_{z \in G_m} |f(z)|\,. \tag{8}$$

The space $A(D)$ endowed with the system of norms (8) is countably normed. The following proposition describes the general form of a linear functional on the space $A(D)$.

THEOREM 5. *For every linear functional $F \in A^*(D)$ there exists a unique function $\varphi(\zeta)$ analytic on a closed simply connected set $\mathbb{C}\backslash D'$, $D' \Subset D$, equal to zero at infinity and such that the value of F at a function $f \in A(D)$ is determined by the identity*

$$F[f] = \frac{1}{2\pi i}\int_l f(\zeta)\varphi(\zeta)\,d\zeta\,. \tag{9}$$

Here l is a simple closed curve lying inside D such that φ is analytic on l and outside l.[3]

PROOF. It follows from the definition of topology by norms (8) that a homogeneous and additive functional $F[f]$ is continuous if and only if there exist a number $m \geq 1$ and a constant C such that the inequality

$$|F[f]| \leq C\|f\|_m \tag{10}$$

holds. Let $F[f]$ be some linear functional on the space $A(D)$, and let $G_m \Subset D$ be a domain corresponding to the norm in (10). Let us choose an intermediate domain D', $G_m \Subset D' \Subset D$. By (10), the functional F can be extended to a linear functional on the space $C(G_m)$ of functions continuous on G_m with the norm (8).

Now let $\zeta \in \mathbb{C}\backslash D'$. Noting that $\dfrac{1}{\zeta - z} \in C(G_m)$, we define the function

$$\varphi(\zeta) = F\Big[\frac{1}{\zeta - z}\Big], \tag{11}$$

where the functional F is applied with respect to the variable z. It is natural to call the function φ the Cauchy-Stieltjes transform of F. By (10), the function $\varphi(\zeta)$ extends to an analytic function on the closed set $\mathbb{C}\backslash D'$. If $\zeta \to \infty$, then $\left\|\dfrac{1}{\zeta - z}\right\|_m \to 0$, and it follows from (10) that $\varphi(\infty) = 0$. Finally, let a simple curve l encircling G_m in the domain D' be chosen close enough to the boundary $\partial D'$ for the function φ to be analytic on l and in the component of the set $\mathbb{C}\backslash l$ containing the infinite point. Then it follows from (11) that

$$\begin{aligned}\frac{1}{2\pi i}\int_l \varphi(\zeta)f(\zeta)\,d\zeta &= \frac{1}{2\pi i}\int_l F\Big[\frac{1}{\zeta - z}\Big]f(\zeta)\,d\zeta \\ &= F\Big[\frac{1}{2\pi i}\int_l \frac{f(\zeta)}{\zeta - z}\,d\zeta\Big] = F[f]\end{aligned}$$

for every function $f \in A(D)$, which proves (9).

Let φ_1 and φ_2 be two functions determining the functional $F[f]$ according to (9). We may assume that the same curve l corresponds to each of them. If $\psi = \varphi_1 - \varphi_2$, then

$$\int_l \psi(\zeta)\zeta^k\,d\zeta = 0\,, \quad k = 0, 1, 2, \dots\,.$$

Since the function ψ is analytic in the exterior of l and $\psi(\infty) = 0$, it follows that $\psi \equiv 0$. Thus, every linear functional is representable by equation (9). The converse statement is obvious. The theorem is proved.

---

[3]In what follows we shall say that the bounded domain whose boundary is l contains all singularities of φ.

Let us apply Theorem 5 to study the completeness of the system of functions $\varphi_n(z) = F(\lambda_n z)$, where $F(z)$ is an entire function and $\{\lambda_n\}$ is a sequence of complex numbers. First results on completeness of a system of functions $\{F(\lambda_n z)\}$ were obtained by A. O. Gelfond in 1937. A. I. Markushevich obtained more complete results using a method close to that described in this section.

THEOREM 6. *Let an entire function of order* ρ

$$F(z) = \sum_{n=0}^{\infty} a_n z^n$$

with all coefficients a_n *nonzero, be of type not exceeding* σ, *and let* $\Lambda = \{\lambda_n\}$ *be a sequence of complex numbers. Then the system* $\varphi_n(z) = F(\lambda_n z)$ *is complete in the disk* $\{z : |z| < R\}$, *where*

$$R^\rho = \frac{1}{\rho\sigma} \max\left(\frac{\overline{\Delta}(\Lambda)}{e}, \underline{\Delta}(\Lambda)\right), \tag{12}$$

and $\overline{\Delta}(\Lambda)$ *and* $\underline{\Delta}(\Lambda)$ *are the upper and lower densities of the sequence* Λ *with respect to the order* ρ.

PROOF. Suppose that the system $\{F(\lambda_n z)\}$ is not complete in the space $A(D)$, $D = \{z : |z| < R\}$. Then by Theorem 5 there exists a function $\psi(z)$ that is analytic outside a disk $\{z : |z| < r\}$, $r < R$, vanishes at infinity, is not equal to zero identically and satisfies

$$\int_{|z|=r} F(\lambda_n z)\psi(z)\, dz = 0\,, \quad n = 0, 1, 2, \dots\,, \quad r < R\,.$$

Let us consider the function

$$\Phi(\lambda) = \int_{|z|=r} F(\lambda z)\psi(z)\, dz\,.$$

This is an entire function vanishing at the points of Λ. Let us estimate its growth. Using

$$|F(\lambda z)| \overset{\text{as}}{\leq} \exp\{(\sigma + \varepsilon)|\lambda|^\rho r^\rho\}\,, \quad \varepsilon > 0\,, \quad r = |z|\,,$$

we obtain

$$\begin{aligned} |\Phi(\lambda)| &\leq \int_{|z|=r} |F(\lambda z)||\psi(z)||dz| \\ &\overset{\text{as}}{\leq} 2\pi r M \exp\{(\sigma + \varepsilon)|\lambda|^\rho r^\rho\}\,, \quad M = \max_{|z|=r} |\psi(z)|\,, \quad \varepsilon > 0\,. \end{aligned}$$

Hence the type of the function Φ does not exceed σr^ρ. If the sequence Λ satisfies $\lim_{n\to\infty} |\lambda_n| = \infty$, then by the inequality $r < R$ and equation (12) it follows from Theorem 3 that $\Phi \equiv 0$. The same statement is a trivial corollary of the uniqueness theorem if Λ has a finite condensation point. Indeed, if

$$\psi(z) = \sum_{n=0}^{\infty} \frac{b_n}{z^{n+1}}\,,$$

then

$$\Phi(\lambda) = \int_{|z|=r} F(\lambda z)\psi(z)\, dz = \sum_{n=0}^{\infty} a_n b_n \lambda^n \equiv 0\,.$$

Since all numbers a_n are different from zero, all numbers b_n must be equal to zero implying $\psi(z) \equiv 0$. This contradiction proves Theorem 6.

REMARK. If $\overline{\Delta}(\Lambda) = \infty$, then the system $\{F(\lambda_n z)\}$ is complete in $A(\mathbb{C})$.

LECTURE 4

Factorization of Entire Functions of Finite Order

4.1. The Weierstrass canonical product

In our study of infinite products we shall assume that an infinite product of entire functions

$$\prod_{n=1}^{\infty} g_n(z)$$

converges at a point z_0 if for some N there exists the limit

$$\lim_{M\to\infty} \prod_{n=N}^{M} g_n(z_0) ,$$

not equal to either zero or infinity. The same infinite product is said to converge uniformly on a set K if for some N the products $\prod_{n=N}^{M} g_n(z)$ tend to a function $h_N(z)$ uniformly with respect to $z \in K$ as $M \to \infty$.

It follows directly from the definition of convergence of infinite products that a general term of a convergent product tends to unity. Hence it is possible after omitting a finite number of factors to define $\log g_n(z)$ using the principal value of logarithm. It follows that (uniform) convergence is equivalent to (uniform) convergence of the series

$$\sum_{n=N}^{\infty} \log g_n(z) .$$

A product is said to be absolutely convergent if the latter series converges absolutely for some N.

Let $\{a_n\}$ be a sequence of complex numbers not equal to zero and such that for some nonnegative integer p the series

$$\sum_{n} |a_n|^{-p-1} < \infty .$$

Let us introduce the infinite product

$$\Pi(z) = \prod_{n} G(z/a_n, p) ,$$

where

$$G(u,p) = \begin{cases} 1-u, & p=0, \\ (1-u)\exp\left[u + \dfrac{u^2}{2} + \cdots + \dfrac{u^p}{p}\right], & p>0 . \end{cases}$$

The functions $G(u,p)$ are called the *Weierstrass primary factors*.

The inequality

$$|\log G(u,p)| \le \sum_{k=p+1}^{\infty} \frac{|u|^k}{k} \le 2|u|^{p+1},$$

evident for $|u| \le 1/2$, implies that the infinite product $\Pi(z)$ converges absolutely and uniformly in every disk $\{z : |z| \le R < \infty\}$. This product is called the *Weierstrass canonical product of genus p*.

4.2. The Hadamard theorem

One of the main theorems of the theory of entire functions is

THEOREM 1 (Hadamard). *An entire function f of finite order ρ may be represented in the form*

$$f(z) = z^m e^{P_q(z)} \prod_{n=1}^{\infty} G\Big(\frac{z}{a_n}, p\Big), \tag{1}$$

where $a_1, a_2, \dots$ are all nonzero roots of the function $f(z)$, $p \le \rho$, $P_q(z)$ is a polynomial in z of degree $q \le \rho$, and m is the multiplicity of the root at the origin.

PROOF. Let us use formula (6) from Lecture 2:

$$\log f(z) = \frac{1}{2\pi}\int_0^{2\pi} \log|f(Re^{i\psi})| \frac{Re^{i\psi}+z}{Re^{i\psi}-z}\,d\psi + \sum_{|a_n|<R} \log\frac{R(z-a_n)}{R^2 - z\bar{a}_n} + iC\,.$$

Without loss of generality we assume here that $m = 0$, i.e., $f(0) \ne 0$. Differentiating this formula $p+1$ times with $p = [\rho]$, we obtain

$$\begin{aligned}[\log f(z)]^{(p+1)} = {} & \frac{(p+1)!}{2\pi}\int_0^{2\pi} \log|f(Re^{i\psi}| \frac{2Re^{i\psi}}{(Re^{i\psi}-z)^{p+2}}\,d\psi \\ & + \sum_{|a_n|<R} \frac{p!\bar{a}_n^{p+1}}{(R^2 - \bar{a}_n z)^{p+1}} - \sum_{|a_n|<R} \frac{p!}{(a_n - z)^{p+1}}\,.\end{aligned}$$

It follows that

$$\begin{aligned}&\Big|[\log f(z)]^{p+1} + \sum_{|a_n|<R} \frac{p!}{(a_n-z)^{p+1}}\Big| \\ &\qquad \le \frac{(p+1)!}{2\pi} \log M_f(R) \frac{4\pi R}{(R-r)^{p+2}} + \frac{p!n(R)}{(R-r)^{p+1}}, \quad r = |z|\,.\end{aligned}$$

The estimates

$$\begin{aligned}\log M_f(R) &\overset{\text{as}}{<} R^{\rho+\varepsilon}\,,\\ n(R) \le \log M_f(eR) &\overset{\text{as}}{<} R^{\rho+\varepsilon}\end{aligned}$$

yield, after passing to the limit as $R \to \infty$,

$$[\log f(z)]^{(p+1)} = -p! \sum_{n=1}^{\infty} \frac{1}{(a_n - z)^{p+1}}\,.$$

Integrating both sides of this indentity along any path joining the points 0 and z and not intersecting cuts from the points $a_1, a_2, \ldots$ to infinity, we obtain

$$\log f(z) - P_q(z) = \sum_{n=1}^{\infty} \left[\log\left(1 - \frac{z}{a_n}\right) + \frac{z}{a_n} + \cdots + \frac{z^p}{pa_n^p}\right], \quad q \leq \rho .$$

Now formula (1) follows, which proves Theorem 1.

Let us remark that the Hadamard theorem was proved with $p = [\rho]$, $q \leq [\rho]$. Representation (1) is possible, generally speaking, with different integers p and q. In what follows, p will denote the smallest integer for which the series $\sum_1^\infty |a_n|^{-p-1}$ converges. With this convention the number q is determined uniquely.

The integer $g = \max(p, q)$ is called the genus of an entire function f. It follows from the Hadamard theorem that the genus of an entire function does not exceed its order.

An entire function of order zero has the form

$$f(z) = Cz^m \prod_{n=1}^{\omega} \left(1 - \frac{z}{a_n}\right), \quad \omega \leq \infty$$

with

$$\sum_{n=1}^{\omega} \frac{1}{|a_n|} < \infty .$$

For $\rho < 1$, by the Hadamard theorem a function $f(z)$ of order ρ has exactly the same form.

An entire function of genus one has the form

$$f(z) = Cz^m e^{az+b} \prod_{n=1}^{\omega} \left(1 - \frac{z}{a_n}\right) e^{z/a_n} .$$

EXAMPLE. The function $f(z) = \dfrac{\sin \pi\sqrt{z}}{\pi\sqrt{z}}$ is entire and of order $\rho = 1/2$. Its zeros are $a_n = n^2$, $n = 1, 2, \ldots$. According to the Hadamard theorem,

$$f(z) = C \prod_{n=1}^{\infty} \left(1 - \frac{z}{n^2}\right),$$

and since $f(0) = 1$, we have $C = 1$. Substituting z^2 instead of z, we obtain

$$\sin \pi z = \pi z \prod_{n=1}^{\infty} \left(1 - \frac{z^2}{n^2}\right)$$

or

$$\sin \pi z = \pi z \prod_{n=-\infty}^{\infty}{}' \left(1 - \frac{z}{n}\right) e^{z/n}.$$

As usual, the prime here means that the factor corresponding to $n = 0$ is omitted.

PROBLEM 1. Let $-\infty < a < b < \infty$. Show that the Fourier transform $F(z)$ of a function $f \in L^2_{[a,b]}$ has an infinite set of zeros (nonreal, generally speaking). The same is true if

$$F(z) = \int_a^b e^{itz} d\sigma(t) , \quad -\infty < a < b < \infty ,$$

where $\sigma(t)$ is a function of bounded variation which is not a step function with a single jump.

PROBLEM 2 (Laguerre). Let $f(z) = e^{-\alpha z^2} g(z)$, where $\alpha \geq 0$ and $g(z)$ is a real entire function of genus $p \leq 1$ with real zeros. Prove that the zeros of the derivative $f'(z)$ are also real and interlace with the zeros of $f(z)$.

HINT. Use the Hadamard theorem and study $f'(z)/f(z)$.

PROBLEM 3. Let $f(z)$ and $g(z)$ be entire functions of order $\rho < 2$ such that

$$f^2(z) + g^2(z) \equiv 1 .$$

Prove that $f(z) = \cos(\alpha z + \beta)$, $g(z) = \sin(\alpha z + \beta)$, $\alpha, \beta \in \mathbb{C}$.

4.3. Estimates for canonical products

Let $\{a_n\}$ be a sequence of complex numbers, $\lim_{n\to\infty} a_n = \infty$, and let $n(r)$ be its counting function. Suppose that for some integer $p \geq 0$ the series $\sum_1^\infty |a_n|^{-p-1}$ converges, and denote

$$\Pi(z) = \prod_{n=1}^{\infty} G\Big(\frac{z}{a_n}, p\Big) .$$

As we proved earlier using the Jensen formula, Section 2.3, the estimate $n(r) \leq \log M_\Pi(er)$ is valid. Now we shall estimate $\log M_\Pi(r)$ from above using $n(r)$. To this end we shall obtain an estimate for the Weierstrass primary factor $G(u, p)$.

LEMMA 1 (the Borel estimate). *For $u \in \mathbb{C}$ the estimates*

$$\log |G(u,p)| \leq A_p \frac{|u|^{p+1}}{1+|u|} , \quad p > 0 , \quad A_p = 3e(2 + \log p) ,$$
$$\log |G(u,0)| \leq \log(1 + |u|)$$

are valid.

PROOF. The latter inequality is evident. Let $p > 0$. If $|u| < p/(p+1)$, then expanding $\log(1-u)$ we obtain

$$\log |G(u,p)| \leq \sum_{n=p+1}^{\infty} \frac{|u|^n}{n} \leq \frac{|u|^{p+1}}{(p+1)(1-|u|)} \leq |u|^{p+1} .$$

If, on the other hand, $|u| > p/(p+1)$, then the inequality $\log(1+|u|) < |u|$ yields

$$\begin{aligned}\log|G(u,p)| &\le 2|u| + \frac{|u|^2}{2} + \cdots + \frac{|u|^p}{p}\\ &= |u|^p\Big(\frac{1}{p} + \frac{1}{p-1}\frac{1}{|u|} + \cdots + \frac{1}{2}\frac{1}{|u|^{p-2}} + 2\frac{1}{|u|^{p-1}}\Big)\\ &\le |u|^p\Big(\frac{p+1}{p}\Big)^{p-1}\Big(2 + \frac{1}{2} + \cdots + \frac{1}{p}\Big) < e(2+\log p)|u|^p\frac{1+|u|}{1+|u|}\\ &= e(2+\log p)\Big(1+\frac{1}{|u|}\Big)\frac{|u|^{p+1}}{1+|u|} < A_p\frac{|u|^{p+1}}{1+|u|}\,,\end{aligned}$$

proving Lemma 1.

THEOREM 2. *Let $\{a_n\}$ be a sequence of complex numbers. If the series*

$$\sum_{n=1}^{\infty}\frac{1}{|a_n|^{p+1}} \tag{2}$$

converges, then the product

$$\Pi(z) = \prod_{n=1}^{\infty} G\Big(\frac{z}{a_n}, p\Big)$$

converges uniformly on every compact set and satisfies the estimate

$$\log|\Pi(z)| \le K_p r^p\Big\{\int_0^r \frac{n(t)}{t^{p+1}}\,dt + r\int_r^\infty \frac{n(t)}{t^{p+2}}\,dt\Big\}\,,$$

where $K_p = (p+1)A_p$, $r = |z|$.

PROOF. First, let $\rho \ge 1$. It follows from the Borel estimate that

$$\begin{aligned}\log|\Pi(z)| &\le A_p\sum_{n=1}^{\infty}\frac{r^{p+1}}{|a_n|^p(r+|a_n|)} = A_p r^{p+1}\int_0^\infty \frac{dn(t)}{t^p(t+r)}\\ &= A_p r^{p+1}\frac{n(t)}{t^p(t+r)}\Big|_0^\infty + A_p r^{p+1}\int_0^\infty\Big[\frac{p}{t^{p+1}(t+r)} + \frac{1}{t^p(t+r)^2}\Big]n(t)\,dt\,.\end{aligned} \tag{3}$$

Since the series (2) converges, by Lemma 1 from the preceding lecture we have

$$\frac{n(t)}{t^{p+1}} \to 0\,, \quad t\to\infty\,,$$

and

$$\begin{aligned}\log|\Pi(z)| &\le A_p r^{p+1}\Big\{\int_0^r + \int_r^\infty\Big\}\Big[\frac{p}{t^{p+1}(t+r)} + \frac{1}{t^p(t+r)^2}\Big]n(t)\,dt\\ &\le K_p r^p\Big\{\int_0^r\frac{n(t)}{t^{p+1}}\,dt + r\int_r^\infty\frac{n(t)}{t^{p+2}}\,dt\Big\}\,, \quad r = |z|\,.\end{aligned}$$

For $p = 0$ the estimate of the canonical product is simplified:

$$\log|\Pi(z)| \le \sum_{n=1}^{\infty} \log\left(1 + \frac{r}{|a_n|}\right) = \int_0^{\infty} \log\left(1 + \frac{r}{t}\right) dn(t)$$
$$= r\int_0^{\infty} \frac{n(t)}{t(t+r)}\, dt \le \int_0^{r} \frac{n(t)}{t}\, dt + r\int_r^{\infty} \frac{n(t)}{t^2}\, dt\,.$$

The theorem is proved.

THEOREM 3 (Borel). *The growth order ρ of a canonical product is equal to the convergence exponent of the sequence of its zeros.*

PROOF. Let p be the smallest integer such that the series

$$\sum_{n=1}^{\infty} \frac{1}{|a_n|^{p+1}}$$

converges, where $\{a_n\}$ is the sequence of zeros of the canonical product $\Pi(z)$, and let ρ_1 be the convergence exponent of the sequence $\{a_n\}$. Then $p \le \rho_1 \le p+1$.

First, let $\rho_1 < p+1$. We choose $\varepsilon > 0$ such that $\rho_1 + \varepsilon < p+1$. Then

$$n(t) \overset{\text{as}}{<} t^{\rho_1+\varepsilon}\,.$$

It follows from the preceding theorem that

$$\begin{aligned}\log M_\Pi(r) &\le K_p r^p\Big\{O(1) + \int_0^r t^{\rho_1+\varepsilon-p-1}\, dt + r\int_r^{\infty} t^{\rho_1+\varepsilon-p-2}\, dt\Big\} \\ &\le K_p r^p\Big\{O(1) + \frac{r^{\rho_1+\varepsilon-p}}{\rho_1+\varepsilon-p} + \frac{r^{\rho_1+\varepsilon-p}}{p+1-\rho_1-\varepsilon}\Big\} \overset{\text{as}}{<} r^{\rho_1+2\varepsilon}.\end{aligned} \tag{4}$$

Now consider the case $\rho_1 = p+1$. We proved in Lemma 1, Section 3.2, that

$$\frac{n(r)}{r^{p+1}}\,, \quad \int_r^{\infty} \frac{n(t)}{t^{p+2}}\, dt$$

tend to zero as $r \to \infty$. Hence it follows from Theorem 2 that

$$\log M_\Pi(r) \overset{\text{as}}{<} \varepsilon r^{p+1} = \varepsilon r^{\rho_1} \tag{5}$$

for any $\varepsilon > 0$.[4]

Thus, in both cases $\rho \le \rho_1$. Comparing this inequality with the corollary to the Jensen formula derived in Section 2.5, we obtain $\rho = \rho_1$. Theorem 3 is proved.

PROBLEM 4. Find a necessary and sufficient conditions for a sequence of complex numbers $\{\alpha_k\}$ to be such that the infinite product

$$\prod_k \frac{\sin \alpha_k z}{\alpha_k z}$$

converge to an entire function. Under what conditions imposed on $\{\alpha_k\}$ this is a function of exponential type?

[4]This is the Poincaré theorem.

LECTURE 5

The Connection between the Growth of Entire Functions and the Distribution of their Zeros

5.1. Functions of noninteger order

THEOREM 1. *The convergence exponent of the zero set of an entire function f of noninteger order is equal to the order of growth of f.*

PROOF. Let f be an entire function of noninteger order ρ, let ρ_1 be the convergence exponent of its zeros , and let $\Pi(z)$ be the canonical product corresponding to the set of zeros of f. According to the Hadamard representation (Theorem 1, Section 4.2), we have

$$f(z) = z^m e^{P_q(z)}\Pi(z), \quad \deg P_q = q \,. \tag{1}$$

Using the Borel theorem (Theorem 3, Section 4.3), we obtain

$$\log M_f(r) \overset{\text{as}}{<} c_1 r^q + r^{\rho_1+\varepsilon}\,, \quad \varepsilon > 0.$$

Hence

$$\log M_f(r) \overset{\text{as}}{<} r^{\lambda+2\varepsilon}\,, \quad \lambda = \max(\rho_1, q)\,,$$

and $\rho \le \lambda$. The opposite inequality is true, since by virtue of Theorem 2, Section 3.2, we have $\rho_1 \le \rho$, and by virtue of the Hadamard theorem $q \le \rho$. The theorem is proved.

THEOREM 2. *If the order ρ of an entire function $f(z)$ is not an integer, then its type σ_f and the upper density of zeros $\overline{\Delta}_f$ simultaneously are equal either to zero, or to infinity, or to positive numbers.*

PROOF. According to Theorem 3, Section 3.2, we have $\overline{\Delta}_f \le e\rho\sigma_f$. To estimate σ_f from above via $\overline{\Delta}_f$ we shall use the bound of a canonical product of genus p proved in Theorem 2, Section 4.3. The inequality

$$n(t) \overset{\text{as}}{<} (\overline{\Delta}_f + \varepsilon)t^\rho\,, \quad \varepsilon > 0\,,$$

yields

$$\log M_\Pi(r) \le K_p r^p \Big\{O(1) + (\overline{\Delta}_f + \varepsilon)\int_0^r t^{\rho-p-1}\,dt + (\overline{\Delta}_f + \varepsilon) r \int_r^\infty t^{\rho-p-2}\,dt\Big\}\,.$$

Since $p < \rho < p+1$, we have

$$\log M_\Pi(r) \overset{\text{as}}{<} C_\rho(\overline{\Delta}_f + \varepsilon) r^\rho\,,$$

and by Hadamard's representation

$$\log M_f(r) \overset{\text{as}}{<} a_0 r^q + C_\rho(\overline{\Delta}_f + \varepsilon)r^\rho \overset{\text{as}}{<} C_\rho(\overline{\Delta}_f + r\varepsilon)r^\rho$$

or

$$\sigma_f \leq C_\rho \overline{\Delta}_f, \tag{2}$$

which proves Theorem 2.

5.2. Functions of integer order

An entire function of integer order ρ may not have zeros at all. It is possible that $\rho = q$, q being the degree of the polynomial in Hadamard's representation, and the order of the canonical product ρ_1 is less than ρ. But another feature of entire functions of integer order is more essential. It turns out that, for an integer order, inequality (2) may fail even for a canonical product. The upper density of the zero set may be finite while the canonical product may be of maximal type.

Consider, for example, the entire functions

$$\sin\frac{\pi}{2}z = \frac{\pi}{2}z\prod_{n=1}^{\infty}\Big(1-\frac{z^2}{4n^2}\Big) = \frac{\pi}{2}z\prod_{n=-\infty}^{\infty}{}'\Big(1-\frac{z}{2n}\Big)e^{z/2n}$$

and

$$\frac{1}{\Gamma(z)} = ze^{\gamma z}\prod_{n=1}^{\infty}\Big(1+\frac{z}{n}\Big)e^{-z/n},$$

where γ is the Euler constant. It is evident that, for both functions, $n(t) \sim t$ and the convergence exponent of zeros is $\rho_1 = 1$. Since the functions differ inessentially from the canonical products, they are of order one. For the former function we have

$$\frac{1}{2}(e^{\frac{\pi}{2}|y|} - e^{-\frac{\pi}{2}|y|}) \leq \Big|\sin\frac{\pi}{2}z\Big| < e^{\frac{\pi}{2}|y|},$$
$$\frac{1}{2}(e^{\frac{\pi}{2}r} - e^{-\frac{\pi}{2}r}) \leq M(r) < e^{\frac{\pi}{2}r}.$$

Hence, $\log M(r) \sim (\pi/2)r$ and $\sigma = \pi/2$. For the latter function, according to the Stirling formula,

$$\log\frac{1}{\Gamma(z)} = -\Big(z-\frac{1}{2}\Big)\log z + z - \frac{1}{2}\log 2\pi + O\Big(\frac{1}{|z|}\Big) = -z(1+o(1))\log z,$$

where the plane is assumed to be cut along the negative real axis, and $|\arg z| < \pi$. Hence

$$\log\frac{1}{|\Gamma(z)|} = -(1+o(1))(\cos\varphi)r\log r, \quad z = re^{i\varphi}, \quad \frac{\pi}{2} < |\varphi| < \pi,$$

and therefore $\log M(r) \geq Cr\log r$. This means that $f = 1/\Gamma$ is of maximal type.

We shall see that the "root of all evil" is the presence of a symmetry in the distribution of zeros of the first function and its absence for the second function.

We remind the reader that p denotes the smallest integer for which the series

$$\sum\frac{1}{|a_n|^{p+1}}$$

converges. By virtue of the Hadamard and Borel theorems from the previous lecture we have $p \le \rho \le p+1$. Two cases are possible if ρ is an integer: either $\rho = p+1$ and the series $\sum |a_n|^{-\rho}$ converges, or $\rho = p$ and the same series diverges.

In what follows we denote by a_ρ the coefficient of z^ρ of the polynomial $P(z)$ in Hadamard's representation.

THEOREM 3 (Lindelöf). *If $\rho = p+1$, then $f(z)$ is an entire function of minimal type for $a_\rho = 0$ and of mean type for $a_\rho \neq 0$.*[5]

PROOF. According to the Hadamard theorem, we have

$$\log|f(z)| \le \operatorname{Re}(a_\rho z^\rho) + \log|\Pi(z)| + O(|z|^{\rho-1}), \quad z \to \infty. \tag{3}$$

It follows from inequality (5), Section 4.3, that if $\rho = p+1$, then

$$\log M_\Pi(r) \overset{\text{as}}{<} \varepsilon r^\rho, \quad \varepsilon > 0, \tag{4}$$

and inequality (3) yields

$$\log M_f(r) \overset{\text{as}}{<} (|a_\rho| + 3\varepsilon) r^\rho. \tag{5}$$

To estimate $\log M_f(r)$ from below we start from the evident relation

$$m(r, \exp(a_\rho z^\rho + \cdots + a_0)) = m\Big(r, \frac{f}{\Pi}\Big), \tag{6}$$

where m is the Nevanlinna proximity function (see Section 2.4). It follows that

$$\frac{|a_\rho|}{2\pi} r^\rho \overset{\text{as}}{<} m(r, \exp(a_\rho z^\rho + \cdots + a_0)) < m(r, f) + m\Big(r, \frac{1}{\Pi}\Big) + \log 2. \tag{7}$$

By virtue of Jensen's formula

$$N(r, 0) = m(r, \Pi) - m\Big(r, \frac{1}{\Pi}\Big) - \log|\Pi(0)|,$$

whence

$$m\Big(r, \frac{1}{\Pi}\Big) \le m(r, \Pi). \tag{8}$$

Taking into account (7), (8) and (4) we obtain

$$\frac{|a_\rho|}{2\pi} r^\rho \overset{\text{as}}{<} m(r, f) + m(r, \Pi) + O(1) \overset{\text{as}}{<} \log M_f(r) + 3\varepsilon r^\rho, \quad \varepsilon > 0. \tag{9}$$

The statement of Theorem 3 follows from (9) and (5).

THEOREM 4 (Lindelöf). *Let $\rho = p$. Set*

$$\delta_f(r) = \Big| a_\rho + \frac{1}{\rho} \sum_{|a_n| < r} a_n^{-\rho} \Big|, \quad \overline{\delta}_f = \limsup_{r\to\infty} \delta_f(r),$$

and $\gamma_f = \max(\overline{\Delta}_f, \overline{\delta}_f)$. Then σ_f and γ_f simultaneously are equal either to zero, or to infinity, or to positive numbers.

[5] A stronger statement $\sigma_f = |a_\rho|$ is true.

PROOF. We shall use the formula

$$[\log f(z)]^{(\rho)} = \frac{\rho!}{2\pi}\int_0^{2\pi} \log|f(Re^{i\psi})| \frac{2Re^{i\psi}}{(Re^{i\psi}-z)^{\rho+1}}\,d\psi + \sum_{|a_n|<R} \frac{(\rho-1)!\overline{a}_n^\rho}{(R^2-\overline{a}_n z)^\rho} - \sum_{|a_n|<R} \frac{(\rho-1)!}{(a_n-z)^\rho}$$

proved in Section 4.2. If we set $z=0$, we obtain

$$[\log f(z)]^{(\rho)}_{z=0} + (\rho-1)! \sum_{|a_n|<R} a_n^{-\rho} = \frac{2\rho!}{2\pi R^\rho}\int_0^{2\pi} \log|f(Re^{i\psi})| e^{-i\rho\psi}\,d\psi + \sum_{|a_n|<R} \frac{(\rho-1)!\overline{a}_n^\rho}{R^{2\rho}}$$

or

$$\left|[\log f(z)]^{(\rho)}_{z=0} + (\rho-1)! \sum_{|a_n|<R} a_n^{-\rho}\right| \le \frac{2\rho!}{R^\rho}\frac{1}{2\pi}\int_0^{2\pi} \big|\log|f(Re^{i\psi})|\big|\,d\psi + (\rho-1)!\frac{n(R)}{R^\rho}. \tag{10}$$

It is easily seen that the logarithm of a primary factor has a root at $z=0$ of multiplicity $\rho+1$. Hence,

$$[\log(\Pi(z)]^{(\rho)}_{z=0} = 0$$

and

$$[\log f(z)]^{(\rho)}_{z=0} = \rho! a_\rho\,. \tag{11}$$

In addition,

$$\frac{1}{2\pi}\int_0^{2\pi} \big|\log|f(Re^{i\psi})|\big|\,d\psi = m(R,f) + m\Big(R,\frac{1}{f}\Big)\,.$$

The Jensen formula yields

$$m\Big(R,\frac{1}{f}\Big) = m(R,f) - N(R,0) + O(1) < m(R,f) + O(1)\,,$$

and

$$\frac{1}{2\pi}\int_0^{2\pi} \big|\log|f(Re^{i\psi})|\big|\,d\psi \le 2m(R,f) + O(1) \le 2\log M_f(R) + O(1)\,. \tag{12}$$

The Jensen formula yields the estimate

$$n(R) \le \log M_f(eR) + O(1)\,. \tag{13}$$

Substituting (11)–(13) in (10) we obtain

$$\left|\rho! a_\rho + (\rho-1)! \sum_{|a_n|<R} a_n^{-\rho}\right| \le 4\frac{\rho!}{R^\rho}\log M_f(R) + (\rho-1)!\,\frac{\log M_f(eR)}{R^\rho} + O\Big(\frac{1}{R^\rho}\Big)$$

or

$$\delta_f(R) \overset{\text{as}}{<} C\,\frac{\log M_f(eR)}{R^\rho}\,,$$

with a constant C independent of the function f. It follows from this inequality that $\overline{\delta}_f \le Ce^\rho\sigma_f$. Since by virtue of (13) we have $\overline{\Delta}_f \le e^\rho\sigma_f$, we obtain finally

$$\gamma_f = \max(\overline{\delta}_f, \overline{\Delta}_f) \le C_1\sigma_f\,. \tag{14}$$

To estimate σ_f from above via γ_f we shall write the Hadamard representation of $f(z)$ in the form

$$f(z) = \exp\Big[\Big(a_\rho + \frac{1}{\rho}\sum_{|a_n|<r} a_n^{-\rho}\Big)z^\rho\Big]\exp P_{\rho-1}(z)$$
$$\times \prod_{|a_n|<r} G\Big(\frac{z}{a_n}, \rho-1\Big) \prod_{|a_n|\ge r} G\Big(\frac{z}{a_n}, \rho\Big)\,, \qquad r = |z|\,,$$

where $P_{\rho-1}$ is a polynomial of degree at most $\rho - 1$. Using the estimate of the primary factor $G(u,p)$ given by Lemma 1, Section 4.3, we obtain

$$\log|f(z)| \le \delta_f(r)r^\rho + A_\rho\Big[r^\rho\int_0^r \frac{dn(t)}{t^{\rho-1}(t+r)} + r^{\rho+1}\int_r^\infty \frac{dn(t)}{t^\rho(t+r)}\Big] + o(r^\rho)\,.$$

Much as in the proof of Theorem 2, Section 4.3, integration by parts yields

$$\log M_f(r) \le \delta_f(r)r^\rho + K_\rho\Big\{r^{\rho-1}\int_0^r \frac{n(t)}{t^\rho}\,dt + r^{\rho+1}\int_r^\infty \frac{n(t)}{t^{\rho+2}}\,dt\Big\} + o(r^\rho)\,.$$

Now we apply the inequality $n(t) \overset{\text{as}}{<} (\overline{\Delta}+\varepsilon)t^\rho$, $\varepsilon > 0$, and obtain

$$\log M_f(r) \le \delta_f(r)r^\rho + 2K_\rho(\overline{\Delta}_f + \varepsilon)r^\rho\,.$$

Therefore,

$$\limsup_{r\to\infty} \frac{\log M_f(r)}{r^\rho} \le \overline{\delta}_f + 2K_\rho\overline{\Delta}_f \le C_\rho\gamma_f\,,$$

with a constant C_ρ, which proves $\sigma_f \le C_\rho\gamma_f$. Together with (14) this proves Theorem 4.

PROBLEM 1 (Valiron). Prove the following statement.

If ρ is not an integer, then convergence of the integral

$$\int^\infty \frac{\log M_f(r)}{r^{\rho+1}}\,dr \tag{15}$$

is equivalent to convergence of the series

$$\sum_{n=1}^\infty \frac{1}{|a_n|^\rho}\,. \tag{16}$$

If ρ is an integer, then convergence of the integral (15) is equivalent to convergence of both the series (16) and the integral

$$\int^\infty \frac{\delta_f(r)dr}{r}\,.$$

LECTURE 6

Theorems of Phragmén and Lindelöf

Let $f(z)$ be an analytic function in a bounded domain D, let ζ be a point of the boundary Γ of this domain, and let $U_\delta(\zeta)$ be the δ-neighborhood of the point ζ. Set

$$\limsup_{z\to\zeta} |f(z)| = \lim_{\delta\to 0} \sup_{z\in U_\delta(\zeta)\cap D} |f(z)| .$$

If the inequality

$$\limsup_{z\to\zeta} |f(z)| \leq M$$

holds at all points of Γ, then we shall say that $|f(z)| \leq M$ on the boundary of the domain D. The *Maximum Principle* for functions analytic in a bounded domain may be stated as follows:

If $|f(z)| \leq M$ on the boundary of a domain M, then $|f(z)| \leq M$ in D.

This statement easily follows from the Maximum Principle in its standard form and compactness of the boundary of D.

6.1. Functions analytic inside an angle

For a function $f(z)$ analytic inside an angle $D = \{z : \alpha < \arg z < \beta\}$ we set

$$M_f(r) = \sup\{|f(re^{i\theta})| : \alpha < \theta < \beta\} .$$

THEOREM 1. *Let D be an angle of opening π/λ, and let $f(z)$ be a function analytic in D satisfying an asymptotic estimate*

$$\log M_f(r) \overset{\text{as}}{<} r^\rho , \tag{1}$$

where $\rho < \lambda$. If $f(z)$ is bounded by a constant M on the sides of D, then $|f(z)| \leq M$ for $z \in D$.

PROOF. Without loss of generality we can assume that $D = \{re^{i\theta} : |\theta| < \alpha\}$, $\alpha = \pi/2\lambda$. Let us choose a number ρ_1 such that $\rho < \rho_1 < \lambda$, and set

$$\varphi_\delta(z) = f(z)e^{-\delta z^{\rho_1}} , \quad \delta > 0 .$$

The asymptotic inequality

$$|\varphi_\delta(z)| \overset{\text{as}}{<} e^{|z|^\rho - \delta|z|^{\rho_1} \cos\rho_1\alpha}$$

holds inside the whole angle D. Since $\rho < \rho_1$ and $\cos\rho_1\alpha > 0$, the inequality

$$|\varphi_\delta(Re^{i\theta})| \leq M , \quad -\alpha < \theta < \alpha,$$

holds for $|z| = R > R_\delta$. Applying the Maximum Principle to the function $\varphi_\delta(z)$ inside the sector $D_R = \{re^{i\theta} : r < R, |\theta| < \alpha\}$, we find that $|\varphi_\delta(z)| \le M$ at an arbitrary point. In other words,

$$|f(z)| \le Me^{\delta|z|^{\rho_1}} .$$

As R tends to infinity we see that this inequality is fullfilled everywhere inside the angle D.

Since $\delta > 0$ is an arbitrary number, we obtain $|f(z)| \le M$ in D, completing the proof of Theorem 1.

PROBLEM 1. Prove that if, for some $\rho < \lambda$, the function $f(z)$ in Theorem 1 satisfies the condition

$$\liminf_{r\to\infty} \frac{\log M_f(r)}{r^\rho} = 0,$$

then its conclusion remains valid.

THEOREM 2. *If a function $f(z)$ analytic inside an angle*

$$D = \left\{z : |\arg z| < \alpha = \frac{\pi}{2\rho}\right\}$$

satisfies the asymptotic inequalities

$$\log M_f(r) \overset{\text{as}}{<} (\sigma + \varepsilon)r^\rho$$

for all $\varepsilon > 0$, and $f(z)$ is bounded on the sides of D by a constant M, then

$$|f(re^{i\theta})| \le Me^{\sigma r^\rho \cos\rho\theta} , \quad re^{i\theta} \in D .$$

PROOF. The function

$$\varphi_\varepsilon(z) = f(z)e^{-(\sigma+\varepsilon)z^\rho}$$

is bounded on a positive ray and on the boundary of D. According to the previous theorem, it is bounded by a constant in each angle $D_+ = \{z : 0 < \arg z < \pi/2\rho\}$, $D_- = \{z : -\pi/2\rho < \arg z < 0\}$. Applying the previous theorem once more, we obtain $|\varphi(z)| \le M$ for $z \in D$, or

$$|f(re^{i\theta})| \le Me^{(\sigma+\varepsilon)r^\rho \cos\rho\theta} , \quad re^{i\theta} \in D , \quad \varepsilon > 0 .$$

The statement of Theorem 2 follows when $\varepsilon \to 0$.

The following corollary from Theorem 2 is frequently used.

THEOREM 3. *If $f(z)$, $z = x + iy$, is an analytic function in the half-plane $\{z : \operatorname{Im} z > 0\}$ such that, for all $\varepsilon > 0$,*

$$M_f(r) \overset{\text{as}}{<} e^{(\sigma+\varepsilon)r},$$

and $|f(x)| \le M$ on the real axis, then

$$|f(x + iy)| \le Me^{\sigma y} . \tag{2}$$

PROOF. If we take $\alpha = \pi/2$ and $\rho = 1$ in Theorem 2 and apply this theorem to $f(-iz)$, we obtain Theorem 3.

REMARK 1. The estimate given by (2) is sharp. It is attained for functions of the form $f(z) = M\gamma e^{-i\sigma z}$, $|\gamma| = 1$. On the other hand, it is easy to verify, introducing the function $f(z)e^{i\sigma z}$, that if the equality is attained in (2) at least at one point, then $f(z) = M\gamma e^{-i\sigma z}$, $|\gamma| = 1$.

REMARK 2. If $f(z)$ is an entire function of exponential type σ, and $|f(x)| \le M$, $-\infty < x < \infty$, then

$$|f(x+iy)| \le Me^{\sigma|y|} \tag{3}$$

in the whole plane.

REMARK 3. If the growth of an entire function $f(z)$ is not higher than of first order and minimal type, and if $|f(z)| \le M$ on the real axis, then $f(z) = \text{const}$.

PROBLEM 2. If $f(z)$ is an entire function of exponential type $\sigma > 0$, and $f(x) \to 0$ as $|x| \to \infty$, then $f(x+iy)e^{-\sigma|y|} \rightrightarrows 0$ as $|x| \to \infty$, uniformly with respect to y.

PROBLEM 3. If $f(z)$ is an entire function of exponential type σ, and

$$|f(x)| \le C(1 + |x|^n) ,$$

then

$$f(z)e^{-\sigma|y|} \le C_1(1 + |z|^n) .$$

If, in addition, $\sigma = 0$, then $f(z)$ is a polynomial.

PROBLEM 4. Prove the following statements.

1. A nonconstant entire function $f(z)$ satisfying the condition

$$\lim_{r\to\infty} \frac{\log M_f(r)}{\sqrt{r}} = 0$$

cannot be bounded on any ray emanating from the origin.

2. If $f(z)$ is a nonconstant entire function of minimal type with respect to the order 1/2, then the function

$$\mu_f(r) = \min\{|f(z)| : |z| = r\}$$

cannot be bounded as $r \to \infty$ (Wiman).

HINT. If $f(z)$ is as stated with zeros $a_1, a_2, \dots$, $f(0) = 1$, and

$$\varphi(z) = \prod_{n=1}^{\infty} \left(1 - \frac{z}{|a_n|}\right) ,$$

then $\mu_f(r) \ge |\varphi(r)|$. Now apply statement 1.

There are many results on the connection between the growth rate of an entire function and the rate of its decrease. We shall mention here the $\cos \pi\rho$-theorem due to Wiman and Valiron stating that

$$\limsup_{r\to\infty} \frac{\log \mu_f(r)}{\log M_f(r)} \ge \cos \pi\rho$$

for an entire function $f(z)$ of order $\rho \le 1$, and the Beurling theorem stating that, for every entire function $f(z)$,

$$\limsup_{r\to\infty} \frac{\log|f(re^{i\theta})|}{\log M_f(r)} \ge -1$$

for every $\theta \in [0, 2\pi]$. Proofs and further results can be found in Beurling [**13**], Kjellberg [**70**], Hayman and Kjellberg [**55**] and in the monograph Essén [**34**].

PROBLEM 5. Prove the following statements.

1. Let $f(z)$ be a bounded analytic function in the right half-plane. If

$$\lim_{x\to+\infty} \frac{\log|f(x)|}{x} = -\infty ,$$

then $f \equiv 0$.

2. Let $f(z)$ be an analytic function in the right half-plane. If

$$|f(z)| \le Me^{-c|z|} , \qquad \operatorname{Re} z \ge 0,$$

for some $c > 0$, then $f \equiv 0$.

HINT. Introduce an auxiliary analytic function $F(z) = f(z)\exp(-\varepsilon z \log z)$, $\operatorname{Re} z > 0$, and apply the first statement.

The Phragmén and Lindelöf theorems proved above treated functions analytic inside an angle.

PROBLEM 6. If $f(z)$ is an analytic function in the strip $\{z : |\operatorname{Im} z| < b\}$, $|f(x \pm ib)| \le M$, and

$$\limsup_{|x|\to\infty} \frac{\log\log\max\{|f(x+iy)| : |y| \le b\}}{|x|} < \frac{\pi}{2b} ,$$

then $|f(z)| \le M$ in the whole strip.

Theorems of a similar type are often used for functions analytic in some other unbounded domains. Such theorems can be found in the monographs by Pólya and Szegö [**111**] (Sect. III, Chap. 5, §6), Evgrafov [**36**] and Tsuji [**123**].

Many theorems of the theory of entire functions remain valid for more general classes of functions.

6.2. Entire functions with values in Banach algebras

A function $\varphi : G \to E$, where G is a domain in $\mathbb{C}$ and E is a Banach space, is called analytic if for all $\lambda \in G$ there exists the derivative

$$\varphi'(\lambda) = \lim_{h\to 0} \frac{\varphi(\lambda+h) - \varphi(\lambda)}{h} , \tag{4}$$

where the limit is considered with respect to the norm in E. It is evident that, for every linear functional $f \in E^*$, the function $f[\varphi(\lambda)]$ is analytic. This remark permits theorems on complex-valued analytic functions to be extended to E-valued analytic functions.

For example, let φ be an entire function with values in E such that $\|\varphi(\lambda)\| \le C$ for all $\lambda \in \mathbb{C}$. For every linear function $f \in E^*$, according to the Liouville theorem, we have $f[\varphi(\lambda)] \equiv \text{const}$. By the Hahn-Banach theorem it follows that $\varphi(\lambda) \equiv \text{const}$. Thus, we have proved the Liouville theorem for E-valued functions.

The growth characteristics for E-valued entire functions are defined in the same way as for complex-valued functions, with the norm in place of the modulus. For example,

$$M_\varphi(r) = \max\{\|\varphi(\lambda)\| : |\lambda| \le r\}$$

for an E-valued entire function $\varphi(z)$. The Phragmén and Lindelöf theorems proved in the previous section remain valid for E-valued functions as well. Indeed, if an abstract function $\varphi(z)$ is analytic in $D = \{z : |\arg z| < \alpha\}$ and satisfies the inequalities

$$\|\varphi(re^{\pm i\alpha})\| \le M ,$$
$$\|\varphi(z)\| \overset{\text{as}}{<} \exp|z|^\rho , \quad \rho < \frac{\pi}{2\alpha} ,$$

and if $f \in B^*$ is a normalized linear functional, then for the scalar analytic function $\Psi(z) = f[\varphi(z)]$ we have

$$|\Psi(re^{\pm i\alpha})| \le M , \quad |\Psi(z)| \overset{\text{as}}{<} \exp|z|^\rho .$$

Applying Theorem 1, we obtain $|f[\varphi(z)]| \le M$ for $z \in D$, and

$$\|\varphi(z)\| = \sup\{|f[\varphi(z)]| : f \in B^*, \ \|f\| = 1\} \le M, \quad z \in D .$$

The same method can be applied for extending other theorems on scalar analytic functions to abstract analytic functions.

Let us also remark that thc formulas expressing the order and type of an entire function via the coefficients of its power expansion

$$f(z) = \sum_{n=0}^{\infty} c_n z^n$$

remain valid if $|c_n|$ is replaced by $\|c_n\|$.

Various theorems on analytic and entire E-valued functions are often used in the theory of Banach algebras (see Gelfand, Raikov, and Shilov [**38**], Bourbaki [**19**], Rudin [**118**], Brudnyi and Gorin [**21**]). We shall describe the simplest examples.

Let B be a Banach algebra with unity e. This means that B is a Banach space and for each pair (x, y) of its elements the product xy is defined which is a bilinear function of x and y, and the inequality $\|xy\| \le \|x\|\|y\|$ holds. In what follows we assume that B is an associative algebra, i.e., $x(yz) = (xy)z$ for all $x, y, z \in B$.

An element $e \in B$ is called the unity of the algebra if $xe = ex = e$ for all $x \in B$.

We remind the reader that the spectrum $\operatorname{spec}(x)$ of an element $x \in B$ is the set of all $\lambda \in \mathbb{C}$ such that $\lambda e - x$ is not invertible in B.

THEOREM 4. *The spectrum of an arbitrary element $x \in B$ is not void.*

PROOF. If the spectrum is void, then the resolvent $(\lambda e - x)^{-1}$ is a B-valued entire function. Its norm tends to 0 as $\lambda \to \infty$. Using the Liouville theorem we conclude that $(\lambda e - x)^{-1} \equiv 0$ which is a contradiction to the identity $\lim_{\lambda\to\infty} \lambda(\lambda e - x)^{-1} = e$.

THEOREM 5 (Le Page). *If $\|xy\| \le C\|yx\|$ for every pair of elements $x, y \in B$ with some constant C, then B is a commutative algebra.*

PROOF. Let us consider the entire function $\varphi(\lambda) = e^{-\lambda x} y e^{\lambda x}$ with

$$e^{\lambda x} = \sum_{k=0}^{\infty} \frac{\lambda^k x^k}{k!} . \tag{5}$$

According to the hypothesis we have

$$\|\varphi(\lambda)\| = \|(e^{-\lambda x} y) e^{\lambda x}\| \leq C \|e^{\lambda x}(e^{-\lambda x} y)\| = C\|y\| ,$$

and by the Liouville theorem $\varphi(\lambda) \equiv \text{const}$. Hence $\varphi'(0) = -xy + yx = 0$, and B is a commutative algebra.

DEFINITION. The value

$$\rho(x) = \sup\{|\lambda| : \lambda \in \operatorname{spec}(x)\}$$

is called the spectral radius of $x \in B$.

THEOREM 6 (I. Gelfand). *The identity*

$$\rho(x) = \lim_{k \to \infty} \|x^k\|^{1/k} \tag{6}$$

holds.

PROOF. Let us consider the analytic B-valued function

$$r_x(\lambda) = \sum_{k=0}^{\infty} \frac{x^k}{\lambda^{k+1}} . \tag{7}$$

The disk of convergence (centered at infinity) of this series coincides with the set

$$\{\lambda : |\lambda| > \limsup_{k \to \infty} \|x^k\|^{1/k}\} .$$

The series converges uniformly inside the disk, and multiplying it by $(x - \lambda e)$ we find $(x - \lambda e) r_x(\lambda) = r_x(x - \lambda e) = e$. Therefore, the function $r_x(\lambda)$ is the resolvent $(\lambda e - x)^{-1}$.

Hence, the convergence disk of series (7) coincides with the largest disk contained in $\overline{\mathbb{C}} \setminus \operatorname{spec}(x)$, which implies that $\rho(x) = \limsup_{k \to \infty} \|x^k\|^{1/k}$. In particular, we find $\rho(x) \leq \|x\|$.

Now let $\lambda \in \operatorname{spec}(x)$. Since $\lambda^n e - x^n = (\lambda e - x) y$, the invertibility of the left-hand side of this identity would imply $(\lambda e - x) y (\lambda^n e - x^n)^{-1} = e$ contradicting the assumption $\lambda \in \operatorname{spec}(x)$. Therefore, $\lambda^n \in \operatorname{spec}(x^n)$ and the inequality $\rho(x^n) \leq \|x^n\|$ implies $|\lambda| \leq \|x^n\|^{1/n}$. Hence,

$$\rho(x) \leq \liminf_{n \to \infty} \|x^n\|^{1/n} ,$$

proving the Gelfand formula (6).

COROLLARY. *The type σ_x of an entire function $e_x(\lambda) = e^{\lambda x}$ is equal to $\rho(x)$.*

THEOREM 7. *Let elements $x, y \in B$ be such that $xy = yx$. Then $\rho(x + y) \leq \rho(x) + \rho(y)$.*

PROOF. Since x and y are commuting, we have $e^{\lambda(x+y)} = e^{\lambda x}e^{\lambda y}$ which follows from the power representation of $e^{\lambda(x+y)}$. It remains to use the inequality for the type of the product of entire functions, which yields

$$\rho(x+y) = \sigma_{e^{\lambda x}e^{\lambda y}} \leq \sigma_{e^{\lambda x}} + \sigma_{e^{\lambda y}} = \rho(x) + \rho(y) ,$$

proving Theorem 7.

6.3. Applications of the Phragmén and Lindelöf theorems to Banach algebras

An element x of a Banach algebra B is called *real* if its spectrum $\operatorname{spec}(x)$ is a real set.

THEOREM 8. *Let every element of a Banach algebra B be representable in the form $w = x + iy$, where x and y are real, and, for every triple x, y, z of real elements, the identity*

$$\|xyz\| = \|yzx\|$$

hold. Then the algebra B is commutative.

PROOF. Let $\mathcal{L}$ be a curve surrounding the spectrum of x. Then

$$e^{\lambda x} = \frac{1}{2\pi i}\int_{\mathcal{L}} e^{\lambda\zeta}(\zeta e - x)^{-1}\,d\zeta \tag{8}$$

and

$$\|e^{\lambda x}\| \leq C\exp\left(\max_{\zeta\in\mathcal{L}} \operatorname{Re}(\lambda\zeta)\right)$$

with a constant C independent of λ. If x is a real element, then, however small $\varepsilon > 0$ be given, we may choose as $\mathcal{L}$ the boundary of a rectangle $\{(\zeta,\eta) : a \leq \zeta \leq b,\ |\eta| \leq \varepsilon\}$. Then for purely imaginary $\lambda = i\mu$ we obtain

$$\|e^{i\mu x}\| \leq C_\varepsilon e^{\varepsilon|\mu|} .$$

As in the proof of the Le Page theorem, let us consider the entire function $\varphi(\lambda) = e^{\lambda x}ye^{-\lambda x}$. Let x be a real element, and λ a real, while μ a nonreal number. Then, using the power expansion (5), it is easy to verify that, with $\mathcal{L}$ sufficiently close to the real axis, the element

$$\frac{1}{2\pi i}\int_{\mathcal{L}} (e^{\lambda\zeta} - \mu)^{-1}(\zeta e - x)^{-1}\,d\zeta$$

is inverse to $e^{\lambda x} - \mu$. Hence, $e^{\lambda x}$ is real for real λ. If y is also a real element, then $\|\varphi(\lambda)\| = \|e^{\lambda x}ye^{-\lambda x}\| = \|y\|$.

If $\lambda = i\mu$, then $\|\varphi(\lambda)\| \leq C'_\varepsilon e^{2\varepsilon|\lambda|}$. Hence, the entire function of exponential type $\varphi(\lambda)e^{2i\varepsilon\lambda}$ is bounded on the imaginary half-axis $\{\lambda : \lambda = i\mu,\ \mu > 0\}$ and on the real axis. By the Phragmén-Lindelöf theorem it is bounded in the entire upper half-plane $\{\lambda : \operatorname{Im}\lambda > 0\}$, and similarly, the function $\varphi(\lambda)e^{-2\varepsilon i\lambda}$ is bounded in $\{\lambda : \operatorname{Im}\lambda < 0\}$. Therefore, $\|\varphi(\lambda)\| \leq C''_\varepsilon e^{2\varepsilon|\lambda|}$, $\varepsilon > 0$, which means that $\varphi(\lambda)$ is of minimal type with respect to the order 1. On the real line we have $\|\varphi(\lambda)\| = \|y\|$. With account taken of Remark 2 of the previous section we obtain $\varphi(\lambda) = \varphi(0) = y$ and $e^{\lambda x}y = ye^{\lambda x}$. Comparing the coefficients at λ in the power expansion we find $xy = yx$.

To complete the proof we notice that the commutativity of real elements implies the commutativity of arbitrary elements. The theorem is proved.

Another application of the Phragmén-Lindelöf theorem is related to commutative Banach algebras. Let us denote by $\mathfrak{M}$ the space of multiplicative linear functionals on a Banach algebra B endowed with the weak topology. Let $x(\cdot)$ be the Gelfand transform of an element x, i.e., the continuous function on $\mathfrak{M}$ defined by $x(M) = M(x)$, $m \in \mathfrak{M}$. Then $\operatorname{spec}(x) = \{z \in \mathbb{C} : z = x(M),\ M \in \mathfrak{M}\}$, cf. Gelfand, Raikov, and Shilov [**38**], Rudin [**118**].

THEOREM 9 (Gelfand). *The unity e of a commutative Banach algebra B is an extreme point of the unit sphere.*

PROOF. [6] Assuming the contrary, there exist elements u, v of the unit sphere such that $2e = u + v$. If $u = e + x$, $x \in B$, then $v = e - x$, and

$$\|e + tx\| = \|e - tx\| = 1\,, \quad -1 \le t \le 1\,.$$

For any nontrivial multiplicative functional $M \in \mathfrak{M}$ we have $M(e) = 1$ and $\|M\| = 1$, and the previous equations with $t = 1$ yield

$$|1 + x(M)| \le 1\,, \quad |1 - x(M)| \le 1.$$

Hence $x(M) = 0$ for every $M \in \mathfrak{M}$. Therefore, $\operatorname{spec}(x) = 0$, and according to the Gelfand formula,

$$\lim_{k\to\infty} \|x^k\|^{1/k} = \rho(x) = 0\,.$$

Let us consider the entire function $e^{\lambda x}$. By the corollary to Theorem 6, its type with respect to the order 1 equals 0. We shall show that the function is bounded on the real line.

It follows from equation (8) that

$$e^{\lambda x} = \lim_{n\to\infty} \left(e + \frac{\lambda x}{n}\right)^n, \quad n \to \infty\,.$$

Hence,

$$\|e^{\lambda x}\| = \lim_{n\to\infty} \left\| \left(e + \frac{\lambda x}{n}\right)^n \right\|.$$

For real λ and $n > |\lambda|$ we have $\|e + \lambda x/n\| = 1$. Hence, $\|(e + \lambda x/n)^n\| \le 1$ and, at last, $\|e^{\lambda x}\| \le 1$.

By the Phragmén-Lindelöf theorem (Remark 3) we obtain $e^{\lambda x} \equiv \text{const}$, which is possible only if $x = 0$. The theorem is proved.

[6] The proof which follows was given by M. Krein.

LECTURE 7

Subharmonic Functions

7.1. Definition and basic properties

A real function $u(z) < +\infty$ is called *subharmonic* in a domain D if at each point $z_0 \in D$ it satisfies two conditions:

a) upper semicontinuity

$$u(z_0) = \lim_{\delta \to 0} \sup_{|z-z_0|<\delta} u(z) \ ;$$

b) the mean-value property

$$u(z_0) \le \frac{1}{2\pi} \int_0^{2\pi} u(z_0 + re^{i\theta})\, d\theta \ , \quad r < \delta(z_0) \ .$$

It is easy to check that the logarithm of modulus of an analytic function is subharmonic. In fact, many properties of the function $\log|f(z)|$ are extendable to the wider class of subharmonic functions.

Many properties of subharmonic functions follow directly from the definition (see, for example, Ronkin [**116**], Hayman and Kennedy [**54**]). Some of them are listed below.

1. *If $\varphi(t)$ is an increasing convex function and $u(z)$ is a subharmonic function, then $\varphi(u(z))$ is subharmonic as well.* In particular, $e^{u(z)}$ is subharmonic. Thus *if $f(z)$ is an analytic function, then for each $\lambda > 0$ the function $|f(z)|^\lambda$ is subharmonic.*

2. *Let $u_1, \dots, u_n$ be subharmonic functions in D. Then the upper envelope*

$$u(z) = \max\left(u_1(z), \dots, u_n(z)\right)$$

is subharmonic in D. In the case of an infinite family of subharmonic functions $\{u_\alpha(z)\}$, locally uniformly bounded from above, the upper envelope need not be upper semicontinuous. However, its upper semicontinuous regularization

$$u^*(z) = \lim_{\delta \to 0} \sup_{|z-\zeta|<\delta} u(\zeta)$$

is subharmonic.

3. *The limit of a decreasing or a uniformly convergent sequence of subharmonic functions is a subharmonic function.*

4. *The sum of finitely many subharmonic functions is a subharmonic function. Moreover, integration with respect to a parameter preserves subharmonicity.*

Namely, let $u(z,p)$, $(z,p) \in D \times G$, be a subharmonic function in D for every $p \in G$ and an upper semicontinuous function in $D \times G$, and let μ_p be a nonnegative measure in G. Then the function

$$u(z) = \int u(z,p)\, d\mu_p$$

is subharmonic in D.

5. The Maximum Principle is valid for subharmonic functions. It may be formulated as follows:

If a subharmonic function $u(z)$ in a domain D attains its maximum value at an interior point $z_0 \in D$, then $u(z) \equiv$ const.

PROOF. If $M = \sup_{z\in D} u(z)$ and $u(z_0) = M$, then by the mean-value property b) we have

$$u(z_0) = \frac{1}{2\pi}\int_0^{2\pi} u(z_0 + re^{i\theta})\, d\theta$$

for small enough $r > 0$. The upper semicontinuity of the function $u(z)$ and the estimate $u(z) \leq M$ yield $u(z_0 + re^{i\theta}) = M$ for $0 \leq \theta \leq 2\pi$, $0 < r < \delta$. Thus, the set of points where $u(z) = M$ is open in D. On the other hand, the upper semicontinuity implies that this set is closed. Hence, $u(z) \equiv M$ for $z \in D$.

This theorem, combined with the lemma on finite covering, yields the following statement.

For $\zeta \in \partial D$ define

$$u(\zeta) = \limsup_{z\to\zeta,\, z\in D} u(z)\ .$$

Then the inequality

$$u(z) \leq \sup_{\zeta\in\partial D} u(\zeta)$$

holds everywhere in D, with the equality valid only if $u(z)$ is constant in D.

PROBLEM 1. Prove the *principle of harmonic majorant*: in order that an upper semicontinuous function $u(z)$, $z \in D$, be subharmonic it is necessary and sufficient that for every subdomain $G \subset D$ and every harmonic function $h(z)$, $z \in G$, satisfying the inequality $u(z) \leq h(z)$ for $z \in \partial G$, the same inequality hold everywhere on G.

6. *Let us define an average*

$$\mathfrak{N}(r, z; u) = \frac{1}{2\pi}\int_0^{2\pi} u(z + re^{i\theta})\, d\theta, \quad r < \operatorname{dist}(z, \partial D)\ ;$$

then

α) $\mathfrak{N}(r, z; u)$ does not decrease as r increases;
β) $\lim_{r\to 0} \mathfrak{N}(r, z; u) = u(z)$.

To prove α), notice that according to the property 4 the function

$$\mathfrak{N}(\zeta, z; u) = \frac{1}{2\pi}\int_0^{2\pi} u(z + \zeta e^{i\theta})\, d\theta$$

is a subharmonic function of ζ and that $\mathfrak{M}(\zeta, z; u) = \mathfrak{M}(|\zeta|, z; u)$. By the Maximum Principle (the property 5) the function $\mathfrak{M}(r, z; u)$ is monotonic in r. To prove β), notice that

$$u(z) \le \mathfrak{M}(r, z; u) \le \max_{|\zeta - z| \le r} u(\zeta)$$

and, by upper semicontinuity,

$$\lim_{r\to 0} \max_{|\zeta - z| \le r} u(\zeta) = u(z) .$$

7. *Each subharmonic function can be represented as a pointwise limit of a decreasing sequence of infinitely differentiable functions.*

To prove this fact, we set

$$u_\varepsilon(z) = \iint u(\omega)\alpha_\varepsilon(z-\omega)\, d\sigma_\omega ,$$

where $d\sigma_\omega$ is the area element, $\alpha_\varepsilon(z) = \varepsilon^{-2}\alpha(\varepsilon^{-1}|z|)$, and $\alpha(t), t \ge 0$, is an infinitely differentiable function supported on $[0, 1]$ and such that

$$2\pi \int_0^1 \alpha(s) s\, ds = 1 .$$

Then

$$\begin{aligned} u_\varepsilon(z) &= \int_0^\varepsilon s\alpha_\varepsilon(s) \int_0^{2\pi} u(z + se^{i\varphi})\, d\varphi\, ds \\ &= 2\pi \int_0^1 s\alpha(s)\mathfrak{M}(\varepsilon s, z; u)\, ds . \end{aligned}$$

Using properties α) and β) of the average $\mathfrak{M}(r)$, we complete the proof.

8. *A twice continuously differentiable function $u(z)$ is subharmonic in a domain D if and only if its Laplacian Δu is nonnegative in D.*

To prove this statement we need an analogue of the Jensen formula from Lecture 2. Let u, v be twice continuously differentiable functions, and let G be a plane domain with the smooth boundary. Then the Green formula is valid:

$$\iint_G (v\Delta u - u\Delta v)\, d\sigma = \int_{\partial G} \left(v\frac{\partial u}{\partial n} - u\frac{\partial v}{\partial n}\right) ds ,$$

where $\partial/\partial n$ is differentiation along the exterior normal. We apply this formula with $G = \{w : \varepsilon < |z - w| < R\}$, $v(w) = \log \dfrac{R}{|z-w|}$, and send ε to zero. We obtain

$$u(z) + \frac{1}{2\pi} \iint_{|z-w|<R} \log \frac{R}{|z-w|} \Delta u(w)\, d\sigma_w = \mathfrak{M}(R, z; u) , \tag{1}$$

proving our assertion.

PROBLEM 2. Prove the following statements:

1. Let us suppose that a function $u(z)$, $z = x + iy$, $a < x < b$ does not depend on y, i.e., $u(z) = \varphi(x)$. For $u(z)$ to be subharmonic it is necessary and sufficient that the function φ be convex.

2. Let $u(z)$ be a subharmonic function in an annulus $R_1 < |z| < R_2$. Then for $R_1 < r < R_2$ the functions $B(r) = \max_{|z|=r} u(z)$ and $\mathfrak{M}(r) = \mathfrak{M}(r, 0; u)$ are convex

functions of $\log r$. The first part of the statement with $u = \log|f|$, f being analytic, is called Hadamard's three-circle theorem.

3. Let a function $u(z)$, $z = re^{i\theta}$, $R_1 < r < R_2$ do not depend on θ, i.e., $u(z) = \psi(r)$. For $u(z)$ to be subharmonic it is necessary and sufficient that ψ be a convex function of $\log r$, i.e.,

$$\psi(r) \leq \frac{\log r - \log r_1}{\log r_2 - \log r_1}\psi(r_2) + \frac{\log r_2 - \log r}{\log r_2 - \log r_1}\psi(r_1)$$

for $R_1 < r < R_2$.

PROBLEM 3. Prove that if a function $u(z)$ is subharmonic in a domain D, and if a function $z(w)$ is analytic in a domain G and with values in D, then the function $v(w) = u(z(w))$ is subharmonic in G.

7.2. The F. Riesz theorem and the Jensen formula

We state here without proof a theorem which is a fundamental fact of the theory of subharmonic functions.

THEOREM 1. *Let $u(z)$ be a subharmonic function in a domain D. Then there exists a unique nonnegative Borel measure μ in D such that $\mu(G) < \infty$ for every subdomain G compactly embedded into D, and $u(z)$ admits the representation*

$$u(z) = \iint_G \log|z - \zeta|\, d\mu_\zeta + h(z) \tag{2}$$

with a function $h(z)$ harmonic in G.

The measure μ is called *the Riesz measure* of the function $u(z)$, and the integral on the right-hand side of (2) is called the *logarithmic potential of μ*. Formula (2) is a generalization of the simple formula

$$\log|f(z)| = \sum_{z_k \in G} \log|z - z_k| + \log\left|\frac{f(z)}{P(z)}\right|,$$

where $f(z)$ is an analytic function in D, $\{z_k\}$ is the set of its zeros in G, $P(z) = (z - z_1)\cdots(z - z_n)$ and the last term on the right-hand side is harmonic in G. In this case the measure μ is a linear combination of Dirac measures supported by the set $\{z_k\}$. If the function $u(z)$ is twice continuously differentiable, then (2) follows directly from (1), and the measure μ has the form

$$d\mu_\zeta = \frac{1}{2\pi}\Delta u(\zeta)\, d\sigma_\zeta .$$

The proof in the general case can be carried out using a careful limit process. Now we shall derive the Jensen formula for subharmonic functions, the Jensen formula of Lecture 2 being a particular case.

THEOREM 2. *Let $u(z)$ be a bounded subharmonic function in a disk $\mathbb{D}_R = \{z : |z| < R\}$, $u(0) \neq -\infty$, and let μ be the Riesz measure of $u(z)$. Then*

$$u(0) + \int_0^R \frac{\mu(t)}{t}\, dt = \mathfrak{M}(R, 0; u) ,$$

where $\mu(t) = \mu(\{z : |z| \leq t\})$.

PROOF. The representation (2) can be written in a disk $\mathbb{D}_r$, $r < R$, in the form

$$u(z) = \iint_{\mathbb{D}_r} \log \left| \frac{r(\zeta - z)}{r^2 - \bar{\zeta} z} \right| d\mu_\zeta + h(z) . \tag{3}$$

Then

$$u(0) = \iint_{\mathbb{D}_r} \log \frac{|\zeta|}{r} \, d\mu_\zeta + h(0) .$$

Since the integrand in (3) vanishes for $|z| = r$, and $h(z)$ is a harmonic function, we obtain

$$\begin{aligned} u(0) &= \int_0^r \log \frac{t}{r} \, d\mu(t) + \mathfrak{N}(r, 0, h) \\ &= -\int_0^r \frac{\mu(t)}{t} \, dt + \mathfrak{N}(r, 0, u) , \end{aligned}$$

proving Theorem 2.

Proofs of the F. Riesz theorem as well as further results on subharmonic functions can be found in the monographs Ronkin [**116**] (short and elementary exposition), Landkof [**78**], Hayman and Kennedy [**54**].

7.3. Phragmén-Lindelöf theorems for subharmonic functions

The Phragmén-Lindelöf theorems proved above for analytic functions are valid for subharmonic functions. The following theorem is similar to Theorem 1 of the previous lecture.

THEOREM 3. *Let D be an angle of opening π/λ, and let $u(z)$ be a function subharmonic in this angle, satisfying an asymptotic estimate*

$$u(z) \overset{\text{as}}{<} |z|^\rho , \quad \rho < \lambda ,$$

and bounded by a constant M on the boundary of the angle. Then $u(z) \le M$ inside the full angle D.

PROOF. Without loss of generality we assume that $D = \{z = re^{i\theta} : |\theta| < \pi/2\lambda\}$ and consider the subharmonic function

$$\omega_\delta(z) = u(z) - \delta |z|^{\rho_1} \cos \rho_1 \theta , \quad \rho < \rho_1 < \lambda,$$

inside the sector $\{|z| < R, \, |\arg z| < \pi/2\lambda\}$. With R tending to infinity, we obtain by the Maximum Principle that $w_\delta(t) \le M$ for an arbitrary fixed z. Passing in this inequality to the limit as $\delta \to 0$, we complete the proof.

Other theorems of Phragmén-Lindelöf type can be derived in a similar way for subharmonic functions.

PROBLEM 4. Prove the following statement of Phragmén-Lindelöf type.

Let $u(z)$ be a subharmonic function in a domain D, and let $u(\zeta) \le M$, $\zeta \in \partial D \backslash E$, $E \subset \partial D$. Assume that there exists a negative harmonic function $h(z)$ in D such that, for every $\delta > 0$,

$$\limsup_{z \to \zeta, \, z \in D} (u(z) + \delta h(z)) \le M$$

at each point $\zeta \in E$. Then $u(z) \le M$ everywhere in D.

For example, if D is a bounded domain, $E = \{\zeta_1, \zeta_2, \dots\}$ is at most countable subset of the boundary ∂D, and $\sup_{z\in D} u(z) < \infty$, then the function

$$h(z) = \sum_{n=1}^{\infty} \frac{1}{2^n} \log|z - \zeta_n| - C$$

will be negative in D for an appropriate constant C. Hence, we conclude that $u(z) \le M$, $z \in D$.

7.4. Logarithmically subharmonic functions

A notion of logarithmically subharmonic function is rather useful.

DEFINITION. A nonnegative function $u(z)$ is called logarithmically subharmonic if the function $v(z) = \log u(z)$ is subharmonic.

For example, if $f(z)$ is analytic, then $|f(z)|^p$, $p > 0$, is a logarithmically subharmonic function.

Let $u_1, \dots, u_n$ be logarithmically subharmonic functions. Evidently, their product $u_1 \cdot \ldots \cdot u_n$ is also a logarithmically subharmonic function. To verify that the finite sum of logarithmically subharmonic functions has the same property, we start with the identity

$$u^2 \Delta \log u = u\Delta u - |\nabla u|^2,$$

which can be easily checked. Here, as usual, ∇u is the gradient of the function u. Using this identity, we obtain

$$\begin{aligned}(u+v)^2 \Delta \log(u+v) &= \Big(1 + \frac{v}{u}\Big) u\Delta u + \Big(1 + \frac{u}{v}\Big) v \Delta v - |\nabla u + \nabla v|^2 \\ &= \Big(1 + \frac{v}{u}\Big)(u^2 \Delta \log u + |\nabla u|^2) + \Big(1 + \frac{u}{v}\Big)(v^2 \Delta \log v + |\nabla v|^2) - |\nabla u + \nabla v|^2 \\ &= \Big(1 + \frac{v}{u}\Big) u^2 \Delta \log u + \Big(1 + \frac{u}{v}\Big) v^2 \Delta \log v + \Big(\frac{v}{u}|\nabla u|^2 - 2(\nabla u, \nabla v) + \frac{u}{v}|\nabla v|^2\Big).\end{aligned}$$

This implies that the sum of two, and hence of an arbitrary finite number, of logarithmically subharmonic functions is logarithmically subharmonic.

A passage to the limit proves that integration with respect to a parameter preserves logarithmical subharmonicity.

PROBLEM 5 (Hardy). Let $f(z)$ be an analytic function in the disk $\{z : |z| < R\}$, and let

$$I(r, f) = \frac{1}{2\pi} \int_0^{2\pi} |f(re^{i\theta})|\, d\theta .$$

Then $I(r, f)$ is an increasing function of r, and $\log I(r, f)$ is a convex function of $\log r$.

THEOREM 4 (Plancherel and Pólya). *Let $f(z)$ be an analytic function in the upper half-plane $\{y > 0\}$, continuous up to the real axis, and let*

$$|f(z)| \overset{\text{as}}{<} e^{(\sigma+\varepsilon)|z|} \tag{3}$$

for an arbitrary $\varepsilon > 0$. If

$$\int_{-\infty}^{\infty} |f(x)|^p\, dx = M < \infty\,, \quad p > 0\,,$$

then

$$\int_{-\infty}^{\infty} |f(x+iy)|^p \, dx \le M e^{p\sigma y}$$

for an arbitrary $y > 0$.

PROOF. Given $N > 0$, the function

$$w_N(z) = \int_{-N}^{N} |f(z+t)|^p \, dt$$

is logarithmically subharmonic in $\mathbb{C}_+$ and bounded on $\mathbb{R}$:

$$w_N(x) \le \int_{-\infty}^{\infty} |f(x+t)|^p \, dt = M \, .$$

Further, by (3),

$$w_N(z) \overset{\text{as}}{<} e^{p(\sigma+\varepsilon)|z|} \, , \quad \varepsilon > 0 \, .$$

The Phragmén-Lindelöf theorem applied to the subharmonic function $\log w_N(z)$ implies that

$$w_N(x+iy) \le M e^{p\sigma y} \, ,$$

or

$$\int_{-N}^{N} |f(x+iy+t)|^p \, dt \le M e^{p\sigma y} \, .$$

Sending N to infinity, we obtain the desired estimate.

REMARK. If $f(z)$ is an entire function of exponential type σ_f such that

$$\int_{-\infty}^{\infty} |f(x)|^p \, dx \le M$$

for some $p > 0$, then the function $f(z)$ is bounded on the real axis.

Indeed, the function $|f(z)|^p$ is subharmonic, and

$$\begin{aligned}
|f(x)| &\le \left\{ \frac{1}{\pi} \iint_{|\zeta|<1} |f(x+\zeta)|^p \, d\sigma_\zeta \right\}^{1/p} \\
&\le \left\{ \frac{1}{\pi} \int_{-1}^{1} d\eta \int_{-\infty}^{\infty} |f(x+\tau+i\eta)|^p \, d\tau \right\}^{1/p} \\
&\le \left\{ \frac{2}{\pi p \sigma} M (e^{p\sigma_f} - 1) \right\}^{1/p} .
\end{aligned}$$

In connection with the Plancherel-Pólya theorem we would like to mention the papers Dzhrbashyan and Avetisyan [**28**] and Luxemburg [**85**].

In the second part of this book we will return to the entire functions belonging to the space $L^p(-\infty,\infty)$ on the real axis.

LECTURE 8

The Indicator Function

8.1. The definition and ρ-trigonometric convexity of the indicator

Let us consider a function $f(z)$ which is analytic inside an angle $D = \{z = re^{i\theta} : \alpha < \theta < \beta\}$ and satisfies the estimate

$$M_f(r) \overset{\text{as}}{<} e^{Ar^\rho} \tag{1}$$

with $M_f(r) = \sup_{\alpha<\theta<\beta} |f(z)|$.

DEFINITION. The function

$$h_f(\theta) = \limsup_{r\to\infty} \frac{\log|f(re^{i\theta})|}{r^\rho}$$

is called the *indicator function* of $f(z)$ with respect to the order ρ.

The indicator function describes the growth of the function $f(z)$ along a ray $\{z : \arg z = \theta\}$.

It follows directly from the definition that the indicator of the product of two functions does not exceed the sum of the indicators of the factors, i.e.,

$$h_{fg}(\theta) \le h_f(\theta) + h_g(\theta) ,$$

and that the indicator of the sum of two functions does not exceed the larger of the two indicators:

$$h_{f+g}(\theta) \le \max\big(h_f(\theta),\ h_g(\theta)\big).$$

For the function

$$f(z) = e^{(A-iB)z^\rho}$$

holomorphic in an angle $\{z = re^{i\theta} : \alpha < \theta < \beta\}$, $\beta - \alpha \le 2\pi$, we have

$$|f(re^{i\theta})| = e^{(A\cos\rho\theta + B\sin\rho\theta)r^\rho} ,$$

and its indicator is equal to

$$H(\theta) = A\cos\rho\theta + B\sin\rho\theta .$$

Such functions are called sinusoidal or ρ-trigonometric. If $0 < \theta_2 - \theta_1 < \pi/\rho$, then the sinusoidal function $H(\theta)$ assuming values h_1 and h_2 at the points θ_1 and θ_2 is unique and can be expressed by the formula

$$H(\theta) = \frac{h_1\sin\rho(\theta_2-\theta) + h_2\sin\rho(\theta-\theta_1)}{\sin\rho(\theta_2-\theta_1)} , \quad \theta_1 \le \theta \le \theta_2 . \tag{2}$$

DEFINITION. A function $K(\theta)$ is called *ρ-trigonometrically convex* on the closed segment $[\alpha, \beta]$ if for $\alpha \leq \theta_1 < \theta_2 \leq \beta$, $0 < \theta_2 - \theta_1 < \pi/\rho$ the equations

$$K(\theta_1) = h_1 , \quad K(\theta_2) = h_2$$

imply the inequality

$$K(\theta) \leq H(\theta) , \quad \theta_1 \leq \theta \leq \theta_2 ,$$

where $H(\theta)$ is a ρ-trigonometric function assuming the values h_1 and h_2 at the points θ_1 and θ_2. A function $K(\theta)$ is called ρ-trigonometrically convex in an open interval if it is ρ-trigonometrically convex on each closed subinterval.

For $\rho = 1$, the corresponding functions are called *trigonometric* and *trigonometrically convex*, respectively.

THEOREM 1. *Let $f(z)$ be a holomorphic function inside an angle, and satisfy inequality* (1). *Then its indicator function h_f with respect to the order ρ is a ρ-trigonometrically convex function.*

PROOF. Let values θ_1 and θ_2 in $[a, b]$ be such that $0 < \theta_2 - \theta_1 < \pi/\rho$, and let $H_\varepsilon(\theta) = A_\varepsilon \cos \rho\theta + B_\varepsilon \sin \rho\theta$ be the ρ-trigonometric function which assumes values $h_f(\theta_j) + \varepsilon$ at θ_j, $j = 1, 2$, $\varepsilon > 0$. Consider the holomorphic function

$$\varphi_\varepsilon(z) = f(z) e^{-(A_\varepsilon - iB_\varepsilon) z^\rho} .$$

We have

$$|\varphi_\varepsilon(re^{i\theta_j})| = |f(re^{i\theta_j})| e^{-H_\varepsilon(\theta_j) r^\rho} \overset{\text{as}}{<} e^{-\frac{\varepsilon}{2} r^\rho} .$$

Hence the function φ_ε is bounded on the rays $\{z : \arg z = \theta_j\}$, $j = 1, 2$, and by the Phragmén-Lindelöf theorem we have

$$|\varphi_\varepsilon(re^{i\theta})| \leq M_\varepsilon , \quad \theta \in [\theta_1, \theta_2] , \quad r > 0 .$$

The latter inequality yields

$$|f(re^{i\theta})| \leq M_\varepsilon e^{r^\rho H_\varepsilon(\theta)}$$

and, according to the definition of the indicator function,

$$h_f(\theta) \leq H_\varepsilon(\theta) .$$

Passing to the limit as $\varepsilon \to 0$, we obtain

$$\text{(3)} \qquad h_f(\theta) \leq H(\theta) , \quad \theta \in [\theta_1, \theta_2] ,$$

where $H(\theta)$ is the ρ-trigonometric function assuming the values $h(\theta_j)$ at θ_j, $j = 1, 2$. The theorem is proved.

Relations (2) and (3) imply the fundamental relation for the indicator function:

$$\text{(4)} \qquad h(\theta_1) \sin \rho(\theta - \theta_2) + h(\theta) \sin \rho(\theta_2 - \theta_1) + h(\theta_2) \sin \rho(\theta_1 - \theta) \leq 0$$

for $\theta_1 < \theta < \theta_2$, $0 < \theta_2 - \theta_1 < \pi/\rho$, which is equivalent to its ρ-trigonometric convexity.

REMARK. If $f(z)$ is an entire function of order ρ, then its indicator h_f is a 2π-periodic ρ-trigonometrically convex function. It is known that for every 2π-periodic ρ-trigonometrically convex function $h(\theta)$ there exists an entire function of order ρ whose indicator coincides with $h(\theta)$. This statement is due to V. Bernstein. The proof of a somewhat more general theorem is given in Levin [**82**].

PROBLEM 1 (Lindelöf). Let a function $f(z)$ analytic in a vertical strip satisfy the estimate

$$|f(x+iy)| = O(|y|^k), \quad |y| \to \infty,$$

with some $K < \infty$. Then the function

$$h_f(x) = \limsup_{|y|\to\infty} \frac{\log|f(x+iy)|}{\log|y|}$$

is convex.

8.2. Properties of trigonometrically convex functions

1. *The maximum of two ρ-trigonometrically convex functions is ρ-trigonometrically convex. Similarly, the upper envelope of a uniformly bounded family of ρ-trigonometrically convex functions is ρ-trigonometrically convex.*

2. *Let $h(\theta)$ be a ρ-trigonometrically convex function in an interval (α, β). If $h(\theta_1) = -\infty$ for some $\theta_1 \in (\alpha, \beta)$, then $h(\theta) \equiv -\infty$ for each $\theta \in (\alpha, \beta)$.*

PROOF. Assertion 1 follows from the definition. Now, in 2, let us prove that $h(\theta) = -\infty$ for each $\theta \in (\alpha, \beta)$ satisfying the condition $\theta_1 < \theta < \theta_1 + \pi/\rho$. We choose a point $\theta_2 \in (\alpha, \beta)$ such that $\theta_1 < \theta < \theta_2 < \theta_1 + \pi/\rho$ and introduce the ρ-trigonometric function $H_\varepsilon(\theta)$ which assumes values $H_\varepsilon(\theta_1) = -1/\varepsilon$, $H_\varepsilon(\theta_2) = \max\{-1/\varepsilon, h(\theta_2)\}$. The ρ-trigonometric convexity of the function h implies that $h(\theta) \le H_\varepsilon(\theta)$, $\theta_1 \le \theta \le \theta_2$. Sending ε to zero, we obtain $h(\theta) = -\infty$. Thus $h(\theta) = -\infty$ for $\theta_1 < \theta < \min(\beta, \theta_1 + \pi/\rho)$. The required assertion now follows for each $\theta \in (\theta_1, \beta)$ and, similarly, for each $\theta \in (\alpha, \theta_1)$.

3. *If a ρ-trigonometrically convex function $h(\theta)$ is bounded, i.e., $|h(\theta)| < K$ for $\theta \in (\alpha, \beta)$, then it is a continuous function of $\theta \in (\alpha, \beta)$, and in each closed subinterval it satisfies a Lipschitz condition.*

PROOF. The proof is based on the fundamental relation (4). We write it in the form

$$\begin{aligned}
&[h(\theta) - h(\theta_1)] \sin\rho(\theta_2 - \theta_1) \\
&\quad\le h(\theta_2)\sin\rho(\theta - \theta_1) + h(\theta_1)[\sin\rho(\theta_2 - \theta) - \sin\rho(\theta_2 - \theta_1)] \\
&\quad= h(\theta_2)\sin\rho(\theta - \theta_1) + 2h(\theta_1)\sin\rho\frac{\theta_1 - \theta}{2}\cos\rho\Big(\theta_2 - \frac{\theta_1 + \theta}{2}\Big),
\end{aligned}$$

which implies

$$h(\theta) - h(\theta_1) \le K_1(\theta - \theta_1), \quad \theta > \theta_1; \quad K_1 = 2K\rho/\sin\rho(\theta_2 - \theta_1).$$

On the other hand,

$$\begin{aligned}
&[h(\theta_2) - h(\theta)] \sin\rho(\theta - \theta_1) \\
&\quad\ge h(\theta_1)\sin\rho(\theta - \theta_2) + h(\theta)[\sin\rho(\theta_2 - \theta_1) - \sin\rho(\theta - \theta_1)] \\
&\quad= h(\theta_1)\sin\rho(\theta - \theta_2) + 2h(\theta)\sin\rho\frac{\theta_2 - \theta}{2}\cos\rho\Big(\theta_1 - \frac{\theta + \theta_2}{2}\Big).
\end{aligned}$$

Fixing θ_1, and sending θ to θ_2, we obtain

$$h(\theta_2) - h(\theta) > -K_1(\theta_2 - \theta)\,, \quad \theta_2 > \theta\,,$$

or after changing the notation,

$$|h(\theta'') - h(\theta')| < K_1|\theta' - \theta''|\,.$$

REMARK. A ρ-trigonometrically convex function does not have to be continuous on a closed segment. Its limit values at the endpoints can be smaller than the values of the function.

PROBLEM 2. Prove that a function $h(\theta)$ is ρ-trigonometrically convex for $\theta \in (\alpha, \beta)$ if and only if the function $u(re^{i\theta}) = r^\rho h(\theta)$ is subharmonic within the angle $D = \{z = re^{i\theta} : \alpha < \theta < \beta,\, r > 0\}$.

PROBLEM 3. Construct a function f analytic within the angle D, continuous up to the bounding rays and satisfying the estimate (1) with the indicator h_f discontinuous at the endpoints of the segment $[\alpha, \beta]$.

4. *Let $h(\theta)$ be a ρ-trigonometrically convex function on the segment $[\alpha, \beta]$. Then*

$$h(\varphi) + h(\varphi + \pi/\rho) \geq 0\,, \quad \alpha \leq \varphi < \varphi + \pi/\rho \leq \beta\,. \tag{5}$$

PROOF. Let us substitute the values $\theta_1 = \varphi + \tau$, $\theta = \varphi + \pi/2\rho$, $\theta_2 = \varphi + \pi/\rho$ into the fundamental relation (4) and pass to the limit as $\tau \to 0$. Using the continuity of the indicator at the interior points of the segment $[\alpha, \beta]$ and the inequality $h(\varphi) \geq h(\varphi + 0)$ for $\varphi = \alpha$, we obtain (5).

5. *If the equality is attained in* (5), *then $h(\theta)$ is a ρ-trigonometric function in the segment $[\varphi, \varphi + \pi/\rho]$.*

PROOF. Let $h(-\pi/2\rho) = h(\pi/2\rho) = 0$. Then the fundamental relation (4) yields $h(\theta) \leq h(0)\cos\rho\theta$. If for some $\theta_0 \in (0, \pi/2\rho)$ the inequality holds, then, applying again the fundamental relation with $\theta_1 = -\theta_0$, $\theta_2 = \theta_0$, $\theta = 0$, we obtain $h(0) < h(0)$. Thus $h(\theta) = h(0)\cos\rho\theta$ everywhere on $[-\pi/\rho, \pi/\rho]$. The general case can be reduced easily to the examined one.

THEOREM 2. *Let $f(z)$ be an analytic function in the angle $D = \{z = re^{i\theta} : \alpha \leq \theta \leq \beta\}$, which satisfies the asymptotic estimate* (1), *and let its indicator with respect to the order ρ be a continuous function on $[\alpha, \beta]$. Then*

$$|f(re^{i\theta})| \leq e^{r^\rho(h(\theta)+\varepsilon)}\,, \quad r > r_\varepsilon\,, \quad \alpha \leq \theta \leq \beta\,. \tag{6}$$

PROOF. We divide the segment $[\alpha, \beta]$ into subintervals with endpoints $\alpha = \theta_0 < \theta_1 < \ldots < \theta_n = \beta$, $\theta_{j+1} - \theta_j < \pi/\rho$. For each subinterval $[\theta_j, \theta_{j+1}]$ we construct the sinusoidal functions $H_j(\theta) = A_j\cos\rho\theta + B_j\sin\rho\theta$ assuming values $h(\theta_j) + \varepsilon/3$ and $h(\theta_{j+1}) + \varepsilon/3$ at the points θ_j and θ_{j+1}, respectively. The segments $[\theta_j, \theta_{j+1}]$ can be chosen small enough that the oscillation of the functions $h(\theta)$ and $H_j(\theta)$ on each of these segments be less than $\varepsilon/3$.

The function

$$\varphi_j(z) = f(z)e^{-(A_j - iB_j)z^\rho}$$

is bounded on the sides of the angle $\theta_j \le \arg z \le \theta_{j+1}$. By the Phragmén-Lindelöf theorem it is also bounded inside the angle. Hence

$$|f(re^{i\theta})| \le C_j e^{H_j(\theta) r^\rho} , \quad \theta_j \le \theta \le \theta_{j+1} ,$$

and for sufficiently large $r_j(\varepsilon)$ and $r > r_j(\varepsilon)$

$$\log |f(re^{i\theta})| < \left[H_j(\theta) + \frac{\varepsilon}{3}\right] r^\rho , \quad \theta_j \le \theta \le \theta_{j+1} .$$

Thus,

$$\log |f(re^{i\theta})| \le [h(\theta) + \varepsilon] r^\rho , \quad \alpha \le \theta \le \beta ,$$

for $r > r(\varepsilon) = \max r_j(\varepsilon)$.

REMARK 1. Similar arguments show that if the indicator of the function $f(z)$ equals $-\infty$ identically, then

$$\frac{\log |f(re^{i\theta})|}{r^\rho} \rightrightarrows -\infty , \quad r \to -\infty ,$$

uniformly on each closed subangle.

REMARK 2. The previous remark implies that if $f(z)$ is an entire function of order ρ and if its indicator with respect to the order ρ equals $-\infty$ at one point, then $f \equiv 0$. This remark is a particular case of the following principle: "No nontrivial entire function which grows not too fast in the complex plane can approach zero too fast as z tends to infinity along any ray."

REMARK 3. There are several different definitions of the order and type of a function $f(z)$ analytic inside an angle $D = \{z = re^{i\theta} : \alpha < \theta < \beta\}$. We shall use the following definitions:

$$\rho_f = \limsup_{r\to\infty} \frac{\log^+ \log^+ M_f(r)}{\log r} ; \quad \sigma_f = \limsup_{r\to\infty} \frac{\log^+ M_f(r)}{r^\rho} .$$

PROBLEM 4. Using the Phragmén-Lindelöf theorem, prove that, for $\rho_f > \pi/(\beta - \alpha)$, the order of growth is simultaneously the order of decrease, i.e.,

$$\limsup_{r\to\infty} \frac{\log |f(re^{i\theta})|}{r^{\rho+\varepsilon}} \equiv 0 , \quad \alpha < \theta < \beta,$$

for each $\varepsilon > 0$.

Other definitions of the order of functions analytic inside an angle can be found in Govorov [**46**], Goldberg and Ostrovskii [**43**], Hayman [**53**], Grishin [**47, 48**].

PROBLEM 5. Let $f(z)$ be a function analytic inside an angle $D = \{z : \alpha < \arg z < \beta\}$ and satisfying estimate (1), and let $h_f(\theta)$ be its indicator with respect to the order ρ. Prove that the indicators of $f(z)$ and $f'(z)$ satisfy the relation $h_{f'}(\theta) \le h_f(\theta)$, $\alpha < \theta < \beta$, where the inequality can hold at the point θ_0 only if $h_f(\theta) \equiv 0$ in some neighborhood of θ_0.

8.3. Applications of properties of the indicator function

THEOREM 3 (Carlson). *Let $f(z)$ be a function analytic and of exponential type in the right half-plane $\{z : \operatorname{Re} z > 0\}$, and let*

$$h_f\Big(\frac{\pi}{2}\Big) + h_f\Big(-\frac{\pi}{2}\Big) < 2\pi\,.$$

If $f(n) = 0$, $n = 0, 1, 2, \ldots$, then $f \equiv 0$.

PROOF. Outside the disks $\{z : |z - n| < \delta\}$ we have

$$|\sin \pi z| > m_\delta e^{\pi|y|}\,.$$

Thus the function $\varphi(z) = f(z)/\sin \pi z$ is analytic in $\{z : \operatorname{Re} z \geq 0\}$ and

$$|\varphi(z)| \overset{\text{as}}{<} \frac{C}{m_\delta} e^{A|z|}$$

outside the same disks. Assuming $\delta < 1/2$, these disks are pairwise disjoint, and by the Maximum Principle the latter inequality holds inside these disks as well. Also, the lower estimate of $\sin \pi z$ implies that

$$h_\varphi(\pm\pi/2) = h_f(\pm\pi/2) - \pi$$

and therefore $h_\varphi(-\pi/2) + h_\varphi(\pi/2) < 0$. By the property 4 of ρ-trigonometrically convex functions it follows that $h_\varphi(\theta) \equiv -\infty$, $-\pi/2 < \theta < \pi/2$. Let us consider a function

$$\varphi_{\alpha,\beta}(z) = \varphi(z) e^{i\alpha z + \beta z}\,, \qquad \alpha \text{ real}, \quad \beta > 0\,.$$

Its indicator equals:

$$h_{\varphi_{\alpha,\beta}}(\theta) = \begin{cases} -\infty, & -\dfrac{\pi}{2} < \theta < \dfrac{\pi}{2}, \\ h_\varphi\Big(\pm\dfrac{\pi}{2}\Big) \mp \alpha, & \theta = \pm\dfrac{\pi}{2}\,. \end{cases}$$

Choosing an appropriate α we can assume that $h_{\varphi_{\alpha,\beta}}(\pm\pi/2) \leq -\eta < 0$. Hence the function $\varphi_{\alpha,\beta}(z)$ is bounded on the imaginary axis by a constant M which does not depend on β. Since $\varphi_{\alpha,\beta}(z)$ is bounded in the right half-plane (this follows from the obvious modification of Theorem 2), we can apply the Phragmén-Lindelöf theorem to arrive at the inequalities

$$|\varphi_{\alpha,\beta}(x)| \leq M$$

and

$$|\varphi(x)| \leq M e^{-\beta x}\,.$$

With β tending to infinity, we obtain $\varphi(x) \equiv 0$ and therefore $f(x) \equiv 0$.

THEOREM 4 (Shilov). *Let $f(x)$ be an infinitely differentiable function on the real axis and let*

$$\sup_{-\infty < x < \infty} |x^p f^{(q)}(x)| \leq C A^p B^q p^{\alpha p} q^{\beta q}\,, \quad p, q = 0, 1, \ldots$$

with some positive A, B, C, α, β, $\alpha + \beta < 1$. Then $f(x) \equiv 0$.

PROOF. The assumptions of the theorem imply that the remainder term in the Taylor formula for the function f tends to zero for every $x \in R$. Hence f can be continued into the whole complex plane as an entire function. Using its Taylor expansion at a point x, we obtain

$$|x^p f(x+iy)| \leq CA^p p^{\alpha p} \sum_q \frac{(B|y|)^q q^{\beta q}}{q!} .$$

By the Stirling formula and Lemma 2 from Section 1.3 we obtain

$$|x^p f(x+iy)| \leq CA^p p^{\alpha p} e^{b|y|^{(1-\beta)^{-1}}} .$$

Hence

$$|f(x+iy)| \leq \frac{Ce^{b|y|^{1/(1-\beta)}}}{\sup\limits_p \left\{ \left(\frac{|x|}{A}\right)^p p^{-\alpha p} \right\}} ,$$

and

$$|f(x+iy)| \leq Ce^{-c|x|^{1/\alpha}+b|y|^{1/(1-\beta)}}$$

with some $c > 0$. Thus f is an entire function of order $\rho \leq 1/(1-\beta)$. Since $\alpha + \beta < 1$, we conclude that

$$\lim_{|x|\to\infty} \frac{\log|f(x)|}{|x|^\rho} = -\infty$$

and hence $f \equiv 0$.

THEOREM 5 (Morgan). *Let $f(t)$, $-\infty < t < \infty$, be a function such that $f(t)e^{A|t|^p}$ is bounded as $|t| \to \infty$ for some $A > 0$ and $p > 1$, and let its Fourier transform $g(x)$ decrease in such a way that for some $B > 0$, $l > 1$ the function $g(x)e^{B|x|^l}$ is bounded as $|x| \to \infty$. If $1/p + 1/l < 1$, then $f(t) \equiv 0$.*

PROOF. For $z \in \mathbb{C}$ we set

$$g(z) = \int_{-\infty}^{\infty} f(t)e^{-izt}\, dt ,$$

and observe that $g(z)$ is an entire function which coincides with the Fourier transform of the function f on the real axis. Further,

$$|g(x+iy)| < C \int_{-\infty}^{\infty} e^{-A|t|^p+|yt|}\, dt .$$

Without loss of generality we may assume that $A > 1/p$ (otherwise, we introduce a function $f(\lambda t)$ instead of $f(t)$). Using the inequality

$$|yt| < \frac{1}{p}|t|^p + \frac{1}{q}|y|^q , \quad \frac{1}{p} + \frac{1}{q} = 1 ,$$

we obtain

$$|g(x+iy)| < Ce^{\frac{1}{q}|y|^q},$$

which means that the order ρ of the function $g(z)$ does not exceed $q < l$. On the other hand, the assumptions of the theorem imply that

$$h_g(0) = \limsup_{x\to+\infty} \frac{\log|g(x)|}{x^\rho} = -\infty .$$

Thus $g(x) \equiv 0$ and hence $f(t) \equiv 0$.

Let us remark that Morgan **[100]** found a precise condition on positive values A and B which guarantees that the assertion of Theorem 5 holds in the case $1/p+1/l = 1$.

Theorem 5 shows that in a study of connection between the decrease of a function and that of its Fourier transform the functions decreasing as fast as e^{-cx^2} play an essential role. The following theorem deals with such functions.

THEOREM 6 (Hardy). *Suppose that, for some nonnegative integer n and for all real x, the following estimates hold:*

$$|f(x)| \le C(1+|x|^n)e^{-x^2/2}\,, \quad |g(x)| \le C(1+|x|^n)e^{-x^2/2}\,,$$

where

$$g(x) = \int_{-\infty}^{\infty} f(t)e^{itx}\,dt$$

is the Fourier transform of f. Then

$$f(x) = e^{-x^2/2}P_n(x)\,, \quad g(x) = e^{-x^2/2}Q_n(x)\,,$$

where P_n and Q_n are polynomials of degrees not exceeding n.

PROOF. The estimate for $f(x)$ implies that

$$g(z) = \int_{-\infty}^{\infty} f(t)e^{itz}\,dt$$

is an entire function, and

$$\begin{aligned} |g(x+iy)| &\le C\int_{-\infty}^{\infty}(1+|t|^n)e^{-t^2/2+yt}\,dt \\ &= C\Big[\int_{|t|\le 4|y|} + \int_{|t|\ge 4|y|}\Big](1+|t|^n)e^{-t^2/2+yt}\,dt\,. \end{aligned} \tag{7}$$

We will denote by C positive constants which depend on n but do not depend on x and y. For the first integral on the right-hand side of (7) we have

$$\begin{aligned} \int_{|t|\le 4|y|}(1+|t|^n)e^{-t^2/2+yt}\,dt &\le C(1+|y|^n)e^{y^2/2}\int_{-4|y|}^{4|y|} e^{-t^2/2+yt-y^2/2}\,dt \\ &= C(1+|y|^n)e^{y^2/2}\int_{-4|y|}^{4|y|} e^{-\frac{1}{2}(t+y)^2}\,dt \\ &\le C(1+|z|^n)e^{y^2/2}\,. \end{aligned} \tag{8}$$

The second integral is bounded since

$$\begin{aligned} \int_{|t|\ge 4|y|}(1+|t|^n)e^{-t^2/2+yt}\,dt &\le 2\int_{4|y|}^{\infty}(1+|t|^n)e^{-t^2/4-(t^2/4-|y|t)}\,dt \\ &\le 2\int_{-\infty}^{\infty}(1+|t|^n)e^{-t^2/4}\,dt \le C\,. \end{aligned} \tag{9}$$

Inserting the estimates (8) and (9) into (7) we obtain

$$|g(x+iy)| \le C(1+|z|^n)e^{y^2/2}\,. \tag{10}$$

Now we observe that $\varphi(z) = e^{z^2/2}g(z)$ is an entire function of order not exceeding two. This function does not grow faster than a polynomial along the coordinate axes. Moreover, the estimate (10) implies that the indicator of $\varphi(z)$ with respect to the order $\rho = 2$ satisfies the estimate

$$h_\varphi(\theta) \le \frac{1}{2}\cos^2\theta\,, \quad 0 \le \theta \le 2\pi. \tag{11}$$

Let us show that $\varphi(z)$ is a polynomial, which will prove the Hardy theorem.

We have $h_\varphi(0) = h_\varphi(\pi/2) \le 0$. By Property 5 of the indicator it follows that

$$h_\varphi(\theta) = k\sin 2\theta\,, \quad 0 \le \theta \le \pi/2\,. \tag{12}$$

Comparing (11) and (12), we conclude that $k\sin 2\theta \le (\cos^2\theta)/2$, or $2k\sin\theta \le (\cos\theta)/2$, $0 \le \theta \le \pi/2$, whence $k \le 0$. Since $\varphi(z)$ has a polynomial bound on the coordinate axes, we obtain by the Phragmén-Lindelöf theorem that $|\varphi(z)| \le C(1 + |z|^n)$ in the first quadrant. Similar estimates are valid in the remaining quadrants. Therefore, $\varphi(z)$ is a polynomial whose degree does not exceed n.

REMARK. If the function $f(x)$ is as stated in Theorem 6, and if $g(x)e^{x^2/2} \to 0$ as $|x| \to \infty$, then we conclude that $Q(x) \to 0$ as $|x| \to \infty$, i.e., $g(x) = f(x) \equiv 0$. This gives a uniqueness theorem.

In Lecture 25, Part III we will present uniqueness theorems generalizing theorems of this section. The recent developments are exposed in Nazarov [**101**].

LECTURE 9

The Pólya Theorem

9.1. Supporting functions of convex sets

We start with the introduction of a notion of *supporting function* $k(\theta)$ of a set $K \subset \mathbb{C}$:

$$k(\theta) = \sup_{z\in K}\{x\cos\theta + y\sin\theta\} = \sup_{z\in K}\{\mathrm{Re}(ze^{-i\theta})\}\,, \qquad \theta \in [0, 2\pi]\,.$$

It is not difficult to prove that the supporting function of a set coincides with the supporting function of its closed convex hull. In what follows we assume that K is a convex compact set.

For each $\theta \in [0, 2\pi]$ the line $l_\theta = \{z : \mathrm{Re}(ze^{-i\theta}) = k(\theta)\}$ is called a supporting line of K. Evidently, it is orthogonal to the ray $\{z : \arg z = \theta\}$, has nonvoid intersection with K, and the set K itself is contained completely in a closed half-plane with the boundary l_θ. The value $|k(\theta)|$ is equal to the length of the segment of the ray $\{z : \arg z = \theta\}$ if $k(\theta) > 0$ or of the ray $\{z : \arg z = \theta + \pi\}$ if $k(\theta) < 0$, cut off by the line l_θ. If $k(\theta)$ is the supporting function of K, then $k(-\theta)$ is the supporting function of $\overline{K}$, where the bar means, as usual, the complex conjugation.

EXAMPLES. The supporting function of the disk $\{z : |z| \le R\}$ is $k(\theta) = R$. The supporting function of a single point $\{z_0 = r_0e^{i\theta_0}\}$ is $k(\theta) = r_0\cos(\theta - \theta_0)$. The latter function is sinusoidal. The converse statement is also true: every sinusoidal function is the supporting function of a set consisting of a single point. The supporting function of the segment $[-id, id]$ is $k(\theta) = d|\sin\theta|$.

THEOREM 1. *The supporting function of a convex compact set is trigonometrically convex. Conversely, every 2π-periodic trigonometrically convex function $k(\theta)$ is the supporting function of some convex compact set K.*

PROOF. Let $k(\theta)$ be the supporting function of a convex compact set K. Then $k(\theta)$ is the upper envelope of a uniformly bounded family of 2π-periodic sinusoidal functions $h(\theta) = \mathrm{Re}(ze^{-i\theta})$, $z \in K$. Hence $k(\theta)$ is trigonometrically convex, which proves the first statement of Theorem 1.

To prove the converse statement, let us consider the set

$$K = \bigcap_{0\le\theta\le\pi} \Pi_\theta\,,$$

where $\Pi_\theta = \{z : \mathrm{Re}(ze^{-i\theta}) \le k(\theta)\}$ is a half-plane, and prove that this set is not empty. Moreover, we shall see that every line $l_\theta = \{z : \mathrm{Re}(ze^{-i\theta}) = k(\theta)\}$ contains points of K. It will show that K is a convex compact set and that l_θ are supporting lines of this compact set.

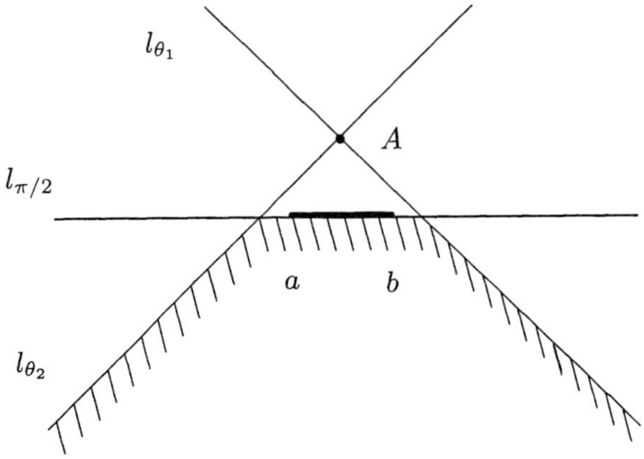

FIGURE 1

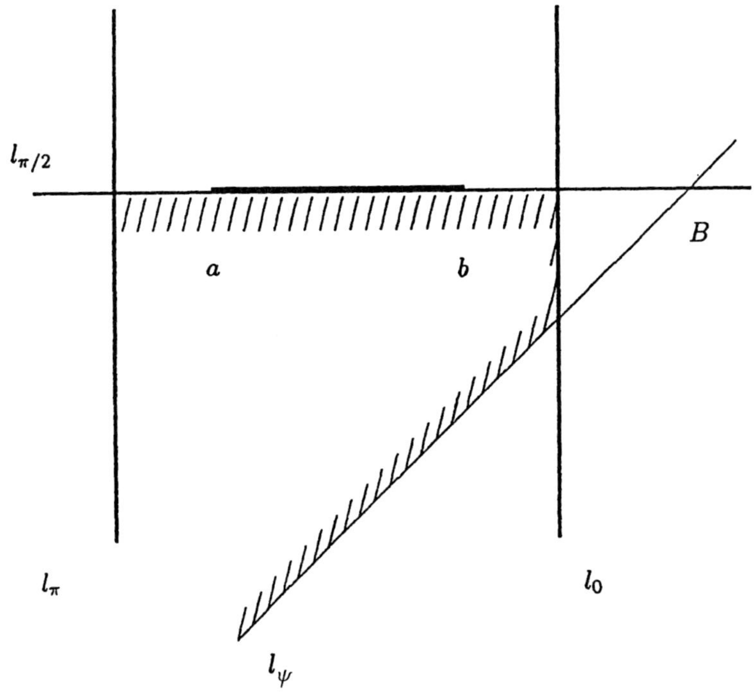

FIGURE 2

Without loss of generality, we fix $\theta = \pi/2$ and prove that $l_{\pi/2}$ contains at least one point which belongs to all half-planes Π_θ. Let us assume that $0 < \theta_1 < \pi/2 < \theta_2 < \pi$, and denote by A the intersection point of the lines l_{θ_1} and l_{θ_2}. Since the function $k(\theta)$ is trigonometrically convex, the point A cannot lie below the line $l_{\pi/2}$ (see Figure 1). Thus, the half-planes Π_{θ_1}, Π_{θ_2} and the line $l_{\pi/2}$ have a common segment which we denote by $[a_{\theta_2}, b_{\theta_1}]$. It may happen that $a_{\theta_2} = b_{\theta_1}$. Setting $a = \sup\{a_{\theta_2} : \pi/2 \leq \theta_2 \leq \pi\}$, $b = \inf\{b_{\theta_1} : 0 \leq \theta \leq \pi/2\}$, we find that $a \leq b$, and hence the segment $[a, b] \subset l_{\pi/2}$ belongs to all half-planes Π_θ, $0 \leq \theta \leq \pi$, and, in particular, to the strip $\{z = x + iy : k(\pi) \le x \le k(0)\}$. Let us show that each half-plane Π_ψ, $-\pi/2 < \psi < 0$ also intersects this segment. Using the definition of trigonometrically convex functions with $\psi = \theta_1 < 0 < \theta_2 = \pi/2$, we find that the intersection point of $l_{\pi/2}$ and l_ψ cannot lie to the left of the intersection point B of the lines $l_{\pi/2}$ and l_0 (see Figure 2). Hence the segment $[a, b]$ is in the half-plane Π_ψ. The case $-\pi < \psi \le -\pi/2$ can be examined using the same arguments, which proves the theorem.

Let F be an entire function of exponential type (EFET). By Theorem 1 from Section 8.1, its indicator h_F is trigonometrically convex, and hence h_F is the supporting function of a convex compact set $I_F \subset \mathbb{C}$. This compact set is called the *indicator diagram* of the function F. It gives a geometrical representation of the growth of F in various directions.

This definition allows us to give, in particular, a simple geometrical interpretation of certain properties of indicators of EFET. For example, let $F(z)$, $G(z)$ be such functions; then the inequality

$$h_{F+G}(\theta) \le \max\{h_F(\theta), h_G(\theta)\}$$

means that the indicator diagram I_{F+G} is contained in the convex hull of the diagrams I_F and I_G.

PROBLEM 1. Prove that if one of the indicator diagrams I_F and I_G can be obtained from the other by a parallel translation, then the indicator diagram of the sum I_{F+G} coincides with the convex hull of the indicator diagrams I_F and I_G.

The sum of sets, $K = K_1 + K_2$, is the set of points $\{z = z_1 + z_2 : z_1 \in K_1, z_2 \in K_2\}$. It is evident that the sum of convex compact sets is a convex compact set. It follows from the definition that the supporting function of the sum of convex compact sets equals $k(\theta) = k_1(\theta) + k_2(\theta)$. Therefore, if $h_{FG}(\theta) = h_F(\theta) + h_G(\theta)$, then $I_{FG} = I_F + I_G$, and conversely, the latter equality implies the former.

9.2. The Borel transform and the Pólya theorem

Let

$$F(z) = \sum_{n=0}^{\infty} \frac{c_n}{n!} z^n$$

be an EFET. It is easy to deduce from the formula for the type of an entire function that

$$\sigma = \sigma_F = \limsup_{n\to\infty} \sqrt[n]{|c_n|}\,.$$

The function

$$f(z) = \sum_{n=0}^{\infty} \frac{c_n}{z^{n+1}}$$

is called the *Borel transform* of the function $F(z)$. By the Hadamard theorem the series converges outside the disk $\{|z| \le \sigma_F\}$ and diverges inside this disk. It is possible that the function $f(z)$ can be analytically continued into the disk $\{z : |z| < \sigma_F\}$. The smallest convex compact set containing all singularities of $f(z)$ is called the *conjugate indicator diagram* of $F(z)$. We denote by $k_F(\theta)$ the supporting function of this compact set.

The following theorem establishes the remarkable connection between the conjugate diagram and the indicator diagram of EFET.

THEOREM 2 (Pólya). *For every EFET $F(z)$ the relation*

$$h_F(\theta) = k_F(-\theta)$$

holds, and hence the conjugate diagram is the reflection in the real axis of the indicator diagram I_F.

PROOF. Let us denote by $\overline{I}$ the conjugate diagram. The proof is based on two integral formulas linking the function $F(z)$ and its Borel transform. The first of them has the form

$$F(z) = \frac{1}{2i\pi} \int_{\partial(\overline{I}+K_\varepsilon)} f(\zeta) e^{\zeta z} \, d\zeta, \tag{2}$$

where K_ε is the disk $\{z : |z| \leq \varepsilon\}$.

Indeed, the integration over the curve $\partial(\overline{I} + K_\varepsilon)$ on the right-hand side may be replaced by the integration over the circle $\partial K_{\sigma+\varepsilon}$ (here σ is the type of the function F). Thus the formula (2) is obtained by the integration of the series

$$e^{\zeta z} f(\zeta) = \sum_{n=0}^{\infty} \frac{c_n}{\zeta^{n+1}} e^{\zeta z} .$$

It follows from (2) that

$$\begin{aligned} |F(re^{i\theta})| &\leq C_\varepsilon \exp\left\{r \max_{\zeta \in \overline{I}+K_\varepsilon} \operatorname{Re}(\zeta z)\right\} \\ &= C_\varepsilon \exp\left\{r\big(k(-\theta) + \varepsilon\big)\right\} . \end{aligned}$$

Hence

$$h(\theta) \leq k(-\theta) . \tag{3}$$

The transform inverse to (2) has the form

$$f(\zeta) = \int_0^\infty F(te^{-i\theta}) e^{-\zeta t e^{-i\theta}} \, d(te^{-i\theta}) \, ; \quad \operatorname{Re}(\zeta e^{-i\theta}) > h(-\theta) . \tag{4}$$

To prove this formula we observe that the inequality

$$|F(te^{-i\theta})| \overset{\text{as}}{<} e^{(h(-\theta)+\varepsilon)t}$$

implies that the integral in (4) converges uniformly in the half-plane $\{\zeta = \xi + i\eta : \xi \cos\theta + \eta \sin\theta \geq h(-\theta) + 2\varepsilon\}$, and hence this integral represents a holomorphic function in the domain $\{\zeta : \operatorname{Re}(\zeta e^{-i\theta}) > h(-\theta)\}$. Let us check that this function coincides with $f(\zeta)$ for $\zeta = re^{i\theta}$, $r > 3\sigma$. Indeed, in this case the integral (4) can be written as

$$\int_0^\infty F(te^{-i\theta}) e^{-rt} e^{-i\theta} \, dt .$$

For the series

$$F(te^{-i\theta}) = \sum_{n=0}^{\infty} \frac{c_n}{n!} t^n e^{-in\theta} ,$$

we have the estimate of the general term

$$\left|\frac{c_n}{n!}\right| \leq \frac{M_F(2t)}{(2t)^n} \overset{\text{as}}{<} \frac{e^{2(\sigma+\varepsilon)t}}{(2t)^n} , \quad n = 0, 1, \dots ,$$

and of the remainder

$$|R_n(t)| \le \sum_{k=n+1}^{\infty} \left|\frac{c_k}{k!}\right| t^k \le \sum_{k=n+1}^{\infty} \frac{1}{2^k} e^{2(\sigma+\varepsilon)t} = \frac{1}{2^n} e^{2(\sigma+\varepsilon)t} .$$

If $r > 3\sigma$ we obtain the estimate

$$\Big| \int_0^\infty F(te^{-i\theta}) e^{-tr} e^{-i\theta}\, dt - \int_0^\infty e^{-tr} e^{-i\theta} \sum_{k=0}^{n} \frac{c_k}{k!} t^k e^{-ik\theta}\, dt \Big| \le \frac{1}{2^n} \int_0^\infty e^{-(\sigma-2\varepsilon)t}\, dt .$$

Integrating termwise, we obtain the equality (4). Thus, $f(\zeta)$ is analytic in the domain $\{\zeta = \xi + i\theta : \xi\cos\theta + \eta\sin\theta > h(-\theta)\}$. Hence, $k(\theta) \le h(-\theta)$. Combining this estimate with (3), we obtain $h(\theta) = k(-\theta)$. The theorem is proved.

REMARK 1. It follows from formula (4) with $\theta = 0$ that the function $f(\zeta)$ coincides with the Laplace transform of the function $F(z)$.

REMARK 2. Let $F(z)$ be a holomorphic function of exponential type inside an angle $\{z : \alpha \le \arg z \le \beta\}$. Then the function $f(\zeta)$ defined by (4) is analytic in the union of half-planes $\{\zeta = \xi + i\eta : \xi\cos\theta + \eta\sin\theta > h(-\theta),\ \alpha \le \theta \le \beta\}$. The complement to this domain is the intersection of closed half-planes $\{\zeta = \xi + i\eta : \xi\cos\theta + \eta\sin\theta \le h(-\theta)\}$, and its boundary contains two rays orthogonal to the rays $\arg\zeta = -\alpha$ and $\arg\zeta = -\beta$. This closed convex set is minimal among all convex sets of such a form containing all singularities of $f(\zeta)$, which is not difficult to prove using the well-known inversion formula for the Laplace transform

$$F(z) = \frac{1}{2i\pi} \lim_{a\to\infty} \int_{a-i\infty}^{a+i\infty} f(\zeta) e^{\zeta z}\, d\zeta .$$

The notions of indicator, indicator diagram, and conjugate diagram can be extended to entire functions with values in a Banach space. To this end, the modulus $|f(re^{i\theta})|$ in the corresponding definitions should be replaced by the norm $\|f(re^{i\theta})\|$.

PROBLEM 2. Prove that the convex hull of the spectrum of an arbitrary element x of a Banach algebra coincides with the indicator diagram of the entire function $e^{\lambda x}$.

EXAMPLE 1. Let $F(z) = \sum_{k=1}^{n} P_k(z) e^{\lambda_k z}$, where λ_k, $k = 1, 2, \dots$, are complex numbers, and $P_k(z)$ are polynomials. The Borel transform of a monomial $z^p e^{\lambda_k z}$ equals $p!(\zeta - \lambda_k)^{-p-1}$. Therefore, the poles at the points $\lambda_k, k = 1, 2, \dots$, are the only singularities of the Borel transform of $F(z)$. The conjugate diagram of $F(z)$ coincides with the convex hull of points $\{\lambda_1, \dots, \lambda_n\}$, and by the Pólya theorem the indicator diagram of $F(z)$ coincides with the convex hull of points $\{\overline{\lambda}_1, \dots, \overline{\lambda}_n\}$.

EXAMPLE 2. Let K be an arbitrary convex compact set. Choose a countable set of points $\{\lambda_k\}$ dense on ∂K, and consider a function

$$f(\zeta) = \sum_{k=0}^{\infty} \frac{c_k}{\zeta - \lambda_k}, \quad \sum |c_k| < \infty .$$

Evidently, the function $f(\zeta)$ is analytic outside K, equals zero at infinity, and cannot be analytically continued through any part of ∂K. Hence K is the indicator diagram of the entire function

$$F(z) = \sum_{k=0}^{\infty} c_k e^{\lambda_k z} .$$

Thus for every closed compact set $K \subset \mathbb{C}$ there exists an entire function of exponential type whose indicator diagram coincides with K.

LECTURE 10

Applications of the Pólya Theorem

10.1. The Paley-Wiener theorem

The following theorem gives a description of the class of entire functions of exponential type that are square integrable on the real axis.

THEOREM 1 (Paley-Wiener). *For a function g to be representable in the form*

$$g(x) = \frac{1}{2\pi}\int_a^b \psi(t)e^{itx}\,dt\,, \quad \psi \in L^2(a,b)\,, \tag{1}$$

it is necessary and sufficient that

a) *it be possible to extend $g(x)$ to the whole complex plane as an EFET;*

b) $g \in L^2(-\infty,\infty)$.

If the interval (a,b) cannot be replaced by a smaller interval, then the segment $[ia, ib]$ of the imaginary axis coincides with the conjugate diagram of $g(z)$.

NECESSITY. By the Fourier-Plancherel theorem, we have $g \in L^2(-\infty,\infty)$ and $\sqrt{2\pi}\|g\|_{L^2(-\infty,\infty)} = \|\psi\|_{L^2(a,b)}$. Further, the function

$$g(z) = \frac{1}{2\pi}\int_a^b e^{itz}\psi(t)\,dt$$

is entire since the integrand is an entire function of $z \in \mathbb{C}$. For $y \geq 0$, we have

$$|g(x+iy)| \leq \frac{1}{2\pi}\int_a^b |\psi(t)|e^{\mathrm{Re}(itz)}\,dt \leq e^{-ay}\frac{1}{2\pi}\int_a^b |\psi(t)|\,dt\,.$$

Similarly, for $y \leq 0$,

$$|g(x+iy)| \leq e^{-by}\frac{1}{2\pi}\int_a^b |\psi(t)|\,dt\,.$$

It follows that $g(z)$ is an EFET and its indicator diagram is contained in the segment $[-ib, -ia]$ of the imaginary axis.

SUFFICIENCY. The Borel transform $\varphi(w)$ of the function g is holomorphic outside the conjugate diagram of g, and, in particular, outside the disk $\{w : |w| \leq \sigma_g\}$. The function φ can be represented using the Laplace transform

$$\varphi(u+iv) = \int_0^\infty g(x)e^{-x(u+iv)}\,dx, \quad u > \sigma_g\,, \tag{2}$$

and, similarly,

$$\varphi(u+iv) = -\int_{-\infty}^{0} g(x)e^{-x(u+iv)}\,dx\,, \quad u < -\sigma_g\,. \tag{3}$$

Since $g \in L^2(-\infty,\infty)$, equations (2) and (3) imply that φ is holomorphic in the half-planes $\{w : \pm\,\mathrm{Re}\, w > 0\}$. Therefore, the conjugate diagram of the function g coincides with a segment $[i\alpha, i\beta]$ of the imaginary axis.

The function $\varphi(u+iv)$ is square integrable on every vertical line which is not the imaginary axis. Let us show that there exist the mean square limits $\operatorname*{l.i.m.}_{u\to\pm 0}\varphi(u+iv)$ equal to

$$\varphi(+0+iv) = \int_0^{\infty} g(x)e^{-ixv}\,dx\,,$$
$$\varphi(-0+iv) = -\int_{-\infty}^{0} g(x)e^{-ixv}\,dx\,,$$

where the integrals are mean square convergent as well. Indeed, by the Plancherel theorem we have

$$\int_{-\infty}^{\infty} |\varphi(+0+iv) - \varphi(u+iv)|^2\,dv = \frac{1}{2\pi}\int_0^{\infty} |g(x)|^2|1-e^{-ux}|^2\,dx \to 0\,, \quad u \searrow 0\,.$$

Similarly, $\varphi(-0+iv) = \operatorname*{l.i.m.}_{u\to -0}\varphi(u+iv)$.

The functions $\varphi(\pm 0 + iv)$ coincide with $\varphi(iv)$ and, consequently, with each other, at imaginary points lying outside the indicator diagram. Thus, if $\psi(v)$ is the Fourier-Plancherel transform of the function $g(x)$, then for such values v we obtain

$$\begin{aligned}\psi(v) &= \int_{-\infty}^{0} g(x)e^{-ixv}\,dx + \int_0^{\infty} g(x)e^{-ixv}\,dx \\ &= \varphi(+0+iv) - \varphi(-0+iv) = 0\,.\end{aligned}$$

The inversion formula implies equation (1) which shows that the indicator diagram of the function $g(z)$ contains the segment $[-ib, -ia]$. The theorem is proved.

10.2. Analytic continuation of a power series

The next application of the Pólya theorem is related to analytic continuation of a power series

$$\varphi(z) = \sum_{n=0}^{\infty} a_n z^n \tag{4}$$

converging in a neighborhood of the origin. First, observe that there exists an EFET $F(w)$ such that $a_n = F(n)$, $n = 0, 1, \ldots$. Indeed, if the series (4) converges for $|z| < r$, then

$$a_n = \frac{1}{2\pi i}\int_{|z|=r} \frac{\varphi(z)}{z^{n+1}}\,dz\,. \tag{5}$$

Choosing a continuous branch of the logarithm $\log z$ for $|\arg z| < \pi$ and setting $\zeta = -\log z$, we obtain $a_n = F(n)$, $n = 0, 1, \dots$, where

$$F(w) = -\frac{1}{2\pi i} \int_{-\log r - i\pi}^{-\log r + i\pi} \varphi(e^{-\zeta}) e^{w\zeta}\, d\zeta$$

is an EFET. Hence the converging series (4) can be represented in the form

$$\varphi(z) = \sum_{n=0}^{\infty} F(n) z^n, \tag{6}$$

where the width of the indicator diagram of the function F along the imaginary axis does not exceed 2π, i.e., $h_F(\pi/2) + h_F(-\pi/2) \leq 2\pi$. If this width is smaller than 2π, then the following theorem asserts that the series (6) may be analytically extended.

THEOREM 2 (Carlson). *If the width of the indicator diagram I_F of the function F along the imaginary axis is less than 2π, then the function φ defined in neighborhood of the origin by the series (6) may be analytically continued into a domain $\mathbb{C}\backslash G$, where $G = \{w = e^{-z},\ z \in \overline{I}_F\}$, and $\varphi(\infty) = 0$.*

Conversely, let a function φ be represented in a neighborhood of the origin by series (4) and also can be continued analytically into the exterior of a set

$$G = \{w : w = e^{-z}\, z \in K\}\,, \tag{7}$$

where K is a convex compact set whose width along the imaginary axis is less than 2π, and let $\varphi(\infty) = 0$. Then there exists an EFET F such that $\overline{I}_F \subset K$ and $a_n = F(n)$, $n = 0, 1, 2, \dots$.

PROOF. To prove the first assertion of the theorem, we use the inversion formula (2) for the Borel transform defined in Section 9.2. Using this formula, we write equation (6) in the form

$$\varphi(z) = \frac{1}{2\pi i} \int_{\partial(\overline{I}_F + K_\varepsilon)} f(\zeta) \sum_{n=0}^{\infty} e^{n\zeta} z^n\, d\zeta,$$

where $\varepsilon > 0$ is chosen so small that the width along the imaginary axis of the compact set $\overline{I}_F + K_\varepsilon$ is less than 2π. For small enough values of $|z|$ the series in the integrand converges and hence

$$\varphi(z) = \frac{1}{2\pi i} \int_{\partial(\overline{I}_F + K_\varepsilon)} \frac{f(\zeta)}{1 - z e^{\zeta}}\, d\zeta\,. \tag{8}$$

The function defined by the integral is analytic outside $G = \{e^{-z},\ z \in \overline{I}_F\}$ and equals zero at infinity. Since the width of $\overline{I}_F$ along the imaginary axis is less than 2π, the set $\overline{\mathbb{C}}\backslash G$ is connected. The first part of the theorem is proved.

Conversely, let all singularities of the function φ represented by series (4) lie in a set G of the form (7), where K is a convex compact set whose width along the imaginary axis is less than 2π, and let $\varphi(\infty) = 0$. Under these assumptions the set $\overline{\mathbb{C}}\backslash G$ is a domain containing the points 0 and ∞. We have

$$a_n = \frac{1}{2\pi i} \int_C \frac{\varphi(z)}{z^{n+1}}\, dz\,, \quad n = 0, 1, \dots\,,$$

where C is a circumference of small radius centered at the origin. Since $\varphi(\infty) = 0$, each of these integrals can be replaced by the integral over an arbitrary contour L which surrounds G and does not surround the origin. Setting

$$F(w) = \frac{1}{2\pi i} \int_L \frac{\varphi(\zeta)}{\zeta^{w+1}}\, d\zeta\,,$$

we obtain $F(n) = a_n$, $n = 0, 1, \ldots$, and

$$F(w) = \frac{1}{2\pi i} \int_\Gamma \varphi(e^{-\zeta}) e^{\zeta w}\, d\zeta\,,$$

where the contour Γ surrounds the compact set K. It follows that F is an EFET and its indicator diagram is contained in K. The theorem is proved.

COROLLARY (Leau, Wigert). *In order that the function φ represented by the series (4) have a singularity at $z = 1$ only, and have a zero at infinity, it is necessary and sufficient that the coefficients of series (4) have the form $a_n = F(n)$, where $F(w)$ is an entire function whose growth does not exceed order one and minimal type.*

PROBLEM 1. Let $\varphi(z) = G(1/(1-z))$, where G is an entire function and $G(0) = 0$. Then the order ρ_G of the function G and the order ρ_F of the entire function $F(w)$ interpolating the coefficients of $\varphi(z)$ are related by the equation $\rho_G = \rho_F/(1-\rho_F)$.

PROBLEM 2. In order that the function

$$\varphi(z) = \sum_{n=0}^{\infty} a_n z^n$$

holomorphic in a neighborhood of the origin have only a finite number of isolated singularities (in whose vicinities the function φ is single-valued) in the disk $|z| < R$, it is necessary and sufficient that

$$a_n = \sum_{k=1}^{N} F_k(n) z_k^{-n} + O\Big(\frac{1}{R^n}\Big)\,, \qquad n \to \infty\,,$$

where $F_k(w)$ are entire functions of minimal exponential type. Moreover, $z_1, \ldots, z_N$ are singularities of $\varphi(z)$ in the disk $\{|z| < R\}$. In order that these singularities be poles, it is necessary and sufficient that the functions $F_k(w)$ be polynomials.

REMARK 1. Equation (8) yields that if the assumptions of the Carlson theorem hold, then the expansion of $\varphi(z)$ in a neighborhood of infinity has the form

$$\varphi(z) = -\sum_{n=1}^{\infty} \frac{F(-n)}{z^n}\,.$$

REMARK 2. If the assumptions of the Carlson theorem hold, then the function $\varphi(z)$ can be continued to infinity along some ray emanating from the origin. If the indicator diagram I_F of the interpolating EFET $F(w)$ is such that the compact set e^{-I_F} does not separate 0 and ∞, then, as before, $\varphi(z)$ can be analytically continued at infinity. Evidently, e^{-I_F} does not separate 0 and ∞ if and only if the sets $I_F + 2m\pi i$, $m \in \mathbb{Z}$, are pairwise disjoint.

PROBLEM 3. Let $F(w)$ be an EFET whose indicator diagram has the width along the imaginary axis less than 2π. Prove that

$$h_F(0) = \limsup_{n\to\infty} \frac{\log|F(n)|}{n}\,.$$

Other applications of the Pólya theorem to problems of analytic continuation may be found in Bieberbach [**16**].

10.3. Analytic functionals

As was stated in Section 3.4, every linear functional F on the space $A(D)$ of analytic functions in a simply connected domain $D \Subset \mathbb{C}$ is defined by the equation

$$F(f) = \frac{1}{2\pi i}\int_l f(\zeta)\varphi(\zeta)\,d\zeta\,,$$

where the function φ determined by F is analytic on the complement of D, $\varphi(\infty) = 0$, and the simple closed curve l surrounds all singularities of φ and lies in D. Thus, the space $A^*(D)$ of linear functionals on $A(D)$ (they are called analytic functionals) is isomorphic to the space $A_0(\mathbb{C}\backslash D)$ of functions analytic on $\mathbb{C}\backslash D$ which are zero at infinity.

The Pólya theorem gives another representation of the space $A^*(D)$ if D is a convex domain. In this case we may assume that compact sets $G_1 \Subset G_2 \Subset \cdots \Subset G_m \Subset \cdots$ exhausting the domain D from the inside are convex and their supporting functions satisfy the condition

$$h_1(\theta) < h_2(\theta) < \cdots < h_m(\theta) < \cdots\,.$$

Then the function $H(\theta) = \lim_{m\to\infty} h_m(\theta)$ is supporting for D. Given a linear functional F, let us introduce the function

$$\Phi(\lambda) = F(e^{\lambda z}) \tag{9}$$

usually called the Fourier-Borel transform of F. According to formulas (8) and (10) from Lecture 3, there exist an integer $m \geq 1$ and a constant C such that

$$|\Phi(\lambda)| \leq C \max_{z\in D_m} |e^{\lambda z}| = C\exp\big(h_m(-\arg\lambda)|\lambda|\big).$$

So $\Phi(\lambda)$ is an EFET and by the Pólya theorem its conjugate diagram is contained in the domain D. Denote by $\varphi(\zeta)$ the Borel transform of the function $\Phi(\lambda)$ and check that

$$F(f) = \frac{1}{2\pi i}\int_l f(\zeta)\varphi(\zeta)\,d\zeta \tag{10}$$

for each function $f \in A(D)$, where l is a simple closed contour in D surrounding all singularities of the function φ.

Indeed,

$$F(e^{\lambda z}) = \Phi(\lambda) = \frac{1}{2\pi i}\int_l e^{\lambda\zeta}\varphi(\zeta)\,d\zeta\,.$$

Differentiating this equation with respect to λ and setting $\lambda = 0$ afterwards, we obtain

$$F(z^k) = \frac{1}{2\pi i}\int_l \zeta^k\varphi(\zeta)\,d\zeta\,.$$

Hence (10) holds for every polynomial. Since polynomials are dense in $A(D)$, equation (10) holds for every function $f \in A(D)$.

Let $\Phi(\lambda)$ be an entire function satisfying an estimate

$$|\Phi(\lambda)| \leq C \exp\left[h_m(-\arg\lambda)|\lambda|\right]$$

for some C and m, and let $\varphi(\zeta)$ be its Borel transform. Then equation (10) defines a linear functional F on $A(D)$ such that $\Phi(\lambda) = F(e^{\lambda z})$. Thus, we have proved

THEOREM 3. *Equation* (9) *defines an isomorphism between the space* $A^*(D)$ *of linear functionals on the space of analytic functions in a convex domain* D *with supporting function* $H(\theta)$, *and the space of entire functions of exponential type whose indicators satisfy the condition*

$$h(\theta) < H(\theta)\ .$$

If $D = \mathbb{C}$, then $H(\theta) \equiv +\infty$, and $A(D)$ is the space of all entire functions endowed with the topology of uniform convergence on each compact set in $\mathbb{C}$. According to Theorem 3, the space $A^*(D)$ can be identified with the space of all EFET.

LECTURE 11

Lower Bounds for Analytic and Subharmonic Functions

11.1. The Carathéodory inequality

For a function $f(z) = u(z) + iv(z)$ analytic in the disk $\{z : |z| \le R\}$ we set $A_f(r) = \max\{u(z) : |z| \le r\}$. It follows from the Maximum Principle for harmonic functions that $A_f(r)$ is a monotonically increasing function of r, and that $|A_f(r)| < M_f(r)$. It appears that for $R > r$ the value $M_f(r)$ can be estimated from above by means of $A_f(R)$.

THEOREM 1 (Carathéodory). *Let $f(z)$ be an analytic function in the disk $\{z : |z| \le R\}$, and let $f(0) = 0$. Then*

$$M_f(r) \le \frac{2r}{R-r} A_f(R) .$$

PROOF. The Schwarz formula, Section 2.1, states that

$$f(z) = \frac{1}{2\pi} \int_0^{2\pi} u(Re^{i\psi}) \frac{Re^{i\psi} + z}{Re^{i\psi} - z} \, d\psi , \quad |z| < R .$$

Further, by the condition $f(0) = 0$, we have

$$0 = \frac{1}{2\pi} \int_0^{2\pi} u(Re^{i\psi}) \, d\psi .$$

Hence

$$f(z) = \frac{1}{2\pi} \int_0^{2\pi} u(Re^{i\psi}) \frac{2z}{Re^{i\psi} - z} \, d\psi .$$

According to the Cauchy theorem

$$\frac{1}{2\pi} \int_0^{2\pi} \frac{d\psi}{Re^{i\psi} - z} = \frac{1}{2\pi i} \int_{|\zeta|=R} \frac{d\zeta}{\zeta(\zeta - z)} = 0 .$$

Thus

$$-f(z) = \frac{1}{2\pi} \int_0^{2\pi} [A_f(R) - u(Re^{i\psi})] \frac{2z}{Re^{i\psi} - z} \, d\psi ,$$

and then

$$|f(z)| \le \frac{1}{2\pi} \int_0^{2\pi} [A_f(R) - u(Re^{i\psi})] \frac{2r}{R-r} \, d\psi = \frac{2r}{R-r} A_f(R) ,$$

proving Theorem 1.

The proven inequality yields *a lower bound for the harmonic function* $u(z)$ *in the disk* $\{z : |z| \le R\}$ provided that $u(0) = 0$:

$$u(z) \ge -\frac{2r}{R-r} \max\{u(Re^{i\psi}) : 0 \le \psi \le 2\pi\}\,, \quad r = |z|\,.$$

It immediately implies

THEOREM 2. *If an analytic function* $f(z)$ *has no zeros in a disk* $\{z : |z| \le R\}$ *and if* $|f(0)| = 1$, *then*

$$\log|f(z)| \ge -\frac{2r}{R-r} \log M_f(R)$$

as $|z| = r < R$. *In particular,* $\log|f(z)| > -2\log M_f(2r)$.

PROBLEM 1. Prove that

$$M_f(r) \le [A_f(R) - \operatorname{Re} f(0)]\frac{2r}{R-r} + |f(0)|\,, \quad r < R$$

for a function analytic in the disk $\{z : |z| \le R\}$.

PROBLEM 2. Let $f(z)$ be an analytic function in the upper half-plane $\{z : \operatorname{Im} z > 0\}$ such that $\operatorname{Im} f(z) > 0$. Prove the estimates

$$\frac{1}{5}|f(i)|\frac{\sin\theta}{r} \le |f(z)| \le 5|f(i)|\frac{r}{\sin\theta}\,, \quad z = re^{i\theta}\,, \quad 0 < \theta < \pi\,, \quad r \ge 1\,.$$

This is the so-called Carathéodory inequality for a half-plane.

HINT. Consider the function

$$F(u) = if\Big(-i\frac{u+1}{u-1}\Big),\ |u| < 1;$$

it satisfies $\operatorname{Re} F(0) \le 0$. Then apply the inequality from the previous problem.

11.2. The Cartan estimate

If the function $f(z)$ has zeros in the disk $\{z : |z| \le R\}$, then

$$\log|f(z)| = \log|P(z)| + \log\left|\frac{f(z)}{P(z)}\right|,$$

where $P(z)$ is a polynomial and the second term is a harmonic function. Therefore, the problem of estimating the function $f(z)$ from below reduces to that for the first term. Evidently, such an estimate is possible only outside some neighborhood of the zeros of the function $f(z)$.

We shall consider a more general problem of estimating from below the logarithmic potential of a finite measure.

THEOREM 3. *Let*

$$u(z) = \iint_{\mathbb{C}} \log|z-\zeta|\,d\mu(\zeta)$$

where μ *is Borel measure,* $\mu(\mathbb{C}) = n < \infty$. *Given* H, $0 < H < 1$, *there exists a system of disks in the complex plane such that*

$$\sum r_j \le 5H\,,$$

where r_j are radii of these disks, and

$$u(z) \geq n \log \frac{H}{e}$$

everywhere outside these disks.

PROOF. Fix $p > 0$. A point $z \in \mathbb{C}$ is said to be p-normal, if $n(t; z) < pt$, $t > 0$, where $n(t; z) = \mu(\{\zeta : |z - \zeta| \leq t\})$. If z is a p-abnormal point, then there exists a number t such that $n(t; z) \geq pt$. Let ρ_z be the l.u.b. of the set of such values t. Since $\mu(\mathbb{C}) < \infty$, the value ρ_z is finite and attained for some t. Indeed, let $t_m \nearrow \rho_z$ and $n(t_m; z) \geq pt_m$. Then

$$n(\rho_z; z) \geq pt_m \to p\rho_z\ , \quad m \to \infty\ .$$

Thus, for every p-abnormal point z there exists a radius ρ_z and an exceptional disk $C_z = \{\zeta : |z - \zeta| < \rho_z\}$. For normal points we set $\rho_z = 0$.

Let $r_1 = \sup\{\rho_z : z \in \mathbb{C}\}$. We will prove that this l.u.b. is attained at some point. Note that for a given $\varepsilon > 0$ one can choose a value R_ε such that the measure μ of the domain $\{z : |z| > R_\varepsilon\}$ is less than ε, and hence $\rho_z \to 0$ as $z \to \infty$. Let $\{z_m\}$ be a sequence such that $\rho_{z_m} \nearrow r_1$. Since the sequence of points z_m lies in some disk $|z| \leq R$, we may assume, without loss of generality, that $z_m \to \zeta$. We have $n(r_1 + \varepsilon; \zeta) \geq pr_1$, and then $n(r_1; \zeta) \geq pr_1$.

We delete from the plane the open exceptional disk C_1 with the center at $\zeta_1 = \zeta$. Similarly, in the remaining part of the plane, the l.u.b. of radii of abnormality is attained at some point ζ_2. We select the corresponding disk C_2. Continuing this construction, we obtain a sequence of exceptional disks $C_1, C_2, \dots$ with centers at $\zeta_1, \zeta_2, \dots$, and radii $r_1 \geq r_2 \geq \cdots$.

Let us show that no point of the plane will be covered by more than five disks C_j. Indeed, let a point z' be covered by disks $C'_1, \dots, C'_k$ with radii $r'_1 \geq r'_2 \geq \cdots \geq r'_k$. Draw vectors from z' to the centers $\zeta'_1, \dots, \zeta'_k$ of these disks. Since the center of each disk lies outside other disks, the angle between each pair of these vectors is larger than $\pi/3$. Thus there are no more than five such vectors.

The disks C_j are exceptional, i.e., $n(r_j; \zeta_j) \geq pr_j$. Therefore,

$$p \sum_j r_j \leq \sum_j n(r_j; \zeta_j) \leq 5\mu(\mathbb{C}) = 5n\ .$$

Choosing $p = n/H$, we obtain

(1)
$$\sum_j r_j \leq 5H.$$

Evidently, if there is an infinite number of disks, then $r_j \to 0$ as $j \to \infty$, and since for each p-abnormal point z there is a radius of abnormality $\rho_z > 0$, every p-abnormal point will be covered by some disk C_j.

It remains to estimate the potential $u(z)$ at an arbitrary normal point z. It is evident that

$$\begin{aligned} u(z) &\geq \iint_{|z-\zeta|\leq 1} \log|z - \zeta|\, d\mu(\zeta) = \int_0^1 \log t\, dn(t; z) \\ &= n(t; z) \log t\Big|_0^1 - \int_0^1 \frac{n(t; z)}{t}\, dt\ , \end{aligned}$$

and since $n(t;z) < pt$, we have

$$u(z) \geq -\int_0^1 \frac{n(t;z)}{t}\,dt\,.$$

In addition, $n(t;z) \leq n = pH$. Therefore,

$$u(z) \geq -\int_0^H \frac{pt}{t}\,dt - \int_H^1 \frac{n}{t}\,dt = -n - n\log\frac{1}{H} = n\log\frac{H}{e}\,,$$

and the theorem is proved.

In particular, if

$$P(z) = \prod_{k=1}^{n}(z - z_k)\,,$$

then the inequality

$$|P(z)| \geq \left(\frac{H}{e}\right)^n \tag{2}$$

holds outside exceptional disks (C_j) with the sum of radii not exceeding $5H$. Due to the Maximum Principle, one can assume that each exceptional disk contains at least one zero z_k. Notice that the estimate (2) is not precise. In the paper Cartan [**24**] (see also Levin [**82,** Chapter 1]) a more precise estimate is proven: for a polynomial $P(z)$ the inequality (2) holds outside disks (C_j) with the sum of radii not exceeding $2H$. In the paper Grishin [**47**] it is proven that in the statement of Theorem 3 one can replace $5H$ by $2H$ as well.[7] The method of proving Theorem 3 presented here is essentially due to L. Ahlfors. This method is frequently used in potential theory (see the monographs Nevanlinna [**102**], Landkof [**78**]) for estimating integral operators with kernels depending on the difference of arguments:

$$\int_X g(|x-\zeta|)\,d\mu(\zeta) = \int_0^\infty g(t)\,d\mu(t;x)\,.$$

In the paper Gorin, Koldobskii [**45**] infinite-dimensional analogs of the Cartan estimate are found.

Ahlfors' method has applications in approximation of a subharmonic function by the logarithm of modulus of entire function. The first general result of such a type was proved by V. S. Azarin. In the paper [**124**] by Yulmukhametov the following theorem is proved.

Let $u(z)$ be an arbitrary subharmonic function of finite order. Then there exists an entire function $f(z)$ such that

$$\big|u(z) - \log|f(z)|\big| = O(\log|z|)\,, \quad |z| \to \infty\,,$$

outside a set of disks (C_j) with finite sum of radii.

---

[7] It seems that the best constant for the sum of radii of exceptional disks is still unknown either for logarithmic potentials or for the logarithm of modulus of monic polynomials, see Hayman [**52,** Problem 4.7].

11.3. Lower bounds for the modulus of an analytic function in a disk

THEOREM 4. *Let $f(z)$ be a function analytic in the disk $\{z : |z| \le 2eR\}$, $|f(0)| = 1$, and let η be an arbitrary small positive number. Then the estimate*

$$\log|f(z)| > -H(\eta)\log M_f(2eR)\,, \quad H(\eta) = \log\frac{15e^3}{\eta}\,,$$

is valid everywhere in the disk $\{z : |z| \le R\}$ except a set of disks (C_j) with sum of radii

$$\sum r_j \le \eta R\,.$$

PROOF. First, we construct the function

$$\varphi(z) = \frac{(2R)^n}{a_1\cdots a_n}\prod_{k=1}^{n}\frac{2R(z-a_k)}{(2R)^2-\bar{a}_k z}\,,$$

$a_1,\dots,a_n$ being the zeros of the function $f(z)$ in the disk $\{z : |z| \le 2R\}$ with account taken, as usual, of their multiplicities. We have $|\varphi(0)| = 1$ and

$$|\varphi(2Re^{i\theta})| = \frac{(2R)^n}{|a_1\cdots a_n|}\,.$$

The function $\psi(z) = f(z)/\varphi(z)$ has no zeros in the disk $|z| \le 2R$, and by Theorem 2 we conclude

$$\begin{aligned}\log|\psi(z)| &\ge -2\log M_\psi(2R) = -2\log M_f(2R) + 2\log\frac{(2R)^n}{|a_1\cdots a_n|}\\ &\ge -2\log M_f(2R) > -2\log M_f(2eR)\end{aligned}$$

for $|z| \le R$. Thus

$$\log|f(z)| \ge -2\log M_f(2eR) + \log|\varphi(z)| \tag{3}$$

for $|z| \le R$. Let us now estimate the second term on the right-hand side of this inequality.

For $|z| \le R$ we have

$$\prod_{k=1}^{n}|(2R)^2-\bar{a}_k z| \le (6R^2)^n\,. \tag{4}$$

Applying Theorem 3 with $H = \eta R/5$, we obtain the inequality

$$\log\prod_{k=1}^{n}|z-a_k| > n\log\frac{\eta R}{5e} \tag{5}$$

everywhere outside the disks (C_j) with the sum of radii not exceeding ηR. Taking into account (4) and (5), we obtain

$$\begin{aligned}\log|\varphi(z)| &= \log\frac{(2R)^{2n}}{|a_1\cdots a_n|} + \log\prod_{k=1}^{n}|z-a_k| - \log\prod_{k=1}^{n}|(2R)^2-\bar{a}_k z|\\ &\ge n\log\frac{\eta R}{5e} - n\log 6R^2 + n\log 2R = n\log\frac{\eta}{15e}\end{aligned}$$

for $|z| \le R$, but outside the exceptional disks (C_j). Using the corollary to the Jensen formula, Section 2.3,

$$n = n(R, f) \le \log M_f(2eR) ,$$

we have

$$\log|\varphi(z)| > -\log M_f(2eR) \log \frac{15e}{\eta} .$$

Inserting this inequality into (3), we obtain

$$\log|f(z)| > -\log M_f(2eR) \log \frac{15e^3}{\eta}$$

for $|z| \le R$, but outside the disks (C_j). The theorem is proved.

Using the Nevanlinna characteristic, we have proved in Section 2.4 the theorem on division of entire functions. Another way of deriving division theorems is based on lower estimates of analytic functions.

THEOREM 5. *Let $f_1(z)$ be an analytic function inside the angle $D = \{z : \alpha < \arg z < \beta\}$, and let $f_2(z)$ be an entire function. Assume that both functions have order ρ and mean type. If the quotient $\varphi(z) = f_1(z)/f_2(z)$ is analytic inside the same angle, and if*

$$|\varphi(Re^{i\alpha})| \overset{\text{as}}{<} e^{AR^\rho} , \quad |\varphi(Re^{i\beta})| \overset{\text{as}}{<} e^{AR^\rho} \tag{6}$$

on the boundary of the angle, then $\varphi(z)$ also has the order ρ and mean type in D.

PROOF. By Theorem 4, we have

$$\log|f_2(z)| \overset{\text{as}}{>} -H(\eta)(\sigma_{f_2} + \varepsilon)(2e)^\rho R^\rho$$

for $|z| \le R$ and outside the exceptional disks (C_j). This implies that the estimate

$$\log|\varphi(z)| \le [H(\eta)(\sigma_{f_2} + \varepsilon)(2e)^\rho + \sigma_{f_1} + \varepsilon]R^\rho = BR^\rho \tag{7}$$

holds for $z \in D$, $|z| \le R$, but outside (C_j).

The exceptional disks (C_j) satisfy the condition

$$\sum_{|z_j|<R} r_j < \eta R ,$$

where z_j are the centers of the disks, and r_j are their radii. Hence, there is a number R_1 lying between R and $(1 - 2\eta)^{-1}R$ such that the circumference $|z| = R_1$ does not intersect the disks (C_j). Using (6) and the Maximum Principle we conclude that the estimate (7) is fulfilled for all R (possibly, with another constant B). The theorem is proved.

LECTURE 12

Entire Functions with Zeros on a Ray

12.1. Asymptotic behavior of canonical products

Let $0 < \lambda_1 \leq \lambda_2 \leq \cdots \leq \lambda_n \leq \cdots$, let $n(r)$ be the counting function of the sequence $\{\lambda_n\}$, and let the limit

$$\Delta = \lim_{r \to \infty} \frac{n(r)}{r^\rho} \tag{1}$$

exist for a noninteger ρ. We consider the canonical product

$$\Pi(z) = \prod_{n=1}^{\infty} G\Big(\frac{z}{\lambda_n}, p\Big) ,$$

where $p = [\rho]$, and

$$G(u, p) = (1-u)e^{u+u^2/2+\cdots+u^p/p} ,$$

and assume that $-\pi < \arg(1-u) < \pi$. For such u the function $\log G(u,p)$ is single-valued in the complex plane cut along the ray $[1, \infty)$.

THEOREM 1. *If a sequence $\{\lambda_n\}$ satisfies (1), then, uniformly with respect to θ, $0 \leq \theta \leq 2\pi$, the asymptotic relation*

$$\Big|\log \Pi(re^{i\theta}) - \frac{\pi\Delta}{\sin \pi\rho} e^{i\rho(\theta-\pi)} r^\rho\Big| \sin\frac{\theta}{2} = o(r^\rho) , \qquad r \to \infty \tag{2}$$

holds in the complex plane cut along the positive ray of the real axis.

PROOF. We have

$$\begin{aligned}\log \Pi(re^{i\theta}) &= \sum_{n=1}^{\infty} \log G\Big(\frac{z}{\lambda_n}, p\Big) = \int_0^\infty \log G\Big(\frac{z}{t}, p\Big)\, dn(t) \\ &= -z^{p+1} \int_0^\infty \frac{n(t)}{t^{p+1}(t-z)}\, dt .\end{aligned}$$

Let us estimate the modulus of the expression

$$\begin{aligned} S &= \log \Pi(re^{i\theta}) + r^{p+1} e^{i(p+1)\theta} \int_0^\infty \frac{\Delta t^\rho dt}{t^{p+1}(t - re^{i\theta})} \\ &= -z^{p+1} \int_0^\infty \frac{n(t) - \Delta t^\rho}{t^{p+1}(t-z)}\, dt .\end{aligned} \tag{3}$$

Because of (1) we obtain

$$
\begin{aligned}
|S| &\le r^{p+1}\int_0^\infty \frac{|n(t)-\Delta t^\rho|}{t^{p+1}|t-re^{i\theta}|}\,dt \\
&< r^{p+1}\int_0^N \frac{|n(t)-\Delta t^\rho|}{t^{p+1}|t-re^{i\theta}|}\,dt + \varepsilon r^{p+1}\int_0^\infty \frac{t^{\rho-p-1}}{|t-re^{i\theta}|}\,dt \qquad (4)\\
&= J_1(r,\theta)+J_2(r,\theta)
\end{aligned}
$$

as $N > N(\varepsilon)$. Then

$$
J_1(r,\theta) = r^{p+1}\int_0^N \frac{|n(t)-\Delta t^\rho|}{t^{p+1}|t-re^{i\theta}|}\,dt = O(r^p) = o(r^\rho)\,, \quad r\to\infty\,, \tag{5}
$$

uniformly with respect to θ, $0\le\theta\le 2\pi$.

To estimate the term J_2 we set $t = r\tau$. Then

$$
\begin{aligned}
J_2(r,\theta) &= \varepsilon r^\rho \int_0^\infty \frac{\tau^{\rho-p-1}}{|\tau - e^{i\theta}|}\,d\tau = \varepsilon r^\rho\Big(\int_0^2 + \int_2^\infty\Big)\frac{\tau^{\rho-p-1}}{|\tau-e^{i\theta}|}\,d\tau \qquad (6)\\
&< \varepsilon r^\rho \Big(\int_0^2 \frac{\tau^{\rho-p-1}}{\sin(\theta/2)}\,d\tau + \int_2^\infty \frac{\tau^{\rho-p-2}}{|1-e^{i\theta}/\tau|}\,d\tau\Big) < \frac{\varepsilon c_\rho r^\rho}{\sin(\theta/2)}\,.
\end{aligned}
$$

Moreover,

$$
\Delta r^{p+1}e^{i(p+1)\theta}\int_0^\infty \frac{t^\rho dt}{t^{p+1}(t-re^{i\theta})} = \Delta r^\rho e^{i(p+1)\theta}\int_0^\infty \frac{t^{\rho-p-1}dt}{t-e^{i\theta}}\,.
$$

The latter integral is easily calculated using residues. It is equal to

$$
-\frac{\pi}{\sin\pi\rho}e^{i(\rho-1-p)\theta - i\pi\rho}\,.
$$

Thus the relations (3)–(6) imply the assertion of the theorem.

PROBLEM 1. Prove asymptotic relation (2) under the assumptions that λ_n are complex numbers such that each angle around the positive ray contains all but finitely many λ_n, $n/\lambda_n^\rho \to \Delta$, $\Delta > 0$, ρ is noninteger, and $\delta\le\theta\le 2\pi-\delta$, $\delta>0$.

REMARK. Taking the real part of both sides of equation (2), we obtain

$$
\log|\Pi(re^{i\theta})| = \frac{\pi\Delta r^\rho\cos\rho(\theta-\pi)}{\sin\pi\rho} + \frac{o(r^\rho)}{\sin(\theta/2)}\,. \tag{2'}
$$

Evidently, this yields

$$
h_\Pi(\theta) = \frac{\pi\Delta}{\sin\pi\rho}\cos\rho(\theta-\pi)\,, \qquad 0<\theta<2\pi\,,
$$

and, by the continuity of the indicator, the latter equation holds for all θ, $0\le\theta\le 2\pi$. By the property of the indicator stated in Theorem 2, Section 8.2, we obtain

$$
\log|\Pi(re^{i\theta})| \overset{\text{as}}{<} \Big[\frac{\pi\Delta}{\sin\pi\rho}\cos\rho(\theta-\pi)+\varepsilon\Big]r^\rho\,.
$$

In what follows, the function $\cos\rho(\theta-\pi)$ is meant to be extended from the interval $\{\theta : 0\le\theta\le 2\pi\}$ as a 2π-periodic function.

THEOREM 2. *If λ_n is a sequence of positive numbers such that $n/\lambda_n \to \Delta$, $n \to \infty$, and if*

$$\Pi(z) = \prod_{n=1}^{\infty} \left(1 - \frac{z^2}{\lambda_n^2}\right),$$

then

$$h_\Pi(\theta) = \pi\Delta|\sin\theta|, \tag{7}$$

and for $\theta \neq 0, \pi$, the limit

$$h_\Pi(\theta) = \lim_{r\to\infty} \frac{\log|\Pi(re^{i\theta})|}{r}$$

exists.

PROOF. The sequence $\{\lambda_n^2\}$ has density Δ with respect to the order $\rho = 1/2$. According to the previous theorem,

$$\log|\Pi(\sqrt{z})| = \pi\Delta R^{1/2} \cos\frac{1}{2}(\varphi - \pi) + \frac{o(R^{1/2})}{\sin(\varphi/2)}, \qquad R \to \infty,$$

where $R^{1/2} = r = |z|$, $\varphi/2 = \theta = \arg z$; i.e.,

$$\log|\Pi(re^{i\theta})| = \pi\Delta|\sin\theta|r + \frac{o(r)}{|\sin\theta|}, \qquad r \to \infty, \qquad 0 < |\theta| < \pi,$$

which completes the proof.

12.2. Theorem on a segment on the boundary of the indicator diagram

THEOREM 3. *Let $f(z)$ be an entire function of exponential type. If f vanishes at a point set $\{\lambda_n\}$ having the density*

$$\Delta = \lim_{n\to\infty} \frac{n}{\lambda_n},$$

then the supporting line of the indicator diagram of $f(z)$, which is orthogonal to the direction $\arg z = 0$, and the indicator diagram itself have a common segment of length at least $2\pi\Delta$.

PROOF. Consider the function

$$\varphi(z) = \frac{f(z)}{\Pi(z)}$$

with

$$\Pi(z) = \prod_{n=1}^{\infty} \left(1 - \frac{z^2}{\lambda_n^2}\right).$$

By Theorem 5 from the previous lecture, $\varphi(z)$ is of exponential type in the upper half-plane.

Theorem 2 implies

$$h_f(\theta) = h_\varphi(\theta) + h_\Pi(\theta) = h_\varphi(\theta) + \pi\Delta|\sin\theta|;$$

i.e., for $|\theta| \le \pi/2$ the supporting function of the indicator diagram I_f coincides with the supporting function of the sum of the indicator diagram I_φ and the segment $[-i\pi\Delta, i\pi\Delta]$. To complete the proof, we observe that the boundary of the convex compact set $I_\varphi + [-i\pi\Delta, i\pi\Delta]$ contains a segment of length at least $2\pi\Delta$ which is parallel to the imaginary axis.

DEFINITION. A domain G bounded by continuous curves $y = a(x)$ and $y = a(x) + d$, $x \in \mathbb{R}$, is called a curvilinear strip of width d.

THEOREM 4. *Assume that all singularities of the Borel transform of an entire function $F(z)$ of exponential type are in a curvilinear strip of width d. If $F(z)$ vanishes at points $\{\lambda_n\}$ and if*

$$\Delta = \lim_{n\to\infty} \frac{n}{\lambda_n} \geq \frac{d}{2\pi},$$

then $F(z) \equiv 0$.

PROOF. Consider the smallest convex compact set K containing all the singularities of the Borel transform of $F(z)$. We see that the vertical line supporting K contains no segment of length $2\pi\Delta$ lying on the boundary of K. Then by the previous theorem we obtain $F(z) \equiv 0$, proving Theorem 4.

Theorems 3 and 4 can be used for the study of the completeness of exponential systems $\{e^{\lambda_n z}\}$.

THEOREM 5. [8] *If $\lambda_n/n \to 1$, $n \to \infty$, then the system of functions $\{e^{\lambda_n z}\}$ is complete in $A(\Omega)$, where Ω is an arbitrary curvilinear strip of width 2π, and is not complete in any simply connected domain which contains a closed segment of length 2π parallel to the imaginary axis.*

PROOF. If a system $\{e^{\lambda_n z}\}$ is not complete, then there exists a function $f(z) \not\equiv 0$ in a neighborhood of the point at infinity satisfying $f(\infty) = 0$, with all its singularities lying in Ω and such that

$$\int_L e^{\lambda_n z} f(z)\, dz = 0,$$

where $L \subset \Omega$ is a simple closed curve surrounding all these singularities. The function

$$\Phi(\lambda) = \int_L e^{\lambda z} f(z)\, dz$$

is an entire function of exponential type, and all singularities of its Borel transform $f(z)$ are in the strip Ω. By Theorem 4, the equations $\Phi(\lambda_n) = 0$, $n = 1, 2, \ldots$ yield $\Phi(\lambda) \equiv 0$, and $f(z) \equiv 0$ giving a contradiction.

Now, let us prove the second assertion of the theorem. Let G be an arbitrary simply connected domain containing a closed segment $\overline{I} = [\overline{a} - i\pi, \overline{a} + i\pi]$. Using (7), we see that the indicator diagram of the function

$$F(\lambda) = e^{\overline{a}\lambda} \prod_{n=1}^{\infty} \left(1 - \frac{\lambda^2}{\lambda_n^2}\right)$$

coincides with the segment $I = [a - i\pi, a + i\pi]$, and the conjugate diagram coincides with the segment $\overline{I} = [\overline{a} - i\pi, \overline{a} + i\pi]$. By the Pólya theorem

$$F(\lambda) = \frac{1}{2\pi i} \int_C e^{\lambda z} f(z)\, dz,$$

[8] This theorem was proved independently by A. F. Leont'ev and the author.

where $C \subset G$ is a simple closed curve surrounding the segment I. A nontrivial functional from $A^*(G)$ corresponding to the function $f(z)$ annihilates all the functions $e^{\lambda_n z}$, and hence this system is not complete.

Final results on completeness of the system of functions $\{e^{\lambda_n z}\}$ in curvilinear strips are established in the papers Malliavin and Rubel [**90**], and Khabibullin [**66**].

Theorem 3 may be used to solve completeness problems for systems of functions in spaces with another topology, for example, in the spaces $L^p(K)$ or $C(K)$ where K is a compact set.

THEOREM 6. *Let K be a rectifiable curve which is the graph of a continuous function defined on a closed segment, and let $\Lambda = \{\lambda_n\}_{n=1}^{\infty}$ be a sequence of complex numbers satisfying the condition*

$$\frac{n}{\lambda_n} \to \Delta > 0\,, \qquad n \to \infty\,.$$

Then the system $E(\Lambda) = \{e^{i\lambda_n z}\}_{n=1}^{\infty}$ is complete in each of the spaces $L^p(K)$, $1 \le p < \infty$, and $C(K)$.

PROOF. For the sake of definiteness, consider the space $C(K)$. If the system $E(\Lambda)$ is not complete in this space, then there exists a nontrivial measure $d\mu(z)$ supported by K and orthogonal to all functions of the system $E(\Lambda)$. Using this measure we construct an entire function of exponential type

$$\Phi(\lambda) = \int_K e^{\lambda z}\, d\mu(z)$$

vanishing at the points of Λ. The Borel transform of this function has the form

$$\varphi(\zeta) = \int_K \frac{d\mu(z)}{\zeta - z}\,,$$

and hence the function $\varphi(\zeta)$ is holomorphic in $\overline{\mathbb{C}} \setminus K$. If this function does not vanish identically, the conjugate diagram of the function $\Phi(\lambda)$ coincides with the convex hull of K or with the convex hull of some part of it. It is evident that such a convex hull has no vertical segment on its boundary, which contradicts Theorem 3. Thus, $\Phi(\lambda) \equiv 0$. This yields

$$\Phi^{(k)}(0) = \int_K z^k\, d\mu(z) = 0\,, \qquad k = 0, 1, 2, \dots\,,$$

and since polynomials are dense in $C(K)$,[9] we obtain that the measure $d\mu$ is orthogonal to the whole space $C(K)$. The theorem is proved.

REMARK 1. Let K be an arbitrary compact set in the complex plane and let $A(K)$ be the closure of polynomials in the uniform norm on this compact set. If Λ and $E(\Lambda)$ are as in Theorem 6, and, for each real a, the diameter of the intersection of K and the vertical line $\operatorname{Re} z = a$ is less than $2\pi\Delta$ (or the intersection is empty), then the system $E(\Lambda)$ is complete in $A(K)$.

It is worth mentioning that by Mergelyan's theorem (Mergelyan [**98**]) the space $A(K)$ coincides with the space of functions continuous on K and holomorphic in the interior of K.

[9] This statement is a rather particular case of a well-known theorem of Mergelyan, see [**98**].

REMARK 2. If there are two points z_1, z_2 in a compact set K such that $\operatorname{Re} z_1 = \operatorname{Re} z_2$ and $|z_1 - z_2| = d > 0$, then the system of functions $E(\Lambda)$, $\Lambda = \{2\pi ki/d\}$, $k \in \mathbb{Z}$ is not complete in $C(K)$.

Indeed, each function of this system assumes equal values at the points z_1 and z_2, and hence the system $E(\Lambda)$ could not be complete in $A(K)$. Evidently, a subset of this system could not be complete as well.

PROBLEM 2. If $a > 0, b > 0$, and an entire function

$$F(\lambda) = \int_{-a}^{b} \cos(\lambda\sqrt{t})\psi(t)\,dt\,, \qquad \psi \in L(-a, b)$$

vanishes on a set $\{\lambda_n\}$, where

$$\frac{n}{\lambda_n} \to \Delta > 0\,, \qquad n \to \infty\,,$$

then $\psi(t) = 0$ a.e. on $(-a, 0)$.

12.3. Lower bound for the canonical product with positive zeros having density

For the canonical product

$$\Pi(z) = \prod_{n=1}^{\infty} G\Big(\frac{z}{\lambda_n}, p\Big)\,, \qquad p < \rho < p+1\,, \qquad \lambda_n > 0\,,$$

with

$$\lim_{r\to\infty} \frac{n(r)}{r^\rho} = \Delta$$

we have established the asymptotic formula

$$\log|\Pi(re^{i\theta})| = \frac{\pi\Delta r^\rho}{\sin\pi\rho}\cos\rho(\theta - \pi) + \frac{o(r^\rho)}{\sin(\theta/2)}\,, \qquad 0 < \theta < 2\pi\,.$$

The latter relation is of no meaning for $\theta = 0$. However, it can be defined more exactly to be valid for $\theta = 0$ as well. To this end, it is necessary to exclude from the complex plane some exceptional disks containing zeros of $\Pi(z)$, i.e., the points λ_n.

DEFINITION. A set of disks (C_j) in the complex plane will be called a C^0-set if

$$\lim_{R\to\infty} \frac{1}{R} \sum_{|z_j|<R} r_j = 0\,,$$

where z_j are the centers of (C_j), and r_j are their radii.

THEOREM 7. *Under the condition*

$$\lim_{t\to\infty} \frac{n(t)}{t^\rho}\,dt = \Delta \tag{8}$$

there exists a C^0-set of disks (C_j) outside which the asymptotic relation

$$\log|\Pi(re^{i\theta})| = \frac{\pi\Delta}{\sin\pi\rho} r^\rho \cos\rho(\theta - \pi) + o(r^\rho)\,, \qquad r \to \infty\,, \tag{9}$$

holds uniformly with respect to θ, $0 \le \theta \le 2\pi$.

PROOF. In the following proof we denote by K various numbers depending on ρ and Δ only. Firstly, we shall prove that, for each small enough value $\varepsilon > 0$, there exists a set of disks $(C_j(\varepsilon))$ with centers z_j and radii r_j such that

$$\limsup_{R\to\infty} \frac{1}{R} \sum_{|z_j|<R} r_j \le \eta(\varepsilon) , \tag{10}$$

where $\eta(\varepsilon) \to 0$ as $\varepsilon \to 0$, and that outside $(C_j(\varepsilon))$ the inequality

$$\left| \log|\Pi(re^{i\theta})| - \frac{\pi\Delta}{\sin\pi\rho} \cos\rho(\theta-\pi) r^\rho \right| \le K\sqrt{\varepsilon}\, r^\rho \tag{11}$$

holds. We have already proved that the inequality

$$\log|\Pi(re^{i\theta})| < \left[\frac{\pi\Delta}{\sin\pi\rho} \cos\rho(\theta-\pi) + \varepsilon\right] r^\rho \tag{12}$$

holds as $r > r_\varepsilon$, and the inequality

$$\log|\Pi(re^{i\theta})| > \left[\frac{\pi\Delta}{\sin\pi\rho} \cos\rho(\theta-\pi) - \varepsilon\right] r^\rho \tag{13}$$

holds as $r > r_\varepsilon$ and $\varepsilon/2 \le \theta \le 2\pi - \varepsilon/2$.

For each natural n we set $R_n = (1+\varepsilon)^n$, $z_0 = R_n e^{i\varepsilon/2}$, and consider the function

$$\Phi(w) = \frac{\Pi(z_0+w)}{\Pi(z_0)} = \frac{\Pi(z)}{\Pi(z_0)} , \qquad z = w + z_0 ,$$

in the disk $|w| \le 2\varepsilon R_n$. We note that the disk $\{|z - z_0| < 2\varepsilon R_n\}$ contains the sector

$$S_n = \left\{ re^{i\theta} : R_{n-1} \le r \le R_n,\ |\theta| \le \frac{\varepsilon}{2} \right\} .$$

Moreover,

$$\max\{|\arg z| : |z - z_0| \le 2\varepsilon R_n\} = \frac{\varepsilon}{2} + \arcsin 2\varepsilon < 3\varepsilon ,$$
$$\max\{|z| : |z - z_0| \le 2\varepsilon R_n\} = R_n(1+2\varepsilon) .$$

By the lower bound of the modulus of a holomorphic function (Theorem 4 of the previous lecture), given $\eta > 0$, the inequality

$$\log|\Phi(w)| > -H(\eta)\log M_\Phi(4\varepsilon e R_n) , \qquad H(\eta) = \log\frac{15e^3}{\eta} ,$$

holds everywhere outside disks $(C_j^{(n)})$ such that

$$\sum_j r_j^{(n)} < 2\varepsilon\eta R_n = 2\varepsilon\eta(1+\varepsilon)^n .$$

Inequalities (12) and (13) imply

$$\begin{aligned} \log M_\Phi(4\varepsilon e R_n) \le \max_{|\theta|\le 3\varepsilon} &\left[\frac{\pi\Delta}{\sin\pi\rho} \cos\rho(\theta-\pi) + \varepsilon\right] (1+2\varepsilon)^\rho R_n^\rho \\ &- \left[\frac{\pi\Delta}{\sin\pi\rho} \cos\rho\left(\frac{\varepsilon}{2}-\pi\right) - \varepsilon\right] R_n^\rho \le K\varepsilon R_n^\rho . \end{aligned}$$

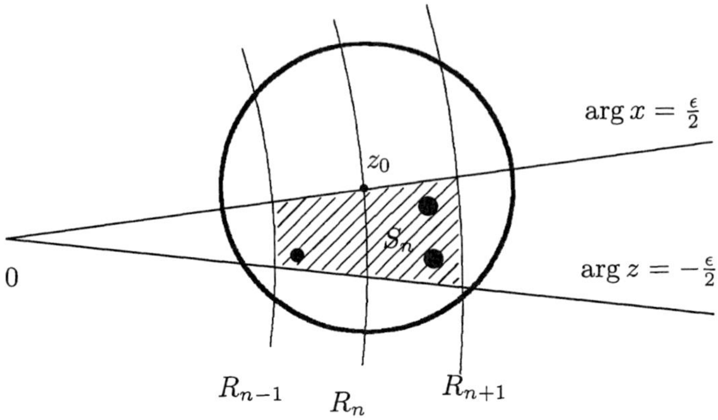

FIGURE 3

Thus, for $|w| \le 2\varepsilon R_n$, but outside the disks $(C_j^{(n)})$,

$$\log|\Phi(w)| > -H(\eta)K\varepsilon R_n^\rho .$$

Setting $w = z - z_0$, we obtain

$$\begin{aligned}\log|\Pi(z)| &= \log|\Pi(z_0)| + \log|\Phi(w)| \\ &> \Big[\frac{\pi\Delta}{\sin\pi\rho}\cos\rho\Big(\frac{\varepsilon}{2}-\pi\Big) - \varepsilon - H(\eta)K\varepsilon\Big]R_n^\rho \\ &> \Big[\frac{\pi\Delta}{\sin\pi\rho}\cos\rho(\theta-\pi) - \big(K_1 + KH(\eta)\big)\varepsilon\Big]r^\rho\end{aligned}$$

for $z = re^{i\theta}$, but outside the exceptional disks $(C_j^{(n)})$.

If $\eta = \eta(\varepsilon)$ is chosen such that

$$H(\eta) = \log\frac{15e^3}{\eta} = \frac{1}{\sqrt{\varepsilon}} ,$$

then

$$\log|\Pi(re^{i\theta})| > \Big[\frac{\pi\Delta}{\sin\pi\rho}\cos\rho(\theta-\pi) - K\sqrt{\varepsilon}\Big]r^\rho \tag{14}$$

with $re^{i\theta} \in S_n$, but outside the set of exceptional disks $(C_j^{(n)})$. Let us now consider the system of disks

$$(C_j(\varepsilon)) = \bigcup_n (C_j^{(n)}) .$$

Let z_j be the centers of these disks and let r_j be their radii. Then for $R_{N-1} \le R < R_N$

$$\sum_{|z_j|<R} r_j \le \sum_{n=0}^{N}\sum_j r_j^{(n)} \le 2\varepsilon\eta\sum_{n=0}^{N}(1+\varepsilon)^n \le 2\eta(1+\varepsilon)^{N+1} \le 2\eta R .$$

Therefore, the system of disks $(C_j(\varepsilon))$ satisfies the condition (10). By estimates (12), (13), and (14), the inequality (11) holds.

In order to complete the proof of Theorem 7 we shall construct an exceptional C^0-set of disks. For this purpose let us choose a sequence $\varepsilon_p \searrow 0$; then $\eta_p = \eta(\varepsilon_p) \searrow 0$. For each p we have found, as described above, an exceptional set of disks $(C_j(\varepsilon_p))$ with centers $z_{j,p}$ and radii $r_{j,p}$. Starting with $R^{(0)} = 1$, we choose a value $R^{(p)} > pR^{(p-1)}$ so that

$$\sum_{|z_j|<R} r_{j,p} < 2\eta_p R$$

as $R \geq R^{(p)}$. We construct the system (C_j) including in it all disks from $(C_j(\varepsilon_p))$ whose centers are in the annulus $R^{(p)} \leq |t| \leq R^{(p+1)}$, $p = 0, 1, 2, \dots$. Then, for $R^{(N)} \leq R < R^{(N+1)}$, we obtain

$$\begin{aligned}
\sum_{|z_j|<R} r_j &= \sum_{p=1}^{N} \sum_{R^{(p-1)} \leq |z_{j,p}| \leq R^{(p)}} r_{j,p} + \sum_{R^{(N)} < |z_{j,N}| < R} r_{j,N} \\
&\leq \sum_{p=1}^{N} 2\eta_{p-1} R^{(p)} + 2\eta_n R \\
&\leq 2\eta_1 \left(\frac{1}{2 \cdot 3 \cdot 4 \cdots N} + \frac{1}{3 \cdot 4 \cdots N} + \cdots + \frac{1}{N} \right) R^{(N)} + 2\eta_{N-1} R^{(N)} + 2\eta_N R \\
&= o(1) R , \qquad N \to \infty .
\end{aligned}$$

Evidently, the asymptotic relation (9) holds outside the disks (C_j). The theorem is proved.

REMARK 1. If

$$\Pi(z) = \prod_{n=1}^{\infty} \left(1 - \frac{z^2}{\lambda_n^2}\right) , \qquad \frac{n}{\lambda_n} \to \Delta ,$$

then, making a change of variables $z^2 = w$, we obtain an entire function of non-integer order $\rho = 1/2$. Applying the latter theorem to this function, we find that outside a C^0-set

$$\log |\Pi(re^{i\theta})| = \pi\Delta |\sin\theta| r + o(r) , \qquad r \to \infty .$$

REMARK 2. If zeros $\{\lambda_n\}$ lie on the ray $\{z : \arg z = \psi\}$, then equation (9) is replaced by

$$\log |\Pi(re^{i\theta})| = \frac{\pi\Delta r^\rho}{\sin \pi\rho} \cos \rho(\theta - \psi - \pi) + o(r^\rho) , \qquad \psi \leq \theta \leq \psi + 2\pi ,$$

valid outside an exceptional C^0-set.

REMARK 3. As an analysis of the proof of Theorem 7 shows, one can replace the assumption (8) by its corollary

$$\log |f(re^{i\theta})| = \frac{\pi\Delta r^\rho}{\sin \pi\rho} \cos \rho(\theta - \pi) + o(r^\rho) , \qquad \delta \leq \theta \leq 2\pi - \delta .$$

PROBLEM 3. Let a sequence of positive integers $\{\lambda_n\}$ has density Δ with respect to the order $\rho = 1$ and let

$$\Pi(z) = \prod_{n=1}^{\infty} G\Big(\frac{z}{\lambda_n}, \rho\Big) .$$

Prove that, outside a C^0-set of disks, the asymptotic relation

$$\log|\Pi(re^{i\theta})| = \Delta \cos\theta r \log r + o(r\log r) , \qquad r \to \infty ,$$

is valid.

PROBLEM 4. Prove that if $f(z)$ is an entire function of minimal type with respect to an order ρ, then

$$\log|f(z)| = o(|z|^{\rho}) , \qquad |z| \to \infty ,$$

everywhere outside a C^0-set.

LECTURE 13

Entire Functions with Zeros on a Ray (Continuation)

13.1. The Valiron theorem

In this section we shall prove theorems which may be regarded as converse to the theorems proved in Lecture 12.

THEOREM 1. *Let $f(z)$ be an entire function of noninteger order ρ with positive zeros, and let, for each $\delta > 0$, the asymptotic relation*

$$\log|f(re^{i\theta})| = \frac{\pi\Delta r^\rho}{\sin\pi\rho}\cos\rho(\theta-\pi) + o(r^\rho) \tag{1}$$

hold uniformly with respect to θ, $\delta \le \theta \le 2\pi-\delta$. Then the limit

$$\lim_{t\to\infty}\frac{n(t)}{t^\rho} = \Delta \tag{2}$$

exists, where $n(t)$ is the zero-counting function of $f(z)$.

PROOF. The asymptotic relation (1) implies that

$$h_f(\theta) = \frac{\pi\Delta}{\sin\pi\rho}\cos\rho(\theta-\pi)$$

for $0<\theta<2\pi$. Since the indicator is a continuous function, the latter equation holds for $\theta = 0$ as well. It yields

$$\log|f(re^{i\theta})| \overset{\text{as}}{<} \left[\frac{\pi\Delta}{\sin\pi\rho}\cos\rho(\theta-\pi)+\varepsilon\right]r^\rho, \qquad \varepsilon>0\,. \tag{3}$$

According to Remark 3 to Theorem 7 of the previous lecture, relations (1) and (3) imply that the asymptotic relation (1) holds everywhere outside an exceptional C^0-set of disks (C_j) containing zeros of $f(z)$.

Let us now choose a number $R>0$ such that the circle $\{z : |z| = R\}$ does not intersect the exceptional disks (C_j). Assume that $f(0) = 1$. Then, by the Jensen formula,

$$\begin{aligned} N(R) &= \int_0^R \frac{n(t)}{t}\,dt = \frac{1}{2\pi}\int_0^{2\pi}\log|f(Re^{i\theta})|\,d\theta \\ &= \frac{\pi\Delta R^\rho}{2\pi\sin\pi\rho}\int_0^{2\pi}\cos\rho(\theta-\pi)\,d\theta + o(R^\rho) = \frac{\Delta}{\rho}R^\rho + o(R^\rho)\,. \end{aligned} \tag{4}$$

If R is a number large enough that the circle $\{z : |z| = R\}$ intersects the exceptional set, then for each $\delta>0$ we can find $h = h(R)$, $0<h<\delta$, such that the circles $|z| = R(1-h)$, $|z| = R(1+h)$, $R > R_0(\delta)$, do not intersect this set. We obtain

$$\frac{\Delta}{\rho}(1-h)^\rho R^\rho + o(R^\rho) \le \int_0^R \frac{n(t)}{t}\,dt \le \frac{\Delta}{\rho}(1+h)^\rho R^\rho + o(R^\rho),$$

which shows that the asymptotic relation (4) holds as $R \to \infty$.

Let us show that equation (4) yields relation (2); this is a standard Tauberian argument from real analysis. Choosing $k > 1$, we easily derive from (4)

$$n(R)\log k \le \int_R^{kR} \frac{n(t)}{t}\,dt = \frac{\Delta}{\rho}(k^\rho - 1)R^\rho + o(R^\rho)\,,$$

and

$$n(kR)\log k \ge \int_R^{kR} \frac{n(t)}{t}\,dt = \frac{\Delta}{\rho}(k^\rho - 1)R^\rho + o(R^\rho)\,.$$

In other words,

$$\frac{n(R)}{R^\rho} \le \frac{\Delta}{\rho}\frac{k^\rho - 1}{\log k} + o(1)\,,$$
$$\frac{n(kR)}{(kR)^\rho} \ge \frac{\Delta}{\rho}\frac{k^\rho - 1}{k^\rho \log k} + o(1)\,.$$

Using these two inequalities, we obtain

$$\frac{\Delta}{\rho}\frac{k^\rho - 1}{k^\rho\log k} \le \liminf_{R\to\infty}\frac{n(R)}{R^\rho} \le \limsup_{R\to\infty}\frac{n(R)}{R^\rho} \le \frac{\Delta}{\rho}\frac{k^\rho - 1}{\log k}$$

for each $k > 1$. Passing to the limit as $k \searrow 1$, we get

$$\liminf_{R\to\infty}\frac{n(R)}{R^\rho} = \limsup_{R\to\infty}\frac{n(R)}{R^\rho} = \Delta\,.$$

The theorem is proved.

In fact, Valiron showed that (2) follows from a far weaker form of (1). To do this we remind the reader that for a function $\varphi(z)$ which is analytic and bounded above in the upper half-plane we have defined $|\varphi(x)|$, $x \in \mathbb{R}$, as

$$|\varphi(x)| = \limsup_{z\to x,\, \operatorname{Im} z > 0} |\varphi(z)|\,.$$

THEOREM 2 (on two constants). *Let $\varphi(z)$ be a bounded analytic function in the half-plane $\{x + iy : y > 0\}$, let $|\varphi(x)| \le M_1$, for $x < 0$, and let $|\varphi(x)| \le M_2$ for $x > 0$. Then*

$$|\varphi(re^{i\theta})| \le M_1^{\theta/\pi} M_2^{1-\theta/\pi} \tag{5}$$

for $0 \le \theta \le \pi$.

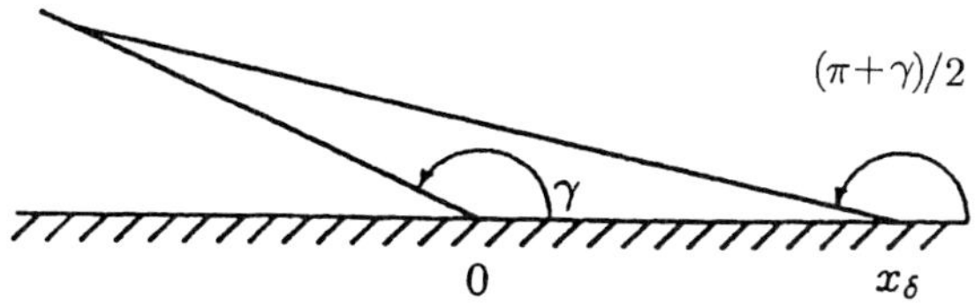

FIGURE 4

PROOF. Let

$$u(z) = \frac{\theta}{\pi}\log M_1 + \left(1 - \frac{\theta}{\pi}\right)\log M_2\ , \qquad z = re^{i\theta}\ .$$

Evidently, $u(z)$ is a harmonic function in the half-plane $y > 0$. The function

$$\omega(z) = \log|\varphi(z)| - u(z)$$

is subharmonic. It is bounded above in the half-plane $\{x + iy\ :\ y > 0\}$ and its boundary values are nonpositive on the real axis. The Phragmén-Lindelöf theorem implies that $\omega(z) \le 0$ for $y > 0$, which is equivalent to (5).

THEOREM 3 (Lindelöf). *Let $\varphi(z)$ be a bounded analytic function in the upper half-plane, and let $|\varphi(x)| \to 0$ as $x \to +\infty$. Then $\varphi(z) \to 0$, $z \to \infty$, uniformly inside the angle $\{z : 0 \le \arg z \le \gamma\}$ for each $\gamma < \pi$.*

PROOF. Indeed, if $|\varphi(z)| \le M$ and if $|\varphi(x)| < \delta$ as $x > x_\delta$, then the inequality

$$|\varphi(z)| \le M^{(1+\gamma\pi^{-1})/2}\delta^{(1-\gamma\pi^{-1})/2}$$

holds inside the angle $\{z\ :\ 0 \le \arg(z - x_\delta) \le (\pi+\gamma)/2\}$. The conclusion of the theorem now follows by inspection of Figure 4.

COROLLARY. *Let $f(z)$ be analytic in the upper half-plane and continuous up to its boundary. Assume that $f(x)$ has distinct limits as $x \to \pm\infty$. Then the function $f(z)$ is unbounded.*

THEOREM 4 (Valiron). *Let the zeros of an entire function $f(z)$ of noninteger order ρ and mean type lie on the positive axis, and let*

$$\log|f(-r)| = \frac{\pi\Delta}{\sin\pi\rho}r^\rho + o(r^\rho)\ , \qquad r \to \infty\ .$$

Then

$$n(t) = \Delta t^\rho + o(t^\rho)\ , \qquad t \to \infty\ .$$

PROOF. According to the Hadamard theorem, Section 4.2, we have

$$\log|f(z)| = \log|\Pi(z)| + \operatorname{Re} P_q(z)\ ,$$

where $\Pi(z)$ is a canonical product, and P_q is a polynomial of degree $q < \rho$. Thus

$$\log|\Pi(-r)| = \frac{\pi\Delta}{\sin\pi\rho}r^\rho + o(r^\rho)\ , \qquad r \to \infty\ . \tag{6}$$

The function $\log \Pi(z)$ may be represented (see Section 12.1) in the form

$$\log \Pi(z) = -z^{p+1} \int_0^\infty \frac{n(t)dt}{t^{p+1}(t-z)}\,, \qquad 0 < \arg z < 2\pi\,, \qquad p = [\rho]\,.$$

Since $f(z)$ is of order ρ and mean type, $n(t) \overset{\text{as}}{<} ct^\rho$, and hence

(7) $$|\log \Pi(z)| < Cr^{p+1} \int_0^\infty \frac{t^{\rho-p-1}}{|t-z|}\, dt < C_\delta r^\rho$$

for $0 < \delta \le \theta \le 2\pi - \delta$. We choose the branch of the function $(-z)^\rho$ in the complex plane cut along the positive axis such that $(-z)^\rho > 0$ on the negative axis, and set

$$\varphi(z) = \frac{\log \Pi(z)}{(-z)^\rho} - \frac{\pi\Delta}{\sin \pi\rho}\,.$$

By (7), the function $\varphi(z)$ is bounded inside the angle $\{z : 0 < \delta \le \arg z \le 2\pi - \delta\}$. Further, the function $\log \Pi(z)$ is real on the negative axis. Together with (6) this implies that $\varphi(-r) \to 0$ as $r \to \infty$. By Theorem 3 we find that $\varphi(z) \to 0$ uniformly inside the angle $\delta \le \arg z \le 2\pi - \delta$ as $z \to \infty$. Hence

$$\log \Pi(re^{i\theta}) = \frac{\pi\Delta r^\rho}{\sin \pi\rho} e^{i\rho(\theta-\pi)} + o(r^\rho)\,, \qquad \delta \le \theta \le 2\pi - \delta\,.$$

According to Theorem 1, this asymptotic relation yields $n(t) = \Delta t^\rho + o(t^\rho)$, which proves Theorem 4.

13.2. Functions of completely regular growth

If zeros $\{\lambda_n\}$ are located on a finite number of rays $\arg z = \psi_k$ with densities Δ_k with respect to t^ρ, ρ being noninteger, then it follows from Theorem 7, Lecture 12, that the canonical product $\Pi(z)$ with zeros at $\{\lambda_n\}$ has the asymptotic behavior

$$\log|\Pi(z)| = \frac{\pi r^\rho}{\sin \pi\rho} \sum_k \Delta_k \cos \rho(\theta - \psi_k - \pi) + o(r^\rho)\,, \qquad \theta - 2\pi < \psi_k \le \theta\,,$$

outside an exceptional C^0-set. The latter sum can be written as a Stieltjes integral:

$$\log|\Pi(z)| = \frac{\pi r^\rho}{\sin \pi\rho} \int_{[0,2\pi]} \cos \rho(\theta - \psi - \pi)\, d\Delta(\psi) + o(r^\rho)\,,$$

where Δ is the measure supported by the points ψ_k, $\Delta(\{\psi_k\}) = \Delta_k$, and, as before, $\cos \rho(\theta - \pi)$ is the 2π-periodic continuation of the function $\cos \rho(\theta - \pi)$ from $[0, 2\pi]$ to the whole axis.

This result can be substantially generalized. To avoid technical difficulties, we restrict ourselves to a narrative exposition omitting the proofs.

For a noninteger ρ, every ρ-trigonometrically convex function $h(\theta)$ may be represented in the integral form

(8) $$h(\theta) = \frac{\pi}{\sin \pi\rho} \int_{[0,2\pi]} \cos \rho(\theta - \psi - \pi)\, d\Delta(\psi)\,,$$

where Δ is a nonnegative Borel measure on $[0, 2\pi)$. Denote by $n_f(r; \psi_1, \psi_2)$ the number of zeros of an entire function $f(z)$ of order ρ in the sector $\{z : |z| \le$

$r,\ \psi_1 \le \arg z < \psi_2\}$. Assume that for each $\psi_1, \psi_2 \in [0, 2\pi)\setminus Q$, where Q is at most countable, there exists the limit

$$\Delta_f(\psi_1, \psi_2) = \lim_{r\to\infty} \frac{n_f(r, \psi_1, \psi_2)}{r^\rho} \tag{9}$$

which is called *the angular density of zeros* of the function $f(z)$. Then the asymptotic relation

$$\log|f(re^{i\theta})| = h_f(\theta) r^\rho + o(r^\rho)\,, \qquad r \to \infty\,, \tag{10}$$

holds everywhere outside an exceptional C^0-set. Here, the indicator $h_f(\theta)$ has the form (8) with the measure Δ coinciding with the angular density of zeros of $f(z)$. In other words, on each semi-interval $[\psi_1, \psi_2)$, $\psi_1, \psi_2 \notin Q$, the measure Δ is defined as $\Delta([\psi_1, \psi_2)) = \Delta_f(\psi_1, \psi_2)$.

For an integer $\rho > 0$, every 2π-periodic ρ-trigonometrically convex function $h(\theta)$ can be represented in the form

$$h_f(\theta) = \int_{[0,2\pi]} (\theta - \psi) \sin\rho(\theta - \psi)\, d\Delta(\psi) + \tau\cos\rho(\theta - \theta_0)\,. \tag{11}$$

In this case, in addition to the existence of the limit (9), the following condition must be imposed: there exists the limit

$$\delta_f = \lim_{R\to\infty} \delta_{f,R} = \lim_{R\to\infty} \Big(a_\rho + \frac{1}{\rho} \sum_{|\lambda_n|\le R} \lambda_n^{-\rho}\Big) \tag{12}$$

(see Lecture 5). The relations (9) and (12) imply that (10) holds outside a C^0-set with the indicator $h_f(\theta)$ defined by (11) with $\delta_f = \tau e^{i\theta_0}$.

The entire functions satisfying (10) are called *the functions of completely regular growth.* The following theorem is valid:

The zeros of an entire function of completely regular growth have an angular density (9) *with*

$$\begin{aligned} &\Delta_f(\psi_1 \pm 0, \psi_2 \pm 0) \\ &\quad = \frac{1}{2\pi\rho}\Big\{h'(\psi_2 \pm 0) - h'(\psi_1 \pm 0) + \rho^2 \int_{\psi_1}^{\psi_2} h(\psi)\, d\psi\Big\}\,, \\ &\qquad\qquad \psi_1 \le \psi_2\,. \end{aligned} \tag{13}$$

For an integer ρ, the relation (12) holds as well.

For $\rho = 1$ (i.e., for entire functions of exponential type) equation (13) has a simple geometrical interpretation. Namely, let us consider the indicator diagram of the function $f(z)$ and draw the supporting lines orthogonal to the directions $\arg z = \psi_1, \psi_2$ (see Figure 5). Let $S = S(\psi_1, \psi_2)$ be the length of the arc of the boundary of the indicator diagram between the supporting points z_1 and z_2 on these lines (for the sake of simplicity, we assume that the indicator diagram has no segment on its boundary). Then the angular density of zeros of the function $f(z)$ of completely regular growth is

$$\Delta_f(\psi_1, \psi_2) = \frac{1}{2\pi} S\,.$$

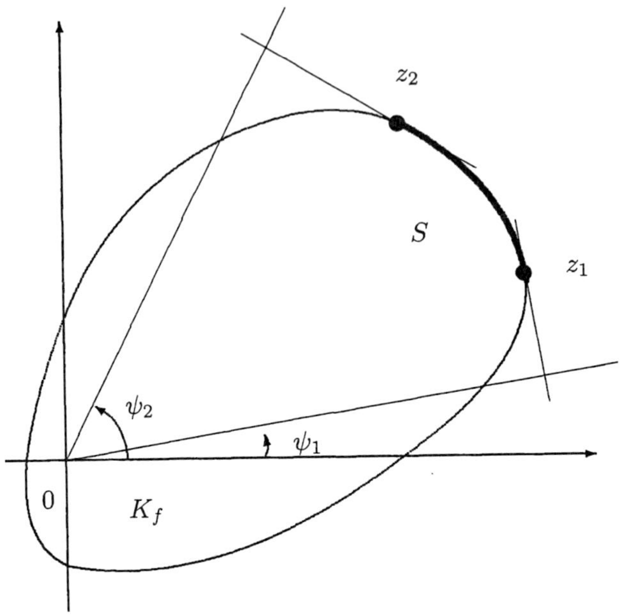

FIGURE 5

The proofs of these results can be found in Chapters II and III of the monograph Levin [**82**], which also contains numerous applications. Further results on functions of completely regular growth can be found in Appendix III to the same monograph. In the recent monograph Ronkin [**117**] the theory of functions of completely regular growth is constructed based on the concept of weak convergence.

V. S. Azarin [**7**] proposed a new approach to the investigation of the asymptotic behavior of entire functions of finite order. This approach is based on the theory of subharmonic functions and allows one to obtain many new results, as well as to simplify proofs of many well-known theorems.

Part II. Entire Functions of Exponential Type

According to the Phragmén-Lindelöf theorem, every function f analytic and of exponential type σ in the upper half-plane $\mathbb{C}_+$ which is bounded by some constant M on the real axis, satisfies the inequality

$$|f(x+iy)| \leq Me^{\sigma y}, \quad y \geq 0 .$$

Therefore, if a sequence of such functions of exponential type not exceeding some σ_0 converges uniformly on the real line, then it converges uniformly in each strip of finite width which is parallel to the real axis. In addition, the limit function is also of exponential type $\sigma \leq \sigma_0$.

A similar effect appears if the uniform norm is replaced by the norm

$$\|f\|_p = \left(\int_{-\infty}^{\infty} |f(x)|^p \, dx\right)^{1/p} .$$

In the second part of the book we study entire functions that are bounded on the real axis in a weaker sense, namely functions of class C.

DEFINITION. By the Cartwright class C we mean the class of all entire functions of exponential type satisfying the inequality

$$\int_{-\infty}^{\infty} \frac{\log^+ |f(t)|}{1+t^2} \, dt < \infty . \tag{0}$$

The functions of this class have completely regular growth (see Lecture 13 in Part I), and moreover, they possess some additional regularity of the growth and distribution of zeros, which is very important for applications. For example, the Fourier transform of a compactly supported function or distribution is an entire function of exponential type satisfying condition (0). The functions of class C also arise in boundary value problems for differential equations where eigenvalues are zeros of such functions. The study of functions of class C is based on two main results: representation of functions of class C in the half-plane, and the Hayman theorem on the estimate from below of the potential of masses located in the closed half-plane. Using this study, we turn to the properties of exponential functions $\{e^{i\lambda_n t}\}$ in the space $L^2(-\pi, \pi)$ and to interpolation problems in some classes of entire functions of exponential type.

LECTURE 14

Integral Representation of Functions Analytic in the Half-plane

14.1. The R. Nevanlinna formula

First, we obtain the Riesz-Herglotz formula. This formula provides a representation of a function that is analytic in the disk $\{w : |w| < R\}$ and has nonnegative real part. The Schwarz formula gives the following representation of such a function $\varphi(w) = u(w) + iv(w)$ in the disk $\{w : |w| < \rho\}$, $\rho < R$:

$$\varphi(w) = \frac{1}{2\pi}\int_{-\pi}^{\pi} \frac{\rho e^{i\theta} + w}{\rho e^{i\theta} - w} u(\rho e^{i\theta})\, d\theta + i \operatorname{Im} \varphi(0)\,. \tag{1}$$

In other words,

$$\varphi(w) = \frac{1}{2\pi}\int_{-\pi}^{\pi} \frac{\rho e^{i\theta} + w}{\rho e^{i\theta} - w}\, d\sigma_\rho(\theta) + i \operatorname{Im} \varphi(0),$$

where σ_ρ is a nonnegative Borel measure on the segment $[-\pi, \pi]$ with identified endpoints. This measure is defined by the relation

$$\sigma_\rho(E) = \int_E u(\rho e^{i\tau})\, d\tau\,.$$

Since $\sigma_\rho([-\pi,\pi]) = u(0) < \infty$, by Helly's theorems one can choose a sequence $\rho_n \uparrow R$ such that there exists the weak limit $\lim_{n\to\infty} \sigma_{\rho_n} = \sigma$, and the representation

$$\varphi(w) = \frac{1}{2\pi}\int_{[-\pi,\pi]} \frac{Re^{i\theta} + w}{Re^{i\theta} - w}\, d\sigma(\theta) + i \operatorname{Im} \varphi(0)$$

holds.

This relation is called the *Riesz-Herglotz formula*. If applied to a function $u(w)$ positive and harmonic in the disk $\{w : |w| < R\}$, this formula gives

$$u(w) = \frac{1}{2\pi}\int_{[-\pi,\pi]} \frac{R^2 - |w|^2}{|Re^{i\theta} - w|^2}\, d\sigma(\theta)\,.$$

Now let $\psi(z) = u(z) + iv(z)$ be a function analytic in the half-plane $\mathbb{C}_+$, and let $v(z) > 0$ for $y > 0$. Setting

$$w = \frac{i-z}{i+z}\,, \quad z = i\,\frac{1-w}{1+w}\,,$$

we obtain the function

$$\varphi(w) = \frac{1}{i}\psi\Big(i\,\frac{1-w}{1+w}\Big)\,,$$

which by the Riesz-Herglotz formula admits the representation

$$\varphi(w) = \frac{1}{2\pi} \int_{[-\pi,\pi]} \frac{e^{i\theta} + w}{e^{i\theta} - w} \, d\sigma(\theta) + i \operatorname{Im} \varphi(0) \, .$$

If the measure $d\sigma$ has a discrete mass $2\pi\sigma$ at the point $\theta = \pi$, one can write this representation in the form

$$\varphi(w) = \frac{1}{2\pi} \int_{(-\pi,\pi)} \frac{e^{i\theta} + w}{e^{i\theta} - w} \, d\sigma(\theta) + \sigma \frac{1 - w}{1 + w} + i \operatorname{Im} \varphi(0) \, ,$$

and the inverse substitution yields the *Nevanlinna formula*

$$\psi(z) = \frac{1}{\pi} \int_{-\infty}^{\infty} \frac{1 + tz}{t - z} \, d\mu(t) + \sigma z + \operatorname{Re} \psi(i) \, , \tag{2}$$

where $(i - t)(i + t)^{-1} = e^{i\theta}$, $t = \tan \theta/2$, and the nonnegative Borel measure μ is defined by the relation $\mu((-\infty, t)) = \frac{1}{2}\sigma((-\pi, 2 \arctan t))$ and is of bounded variation on the whole real axis. If

$$\nu(E) = \int_E (1 + t^2) \, d\mu(t) \, ,$$

then relation (2) can be written in the form

$$\psi(z) = \frac{1}{\pi} \int_{-\infty}^{\infty} \left[\frac{1}{t - z} - \frac{t}{1 + t^2} \right] d\nu(t) + \sigma z + \operatorname{Re} \psi(i) \, , \tag{3}$$

where $\sigma \geq 0$, and

$$\int_{-\infty}^{\infty} \frac{d\nu(t)}{1 + t^2} < \infty \, . \tag{4}$$

Taking the imaginary part of (3) we obtain

$$v(z) = \frac{y}{\pi} \int_{-\infty}^{\infty} \frac{d\nu(t)}{|t - z|^2} + \sigma y \, . \tag{5}$$

Thus, the following theorem is proved.

THEOREM 1. *Every function $v(z)$ nonnegative and harmonic in the half-plane $\mathbb{C}_+$ admits representation* (5), *where ν is a nonnegative Borel measure satisfying condition* (4) *and $\sigma \geq 0$.*

One can easily obtain an explicit expression for the measure ν. Indeed, if $y > 0$, then integrating both sides of (5) with respect to x we obtain

$$\int_\alpha^\beta v(x + iy) \, dx = \frac{1}{\pi} \int_{-\infty}^{\infty} \left[\arctan \frac{x - t}{y} \Big|_\alpha^\beta \right] d\nu(t) + (\beta - \alpha) \sigma y \, .$$

The quantity $\arctan \frac{x - t}{y} \Big|_\alpha^\beta$ is equal to the angle at which the segment $[\alpha, \beta]$ is viewed from the point $z = x + iy$. Therefore, if $\nu(t)$ is continuous at the points α and β, we have

$$\lim_{y \downarrow 0} \int_\alpha^\beta v(x + iy) \, dx = \nu(\beta) - \nu(\alpha) \, . \tag{6}$$

This relation remains true for arbitrary α and β if $\nu(t)$ is defined as $\nu(t) = [\nu(t+0)+\nu(t-0)]/2$ at the jump points of the function $\nu(t)$. It follows from relation (6) that the function v uniquely defines both the measure μ and the number σ.

REMARK. Suppose that the function $v(z)$ in Theorem 1 has continuous boundary values $v(x)$ on the real axis. Passing to the limit under the integral sign in (6) for each α and β, we have

$$\int_\alpha^\beta v(t)\,dt = \nu(\beta) - \nu(\alpha)\,,$$

or

$$\nu(E) = \int_E v(t)\,dt\,.$$

In the general case we denote $v(x) = \liminf_{y\to 0} v(x+iy)$,[10] and applying the Fatou Lemma obtain the inequality

$$\int_\alpha^\beta v(t)\,dt \le \nu(\beta) - \nu(\alpha)$$

which can be written in the form

$$\nu(E) \ge \int_E v(t)\,dt\,.$$

COROLLARY 1. *If the function $v(t)$ is positive and harmonic in $\mathbb{C}_+$, then*

$$\int_\infty^\infty \frac{v(t)}{1+t^2}\,dt < \infty\,.$$

COROLLARY 2. *Let $f(z)$ be an analytic function in $\mathbb{C}_+$ which is nonvanishing and continuous up to the real axis, and let $\log|f(z)| \le 0$, $z \in \mathbb{C}_+$. Then*

$$\log|f(z)| = \frac{y}{\pi}\int_{-\infty}^{\infty} \frac{\log|f(t)|}{|t-z|^2}\,dt + \sigma y\,, \quad \sigma \le 0\,.$$

The example

$$f(z) = \frac{z^2}{(i+z)^2}\exp\Big(-\frac{i}{z}\Big)$$

shows that the condition that $f(z)$ is nonvanishing cannot be omitted.

14.2. Representation of a function $f(z)$ analytic in the half-plane such that $\log|f(z)|$ admits a positive harmonic majorant

Now, we shall consider functions that may have zeros in the half-plane.

---

[10] The famous Fatou theorem asserts that the limit exists for a.e. $x \in \mathbb{R}$; see for example Nevanlinna [**102**] or Koosis [**71**].

THEOREM 2. *Let a function $f(z)$ be analytic in the half-plane $\mathbb{C}_+$, let*

$$\log|f(z)| \le u(z)$$

where $u(z)$, $z \in \mathbb{C}_+$, is a positive harmonic function, and let $\{a_k\}_1^\infty$ be the sequence of all zeros of $f(z)$. Then

$$\sum_{k=1}^{\infty} \frac{\operatorname{Im} a_k}{1+|a_k|^2} < \infty . \tag{7}$$

PROOF. If $v(z)$ is the harmonic conjugate to $u(z)$, then the function

$$\varphi(z) = f(z)e^{-u(z)+iv(z)}$$

is analytic in $\mathbb{C}_+$, has the same zeros as $f(z)$, and satisfies the estimate $|\varphi(z)| \le 1$. Without loss of generality we may assume $\varphi(i) \neq 0$. Consider the sequence of functions

$$\Pi_N(z) = \prod_{k=1}^{N} \frac{1-z/\overline{a}_k}{1-z/a_k} . \tag{8}$$

By the Phragmén-Lindelöf theorem we have $|\varphi(z)\Pi_N(z)| \le 1$, $z \in \mathbb{C}_+$. Since the absolute value of each factor in the product (8) is greater than 1 for $z \in \mathbb{C}_+$, we have that $|\Pi_N(i)|^2$ is an increasing bounded sequence. Hence,

$$\sum_{k=1}^{\infty} \left(\left| \frac{1-i/\overline{a}_k}{1-i/a_k} \right|^2 - 1 \right) < \infty .$$

Finally, we have

$$\left| \frac{1-i/\overline{a}_k}{1-i/a_k} \right|^2 - 1 = \frac{|\overline{a}_k - i|^2 - |a_k - i|^2}{|a_k - i|^2} \ge \frac{4 \operatorname{Im} a_k}{1+|a_k|^2} .$$

This completes the proof of the theorem.

REMARK. Under the assumptions of Theorem 1 we have

$$\sum_{|a_k| \ge \varepsilon} \left| \operatorname{Im} \frac{1}{a_k} \right| < \infty . \tag{9}$$

In particular, if the function $f(z)$ does not vanish in a neighborhood of the point $z = 0$, then condition (7) takes the form

$$\sum_{k=1}^{\infty} \left| \operatorname{Im} \frac{1}{a_k} \right| < \infty . \tag{10}$$

COROLLARY. *If a function $f(z)$ is analytic and bounded in $\mathbb{C}_+$ and vanishes on a sequence $\{a_k\}_0^\infty$ such that*

$$\sum_{|a_k| \ge \varepsilon} \left| \operatorname{Im} \frac{1}{a_k} \right| = \infty ,$$

then $f(z) \equiv 0$.

PROBLEM. Prove the Müntz theorem on the completeness of the system $\{t^{a_n}\}_0^\infty$ in $C[0,1]$:

Let $\{a_n\}$ be a sequence of numbers $0 = a_0 < a_1 < \cdots$. For the system of powers $\{t^{a_n}\}$ to be complete in the space $C[0,1]$ it is necessary and sufficient that $\sum_1^\infty a_n^{-1} = \infty$.

HINT. To prove the necessity, represent the function

$$S(x) = \frac{1}{1+x^2} \prod_{k=1}^{\infty} \frac{1 - x/a_k}{1 + x/a_k}$$

in the form

$$S(x) = \int_0^1 t^x f(t)\, dt\,.$$

To obtain a representation of functions satisfying the conditions of Theorem 2 we need the following lemma:

LEMMA 1. *Let a sequence $\{a_k\} \subset \mathbb{C}_+$ satisfy condition* (7). *Then the series*

$$v(z) = \sum_{k=1}^{\infty} \log \left| \frac{z - a_k}{z - \overline{a}_k} \right|$$

converges uniformly on each compact set in $\mathbb{C}_+$.

PROOF. We have

$$\begin{aligned} v(z) &= \frac{1}{2} \sum_{k=1}^{\infty} \log \left| \frac{z - a_k}{z - \overline{a}_k} \right|^2 = \frac{1}{2} \sum_{k=1}^{\infty} \log \left[1 - \frac{|z - \overline{a}_k|^2 - |z - a_k|^2}{|z - \overline{a}_k|^2} \right] \\ &= \frac{1}{2} \sum_{k=1}^{\infty} \log \left[1 - \frac{4y \operatorname{Im} a_k}{|z - \overline{a}_k|^2} \right]. \end{aligned} \tag{11}$$

Since, for $\operatorname{Im} \zeta \le 0$, we have $|\zeta - i|^2 \ge |\zeta|^2 + 1$, it follows that

$$\sup_{\operatorname{Im} \zeta \le 0} \frac{1 + |\zeta|^2}{|\zeta - z|^2} \le \sup_{\operatorname{Im} \zeta \le 0} \left| \frac{\zeta - i}{\zeta - z} \right|^2 .$$

The right-hand side of this inequality is bounded when z belongs to a compact set in $\mathbb{C}_+$. Taking into account the inequality

$$\frac{1}{|z - \overline{a}_k|^2} \le \frac{1}{1 + |a_k|^2} \cdot \sup_{\operatorname{Im} \zeta \le 0} \left| \frac{\zeta - i}{\zeta - z} \right|^2$$

and condition (7) we see that the series

$$\sum_{k=1}^{\infty} \frac{\operatorname{Im} a_k}{|z - \overline{a}_k|^2}$$

converges uniformly on each compact set in $\mathbb{C}_+$. This implies the uniform convergence of series (11) and, finally, the statement of the lemma.

The function

$$\prod_{k=1}^{\infty} \frac{1 - z/a_k}{1 - z/\overline{a}_k}$$

is called the *Blaschke product* for the upper half-plane, and relation (7), which, in the case $|a_k| \geq \varepsilon > 0$, provides the convergence of the product, is called the *Blaschke condition*.

THEOREM 3. *Let the function $f(z)$ be analytic in the half-plane $\mathbb{C}_+ = \{z : \operatorname{Im} z > 0\}$ and let the function $\log|f(z)|$ have a positive harmonic majorant $u(z)$ in $\mathbb{C}_+$. Then*

$$\log|f(z)| = \sum_{k=1}^{\infty} \log\left|\frac{z-a_k}{z-\bar{a}_k}\right| + \frac{y}{\pi}\int_{-\infty}^{\infty}\frac{d\nu_1(t)}{|t-z|^2} - \frac{y}{\pi}\int_{-\infty}^{\infty}\frac{d\nu_2(t)}{|t-z|^2} + \sigma y\,. \tag{12}$$

Here $\{a_k\}$ is the zero set of the function $f(z)$ in $\mathbb{C}_+$, σ is a real number, and the ν_i are nonnegative measures such that

$$\int_{-\infty}^{\infty}\frac{d\nu_i(t)}{1+t^2} < \infty\,, \quad i = 1, 2\,. \tag{13}$$

PROOF. Let $v(z)$ be the harmonic conjugate to $u(z)$. The function

$$\varphi(z) = f(z)e^{-u(z)-iv(z)}$$

is analytic in $\mathbb{C}_+$, has the same zero set and satisfies the bound $|\varphi(z)| \leq 1$, $z \in \mathbb{C}_+$. By Lemma 1 the series

$$w(z) = \sum_{k=1}^{\infty} \log\left|\frac{z-a_k}{z-\bar{a}_k}\right|$$

converges in $\mathbb{C}_+$. Let us denote $\gamma_\infty(z) = \log|\varphi(z)| - w(z)$ and prove that $-\gamma_\infty$ is a nonnegative and harmonic function in $\mathbb{C}_+$.

Indeed, for each $n \geq 1$ the function

$$\gamma_n(z) = \log|\varphi(z)| - \sum_{k=1}^{n}\log\left|\frac{z-a_k}{z-\bar{a}_k}\right| = \log\left|\varphi(z)\prod_{k=1}^{n}\frac{z-\bar{a}_k}{z-a_k}\right|$$

is subharmonic and bounded above in $\mathbb{C}_+$. Since $\gamma_n(z) \leq 0$ on the boundary of the half-plane, the Phragmén-Lindelöf theorem for subharmonic functions (see Lecture 7) yields $\gamma_n(z) \leq 0$ for all $z \in \mathbb{C}_+$. As $n \to +\infty$ we obtain $\gamma_\infty(z) \equiv \log|\varphi(z)| - w(z) \leq 0$, $z \in \mathbb{C}$. It is obvious that the function $\gamma_\infty(z)$ is harmonic in $\mathbb{C}_+$. Therefore, we have

$$\log|f(z)| = u(z) + \gamma_\infty(z) + w(z)\,, \tag{14}$$

and Theorem 1 gives representation (12), as well as inequalities (13), completing the proof of Theorem 3.

REMARK. Let a function $f(z)$ be analytic in $\overline{\mathbb{C}}_+$ and let $\log|f(z)|$ have a positive harmonic majorant in $\mathbb{C}_+$. Then $\log|f(t)|\,dt = d\nu_1(t) - d\nu_2(t)$ where ν_1, ν_2 are the measures from representation (12), and hence, this representation takes the form:[11]

$$\log|f(z)| = \frac{y}{\pi}\int_{-\infty}^{\infty} \frac{\log|f(z)|}{|t-z|^2}\,dt + \sigma y + \sum_{k=1}^{\infty} \log\left|\frac{z-a_k}{z-\overline{a}_k}\right| .$$

In addition, relation (13) yields

$$\int_{-\infty}^{\infty} \frac{|\log|f(t)||}{1+t^2}\,dt < \infty$$

or, equivalently,

$$\int_{-\infty}^{\infty} \frac{\log^+|f(t)|}{1+t^2}\,dt < \infty , \qquad \int_{-\infty}^{\infty} \frac{\log^-|f(t)|}{1+t^2}\,dt < \infty .$$

The following uniqueness theorem is a direct consequence of the above Remark.

THEOREM 4 (Carleman). *If $f(z)$ is analytic and bounded in $\overline{\mathbb{C}}_+$[12] and if*

$$\int_{-\infty}^{\infty} \frac{\log^-|f(t)|}{1+t^2}\,dt = \infty ,$$

then $f(z) \equiv 0$.

14.3. Application to the theory of quasianalytic classes

Let a sequence $\{m_n\}_0^\infty$ of positive numbers be given. Denote by C_{m_n} the class of all infinitely differentiable functions on the real axis satisfying the inequalities

$$(15) \qquad |f^{(n)}(t)| \le k^n m_n , \quad -\infty < t < \infty , \quad n = 0, 1, 2, \dots ,$$

with some constant k independent of n and t. It is evident that this class is linear.

The class C_{m_n} is said to be (Δ)-*quasianalytic* if, for each $t_0 \in \mathbb{R}$, the unique function $f \in C_{m_n}$ which satisfies the relations $f^{(n)}(t_0) = 0$, $n = 0, 1, \dots$, is $f(t) \equiv 0$.

It is easy to see, estimating the remainder term in the Taylor series, that if $m_n = n!$, then the functions of class C_{m_n} are analytic on the real axis. Hadamard posed the problem to find a criteria for a class C_{m_n} to be (Δ)-quasianalytic. It is clear that the slower the growth of a sequence $\{m_n\}$ is, the more likely are the chances for the class C_{m_n} to be (Δ)-quasianalytic. The growth of a sequence $\{m_n\}$ can be measured by means of the function

$$(16) \qquad T(r) = \sup_{n \ge 0} \frac{r^n}{m_n} .$$

This function was introduced by A. M. Ostrowski. Since the function $\log T(r)$ is the upper envelope of the family of functions $\{n \log r - \log m_n\}_{n=0}^\infty$ that are linear with respect to $\log r$, it is convex with respect to $\log r$. It follows that $\log T(r)$ is a continuous function. It is evident that the slower is the growth of the sequence $\{m_n\}$, the faster is the growth of $T(r)$.

[11] This representation was obtained by R. Nevanlinna in 1925.

[12] Taking into account Fatou's theorem (see footnote 10), one can assume that $f(z)$ is analytic and bounded in $\mathbb{C}_+$.

THEOREM 5 (Carleman, A. Ostrowski). *For the class C_{m_n} to be (Δ)-quasi-analytic, it is necessary and sufficient that*

$$\int_1^\infty \frac{\log T(r)}{r^2} = \infty . \tag{17}$$

PROOF. Without loss of generality we may assume $k = 1$ in inequality (15); otherwise we may consider the function $f(k^{-1}t)$. Introducing the function $f(t+t_0)$, we may suppose

$$f^{(n)}(0) = 0 , \quad n = 0, 1, 2, \ldots . \tag{18}$$

To prove the sufficiency of condition (17) we redefine the function $f(t)$ setting $f(t) = 0$ for $t < 0$. The resulting function belongs to the class C_{m_n}. Let $F(z)$ be its Fourier transform:

$$F(z) = \int_0^\infty f(t)e^{izt}\, dt , \quad \operatorname{Im} z > 0 . \tag{19}$$

Integrating by parts we have

$$F(z) = \frac{(-1)^n}{(iz)^n} \int_0^\infty f^{(n)}(t)e^{itz}\, dt , \quad \operatorname{Im} z > 0 ,$$

whence

$$|F(z)| \le \frac{m_n}{|z|^n} \int_0^\infty e^{-ty}\, dt = \frac{1}{y}\frac{m_n}{|z|^n} \le \frac{1}{yT(|z|)} , \quad y = \operatorname{Im} z .$$

The function $F(z+i)$ is bounded and analytic in $\overline{\mathbb{C}}_+$, and we can apply Carleman's uniqueness theorem. Condition (17) yields

$$\int_{-\infty}^\infty \frac{\log^- |F(x+iy)|}{1+x^2}\, dx = \infty$$

and, hence, $F(z) \equiv 0$. Taking the inverse transform in (19) we obtain $f(t) \equiv 0$ proving the sufficiency of condition (17).

Now let

$$\int_1^\infty \frac{\log T(r)}{r^2}\, dr < \infty .$$

Consider the function

$$u(z) = -\frac{y}{\pi} \int_{-\infty}^\infty \frac{\log[(1+t^2)T(|t|)]}{(t-x)^2 + y^2}\, dt .$$

It is harmonic in $\mathbb{C}_+$ and continuous in the closed half-plane $\overline{\mathbb{C}}_+$. Since $T(r) \ge 1/m_0$, the function $u(z)$ is bounded from above in $\mathbb{C}_+$. If $v(z)$ is its harmonic conjugate, then the function

$$F(z) = e^{u(z)+iv(z)}$$

is bounded and analytic in $\mathbb{C}_+$ and satisfies the estimate

$$|F(x)| \le \frac{1}{(1+x^2)T(|x|)}$$

on the real axis. By the Lindelöf theorem (Theorem 3, Lecture 13) we have $F(z) \to 0$, $|z| \to \infty$, $z \in \mathbb{C}_+$, and hence the Jordan Lemma implies that the Fourier transform

$$f(t) = \frac{1}{\pi} \int_{-\infty}^{\infty} F(x) e^{-ixt}\, dx$$

vanishes for $t < 0$. On the other hand, $f(t)$ is not identically zero and

$$f^{(n)}(t) = \frac{(-1)^n}{\pi} \int_{-\infty}^{\infty} x^n F(x) e^{-ixt}\, dx\,.$$

Hence

$$|f^{(n)}(t)| \le \frac{1}{\pi} \int_{-\infty}^{\infty} \frac{|x|^n dx}{(1+x^2)T(|x|)} \le \frac{1}{\pi} \int_{-\infty}^{\infty} \frac{|x|^n dx}{(1+x^2)|x|^n/m_n} = m_n\,.$$

Thus $f \in C_{m_n}$, which completes the proof of the theorem.

For other theorems on (Δ)-quasianalyticity see the monographs by Carleman [**22**] and Mandelbrojt [**91, 93**]. An abstract operator version of the Hadamard Δ-quasianalyticity problem can be found in Lyubich and Tkachenko [**88**].

LECTURE 15

The Hayman Theorem

We consider the asymptotic behavior of a function in $\mathbb{C}_+$ which is represented by a more general formula than (12) of Section 14.2. Namely, we shall study functions of the form

$$v(z) = \iint_{\mathbb{C}_+} \log\left|\frac{\zeta - \bar{z}}{\zeta - z}\right| d\mu(\zeta) + \frac{y}{\pi}\int_{-\infty}^{\infty} \frac{d\nu(t)}{(t-x)^2 + y^2}, \tag{1}$$

where $d\mu(\zeta)$ and $d\nu(t)$ are nonnegative Borel measures such that

$$\iint_{\mathbb{C}_+} \frac{\operatorname{Im}\zeta}{1+|\zeta|^2}\, d\mu(\zeta) < \infty, \qquad \int_{\mathbb{R}} \frac{d\nu(t)}{1+t^2} < \infty.$$

A set of disks $(C_j)_1^\infty$, centered at points z_j of the upper half-plane and of radii ρ_j is called a set of finite view if

$$\sum_{j=1}^{\infty} \frac{\rho_j}{r_j} < \infty, \tag{2}$$

where $r_j = |z_j|$.

THEOREM 1 (Hayman). *The asymptotic relation*

$$v(z) = o(|z|), \qquad |z| \to \infty, \tag{3}$$

holds everywhere in $\mathbb{C}_+$ outside some exceptional set of disks of finite view.

PROOF. The proof of this theorem is similar to that of the Cartan estimate (see Section 11.2). First, to transform representation (1), we define the measure $dm(\zeta)$ and the kernel $K(z,\zeta)$ by the relations

$$dm(\zeta) = \begin{cases} \dfrac{\eta\, d\mu(\zeta)}{1+|\zeta|^2}, & \eta > 0, \\[2ex] \dfrac{d\nu(\xi)}{\pi(1+|\xi|^2)}, & \eta = 0, \end{cases}$$

and

$$K(z,\zeta) = \begin{cases} \dfrac{1+|\zeta|^2}{\eta} \log\left|\dfrac{\zeta-\bar{z}}{\zeta - z}\right|, & \eta > 0, \\[2ex] (1+|\xi|^2)\dfrac{y}{(\xi - x)^2 + y^2}, & \eta = 0, \end{cases}$$

where $z = x + iy$, $\zeta = \xi + i\eta$. The function $v(z)$ now can be represented in the form

$$v(z) = \iint_{\overline{\mathbb{C}}_+} K(z,\zeta)\, dm(\zeta),$$

where

$$m(\overline{\mathbb{C}}_+) = \iint_{\mathbb{C}_+} dm(\zeta) < \infty \, .$$

Since $v(z) \geq 0$ for $z \in \mathbb{C}_+$, it is sufficient to obtain the estimate of $v(z)$ from above. It is easily seen that, for the kernel $K(z,\zeta)$,

$$K(z,\zeta) \leq 2\frac{1+|\zeta|^2}{y} \log \frac{3y}{|\zeta - z|} \, , \qquad |\zeta - z| \leq \frac{y}{2} \, , \tag{4}$$

and[13]

$$K(z,\zeta) \leq 4\frac{1+|\zeta|^2}{|\zeta - z|^2} y \, , \qquad y > 0 \, . \tag{5}$$

At the next step, we fix small $\varepsilon > 0$, choose a number R_ε large enough for the inequality

$$\iint_{|\zeta| \geq R_\varepsilon} dm(\zeta) \leq \frac{\varepsilon}{4} \tag{6}$$

to be valid, define the measure $m^{(\varepsilon)}$ by the relation

$$m^{(\varepsilon)}(E) = m(E \cap \{z \in \mathbb{C}_+ \, , \, |z| > R_\varepsilon\}) \, ,$$

and introduce the function

$$u^{(\varepsilon)}(z) = \iint_{\overline{\mathbb{C}}_+} K(z,\zeta) \, dm^{(\varepsilon)}(\zeta) \, .$$

A point $z \in \overline{\mathbb{C}}_+$ is called ε-*normal* if the inequality

$$\iint_{|\zeta - z| \leq t} dm^{(\varepsilon)}(\zeta) < \frac{\varepsilon t}{|z|}$$

holds for all $t > 0$. Otherwise z is called ε-*abnormal*. It is obvious that the inequality holds for $t > |z|/4$. Therefore, if a point z is ε-abnormal, and if ρ_z is the least upper bound of the values t for which

$$\iint_{|\zeta - z| \leq t} dm^{(\varepsilon)} \geq \frac{\varepsilon t}{|z|} \, , \tag{7}$$

then ρ_z admits the estimate

$$\rho_z \leq \frac{1}{4}|z| \, . \tag{8}$$

As in the proof of the Cartan estimate (Section 11.2), inequality (7) holds for $t = \rho_z$ as well. Now, we shall prove that, for sufficiently large $|z|$, at each ε-normal point z the inequality

$$u^{(\varepsilon)}(z) \leq C\varepsilon|z| \, , \tag{9}$$

holds with C independent of both ε and z. To this end, let us consider the sequence of half-annuli

$$A_n = \{z : 2^n \leq |z| \leq 2^{n+1}\} \cap \overline{\mathbb{C}}_+ \, .$$

[13] Compare with (11) from the preceding lecture.

For a fixed $n > 0$, choose the disk $C_z^{(n)}$ centered at an ε-abnormal point $z \in A_n$ and having the largest radius $\rho_1^{(n)}$ among all ε-abnormal points in A_n (such points exist, see Section 11.2). Excluding the interior of this disk, we choose the point $z_2^{(n)}$ in the remaining set with the largest possible radius $\rho_2^{(n)}$ of abnormality and continue the process. The radii $\rho_j^{(n)}$ of the constructed disks $C_{z_j}^{(n)}$ form a sequence which is either finite or decreases to zero, and all abnormal points in A_n belong to the set $\bigcup_j C_{z_j}^{(n)}$.

Repeating the arguments used in the proof of the Cartan estimate, we prove that each point $z \in A_n$ belongs to at most five exceptional disks. Therefore,

$$\varepsilon \sum_j \frac{\rho_j^{(n)}}{r_j^{(n)}} \le 5 \int_{2^n - \rho_1^{(n)} \le |\zeta| \le 2^{n+1} + \rho_1^{(n)}} dm^{(\varepsilon)}(\zeta) \, . \tag{10}$$

Since $r_1^{(n)} \le 2^{n+1}$, then by (8) we have $\rho_1^{(n)} \le 2^{n-1}$. Therefore, all exceptional disks that correspond to the half-annulus A_n belong to the annulus

$$\{z : 2^{n-2} \le |z| \le 2^{n+2}\} \, .$$

Then, according to (10), the union over all n of all such disks has a finite view:

$$\sum_n \sum_j \frac{\rho_j^{(n)}}{r_j^{(n)}} \le \frac{25}{\varepsilon} \iint_{\overline{\mathbb{C}}_+} dm^{(\varepsilon)}(\zeta) < \infty \, .$$

Let us estimate the function $u^{(\varepsilon)}(z)$ at ε-normal points. By virtue of (4) we have

$$\begin{aligned} \iint_{\delta \le |\zeta - z| \le y/2} K(z,\zeta) dm^{(\varepsilon)}(\zeta) &\le 2 \iint_{\delta \le |\zeta - z| \le y/2} \frac{1 + |\zeta|^2}{y} \log \frac{3y}{|\zeta - z|} \, dm^{(\varepsilon)}(\zeta) \\ &\le 6 \frac{1 + |z|^2}{y} \int_\delta^{y/2} \log \frac{3y}{t} \, dm_z^{(\varepsilon)}(t) \\ &= 6 \frac{1 + |z|^2}{y} \left[m_z^{(\varepsilon)}(t) \log \frac{3y}{t} \Big|_\delta^{y/2} + \int_\delta^{y/2} \frac{m_z^{(\varepsilon)}(t)}{t} \, dt \right] , \end{aligned}$$

where

$$m_z^{(\varepsilon)}(t) = \iint_{|z - \zeta| \le t} dm^{(\varepsilon)}(\zeta) \, .$$

If z is an ε-normal point, then

$$m_z^{(\varepsilon)}(t) \le \frac{\varepsilon t}{|z|} \, ,$$

and, as $\delta \to 0$, we have

$$v_1(z) \stackrel{\text{def}}{=} \iint_{|\zeta - z| \le y/2} K(z,\zeta) \, dm^{(\varepsilon)}(\zeta) \le C_1 \varepsilon |z| \, .$$

Using estimate (5) for

$$v_2(z) \stackrel{\text{def}}{=} \iint_{y/2 \le |\zeta - z| \le r/2} K(z,\zeta) \, dm^{(\varepsilon)}(\zeta) \, ,$$

with $r = |z|$, we obtain

$$v_2(z) \le 4 \iint_{y/2 \le |\zeta - z| \le r/2} \frac{1+|\zeta|^2}{|\zeta - z|^2} y\, dm^{(\varepsilon)}(\zeta) \le 12(1+r^2)y \int_{y/2}^{r/2} \frac{dm^{(\varepsilon)}(t)}{t^2}$$
$$= 12(1+r^2)y \left[\frac{m_z^{(\varepsilon)}(t)}{t^2} \Big|_{y/2}^{r/2} + 2\int_{y/2}^{r/2} \frac{m_z^{(\varepsilon)}(t)}{t^3}\, dt \right],$$

and using the inequality $m_z^{(\varepsilon)}(t) \le \dfrac{\varepsilon t}{r}$ again we have

$$v_2(z) \le 12(1+r^2)y \left[\frac{4m_z^{(\varepsilon)}(r/2)}{r^2} + \frac{\varepsilon}{r} \int_{y/2}^{r/2} \frac{dt}{t^2} \right] \le C_2 \varepsilon r .$$

Further, let

$$v_3(z) \stackrel{\text{def}}{=} \iint_{|\zeta - z| \ge r/2} K(z,\zeta) dm^{(\varepsilon)}(\zeta) .$$

Then

$$v_3(z) \le 4y \iint_{|\zeta - z| \ge r/2} \frac{1+|\zeta|^2}{|\zeta - z|^2} dm^{(\varepsilon)}(\zeta)$$
$$\le 16\frac{y}{r^2} \iint_{|\zeta - z| \ge r/2} dm^{(\varepsilon)}(\zeta) + 2y \iint_{|\zeta - z| \ge r/2} \left| \frac{\zeta}{\zeta - z} \right|^2 dm^{(\varepsilon)}(\zeta) .$$

Since in this case $\left| \dfrac{\zeta}{\zeta - z} \right| \le 3$ and since (6) holds, we have

$$v_3(z) \le C_3 y \int_{|\zeta - z| \ge r/2} dm^{(\varepsilon)}(\zeta) \le C_3 \varepsilon y .$$

Combining the estimates of the functions $v_j(z)$, $j = 1, 2, 3$, we arrive at relation (3).

To complete the proof of the Hayman Theorem let us represent the function $v(z)$ in the form

$$v(z) = \iint_{|\zeta| \le R_\varepsilon} K(z,\zeta)\, dm(\zeta) + u^{(\varepsilon)}(z) .$$

The integral here is $O\Big(\dfrac{1}{|z|}\Big)$. Therefore relation (9) yields that the inequality

$$|v(z)| \le C\varepsilon |z|$$

holds for all sufficiently large $|z|$ if z does not belong to a set of finite view, C being independent both of ε and z.

Let us choose sequences $\{\varepsilon_p\}_1^\infty$, $\varepsilon_p \downarrow 0$ and $\{R_p\}_1^\infty$, $R_p \uparrow \infty$ so that for each p, the condition

$$\sum_j \frac{\rho_{j,p}}{r_{j,p}} \le \frac{1}{2^p}$$

holds for exceptional disks centered in the exterior of the half-disk

$$K_{R_p} = \{z : |z| \le R_p, \operatorname{Im} z \ge 0\} ,$$

and the estimate

$$v(z) \le C\varepsilon_p |z|$$

holds for all $z \notin K_{R_p}$ such that z does not belong to the union of exceptional disks. Finally, let us include into the exceptional set all disks centered in the half-annulus

$$\{z : R_p \le |z| \le R_{p+1}\,, \operatorname{Im} z \ge 0\}$$

that correspond to the value ε_p. Then we have

$$\sum_p \sum_j \frac{\rho_{j,p}}{r_{j,p}} \le 2$$

and

$$v(z) = o(|z|)$$

outside the exceptional set. This completes the proof of the theorem.

We would like to mention that Hayman [**50**] proved the asymptotic relation (3) for $|z| \notin E$ where $E \subset \mathbb{R}_+$ is a set of finite logarithmic length, i.e.,

$$\int_D \frac{dx}{x} < \infty. \tag{11}$$

A version of the theorem given here was obtained by Azarin. It is evident that condition (11) follows from (2). There are many works containing further generalizations and refinements of the Hayman theorem. We mention here the papers Azarin [**5**] and Grishin [**47**], and the monograph Essén [**35**].

LECTURE 16

Functions of Class C and their Applications

16.1. Properties of functions of class C

First, we remind the reader that, an entire function $f(z)$ belongs to the class C if

a) $f(z)$ is an entire function of exponential type;

b) $\displaystyle\int_{-\infty}^{\infty} \frac{\log^+ |f(t)|}{1+t^2}\, dt < \infty.$

THEOREM 1 (M. G. Krein). *For an entire function $f(z)$ to belong to the class C it is necessary and sufficient that the function $\log|f(z)|$ have positive harmonic majorants in the upper and lower half-planes $\mathbb{C}_+$ and $\mathbb{C}_-$.*

SUFFICIENCY. By virtue of Theorem 3, Section 14.3, the existence of a positive harmonic majorant in the half-plane $\mathbb{C}_+$ yields the representation

$$\log|f(z)| = \sum_{k=1}^{\infty} \log\left|\frac{z-a_k}{z-\bar{a}_k}\right| + \frac{y}{\pi}\int_{-\infty}^{\infty} \frac{d\nu(t)}{|t-z|^2} + \sigma_+ y$$

with

$$\sum_k \frac{\operatorname{Im} a_k}{1+|a_k|^2} < \infty\,, \qquad \int_{-\infty}^{\infty} \frac{d|\nu|(t)}{1+t^2} < \infty\,.$$

Applying the Hayman theorem, we obtain

$$\log|f(z)| = \sigma_+ y + o(|z|)\,, \qquad y \ge 0\,, \tag{1}$$

and, similarly,

$$\log|f(z)| = \sigma_- y + o(|z|)\,, \qquad y \le 0\,, \tag{2}$$

everywhere outside the union of exceptional disks $\{C_j\}$ that form a set of finite view. Outside these disks we have

$$|f(z)| \overset{\text{as}}{<} e^{(\sigma+\varepsilon)|z|}\,, \qquad \sigma = \max\{\sigma_+, \sigma_-\}\,, \tag{3}$$

for each $\varepsilon > 0$, and by the Maximum Principle this inequality still holds inside the disks C_j. Condition b) is a direct consequence of the Remark to Theorem 3, Section 14.3.

NECESSITY. Let $f(z) \in C$. Condition b) yields the convergence of the integral

$$u(z) = \frac{y}{\pi} \int_{-\infty}^{\infty} \frac{\log^+ |f(t)|}{|t-z|^2}\, dt , \qquad y > 0 ,$$

which defines a positive harmonic function in $\mathbb{C}_+$ with boundary values on the real axis coinciding with $\log^+ |f(t)|$. Let $v(z)$ be the harmonic conjugate to $u(z)$ in $\mathbb{C}_+$. Then the function

$$\varphi(z) = f(z) e^{-u(z)-iv(z)}$$

is analytic and of exponential type in the half-plane $\mathbb{C}_+$, and $|\varphi(x)| \leq 1$, $-\infty < x < \infty$. By the Phragmén-Lindelöf theorem we have

$$|\varphi(x+iy)| \leq e^{\sigma y} , \qquad y \geq 0 ,$$

where σ is the type of the function $f(z)$. Hence

$$\log |f(z)| \leq \frac{y}{\pi} \int_{-\infty}^{\infty} \frac{\log^+ |f(t)|}{|t-z|^2}\, dt + \sigma y ,$$

i.e., the function $\log |f(z)|$ has a positive harmonic majorant in the upper half-plane. Similarly, we prove the existence of a positive harmonic majorant in the lower half-plane, completing the proof of Theorem 1.

Together with the Hayman theorem this yields the following important result related to functions of class C.

THEOREM 2. *Every function $f(z)$ of class C satisfies the following relations:*

$$\log |f(z)| = \sigma_+ y + o(|z|) , \qquad y \geq 0 , \tag{4}$$

$$\log |f(z)| = \sigma_- y + o(|z|) , \qquad y \leq 0 , \tag{5}$$

everywhere outside a system of exceptional disks of finite view.

REMARK. The arguments used in the proof of Theorem 1 can be used to obtain statements similar to Theorems 1 and 2 for functions analytic and of exponential type in the closed upper half-plane for which the integral

$$\int_{-\infty}^{\infty} \frac{\log^+ |f(t)|}{1+t^2}\, dt < \infty$$

converges. A conformal mapping of the upper half-plane onto the unit disk transforms this class into the *Nevanlinna class*, i.e., class of functions $f(z)$ analytic in the unit disk and such that $\log |f(z)|$ admits a positive harmonic majorant.

If the function f has only real zeros, then the sufficient condition in the previous theorem yields the following result.

THEOREM 3 (M.G. Krein). *If $f(z)$ is an entire function, and if the representation*

$$\frac{1}{f(z)} = \sum_n \frac{c_n}{z - \lambda_n} \tag{6}$$

holds with some real λ_n's satisfying

$$\sum \frac{|c_n|}{1+|\lambda_n|} < \infty ,$$

then $f(z)$ belongs to the class C.

PROOF. Let us use the representation

$$\frac{1}{f(z)} = \varphi_1(z) - \varphi_2(z) + i[\varphi_3(z) - \varphi_4(z)] ,$$

where the functions $\varphi_j(z)$ are represented by series of the form (6) with positive coefficients c_n. Since $\operatorname{Im} \varphi_j(z) < 0$ for $\operatorname{Im} z > 0$, we have $|\varphi_j(z) - i| > 1$ implying that each function

$$u_j(z) \stackrel{\text{def}}{=} \log |\varphi_j(z) - i|$$

is positive and harmonic in $\mathbb{C}_+$. Using the inequality

$$\log^+ \left| \sum_{k=1}^{m} a_k \right| \le \sum_{k=1}^{m} \log^+ |a_k| + \log m ,$$

we obtain

$$\log^+ \left| \frac{1}{f(z)} \right| \le \sum_{j=1}^{4} \log^+ |\varphi_j(z)| + \log 4 .$$

Since $|\varphi_j(z)| \le |\varphi_j(z) - i|$, we have

$$\log^+ \left| \frac{1}{f(z)} \right| \le \sum_{j=1}^{4} \log^+ |\varphi_j(z) - i| + \log 4 = \sum_{j=1}^{4} u_j(z) + \log 4 .$$

Hence, the function $-\log |f(z)|$ admits a positive harmonic majorant in the upper half-plane, and the same is true for the lower half-plane. By the Hayman theorem, asymptotic relations (1) and (2) hold outside a set of finite view and applying the Maximum Principle we obtain asymptotic inequality (3) everywhere in $\mathbb{C}$. This proves that $f(z)$ is an entire function of exponential type. By virtue of the Remark to Theorem 3, Section 14.3, we have

$$\int_{-\infty}^{\infty} \frac{\log^+ |f(t)|}{1+t^2}\, dt = \int_{-\infty}^{\infty} \frac{\log^- |1/f(t)|}{1+t^2}\, dt < \infty .$$

Hence, the function $f(z)$ belongs to the class C. This completes the proof.

It should be mentioned that Theorem 3 still remains true if the function $1/f(z)$ is representable in the form

$$\text{(6a)} \qquad \frac{1}{f(z)} = \sum_n c_n \left[\frac{1}{z-\lambda_n} + \frac{1}{\lambda_n} + \frac{z}{\lambda_n^2} + \cdots + \frac{z^{p-1}}{\lambda_n^p} \right] ,$$

where p is a positive integer, λ_n's are real numbers, and

$$\sum_n \left| \frac{c_n}{\lambda_n^{p+1}} \right| < \infty .$$

Indeed, we can write the previous representation in the form

$$\frac{1}{z^p f(z)} = \sum_n \frac{c_n}{\lambda_n^p (z - \lambda_n)}$$

which reduces the problem to the case already considered.

REMARK. In articles [**76, 77**] by Krein, Theorems 1 and 3 were applied for studying the theory of selfadjoint operators in Hilbert space and for a description of spectra of boundary value problem for the Sturm-Liouville operator. A description of the class of entire functions admitting representation (6) (or (6a)) was given in Ostrovskii [**106**].

In Part III of this book we prove the Matsaev theorem which gives a generalization of Theorem 3. Other generalizations of the Krein Theorem may be found in Chapter VI of Goldberg and Ostrovskii [**43**].

It follows from Theorem 2 that if $f(z)$ belongs to the class C, then the limit

$$\lim_{r\to\infty} \frac{\log|f(re^{i\theta})|}{r} = \begin{cases} \sigma_+ \sin\theta\,, & 0 \le \theta \le \pi \\ \sigma_- |\sin\theta|\,, & \pi \le \theta \le 2\pi \end{cases} \tag{7}$$

exists for almost all $\theta \in [0, 2\pi]$. Indeed, for any $\varepsilon > 0$, one can choose sufficiently large R_ε such that the sum of openings of the angles at which the exceptional disks C_j centered outside the disk $\{z : |z| < R_\varepsilon\}$ are viewed, is less than ε. Relation (7) holds for all θ such that the ray $\arg z = \theta$ does not belong to these angles. Since ε is an arbitrary small number, the limit in (7) exists almost everywhere in $[0, 2\pi]$.

THEOREM 4 (on the sum of indicator functions). *If $f_1(z)$ is an entire function of exponential type and $f_2(z)$ belongs to the class C, then*

$$h_{f_1 f_2}(\theta) = h_{f_1}(\theta) + h_{f_2}(\theta)\,, \tag{8}$$

and hence, the indicator diagram of the product $f_1 f_2$ is the sum of the indicator diagrams of $f_1(z)$ and $f_2(z)$.

PROOF. On a dense set of rays we have

$$h_{f_1 f_2}(\theta) = \limsup_{r\to\infty} \frac{\log|f_1(re^{i\theta})|}{r} + \lim_{r\to\infty} \frac{\log|f_1(re^{i\theta})|}{r}\,,$$

i.e., $h_{f_1 f_2}(\theta) = h_{f_1}(\theta) + h_{f_2}(\theta)$. The statement of the theorem follows since the indicators are continuous functions.

Using the same arguments as in the Remark to Theorem 2 we obtain the following statement:

REMARK. If $f_1(z)$, $f_2(z)$ are functions analytic and of exponential type in $\mathbb{C}_+$, and if

$$\int_{-\infty}^{\infty} \frac{\log^+|f_2(t)|}{1+t^2}\, dt < \infty\,,$$

then relation (8) holds for $0 < \theta < \pi$.

It follows from Theorem 2 that the functions of class C are of completely regular growth. The functions of completely regular growth were described in Section 13.2. The theorem on the sum of indicators still remains true if $f_2(z)$ is an arbitrary entire function of completely regular growth. If no assumption on the regularity of growth of entire functions $f_1(z)$ and $f_2(z)$ is made, one can claim only the trivial inequality $h_{f_1 f_2}(\theta) \leq h_{f_1}(\theta) + h_{f_2}(\theta)$. There exist examples of entire functions for which $h_{f_1 f_2}(\theta) < h_{f_1}(\theta) + h_{f_2}(\theta)$ at every $\theta \in [0, 2\pi]$.

Property (8) is characteristic for functions of completely regular growth. It is proved by Azarin [**6**] that if $f_2(z)$ is a fixed entire function and relation (8) holds for an arbitrary entire function $f_1(z)$, then $f_2(z)$ is of completely regular growth.

16.2. The Titchmarsh convolution theorem and a problem of Gelfand

It is easy to see that the class C contains entire functions of exponential type bounded on the real axis. In particular, such is the Fourier transform F of an integrable function f with compact support. The Fourier transform F is an entire function of exponential type, and its conjugate diagram, after being rotated by the angle $\pi/2$, coincides with the supporting segment of f, that is, with the smallest segment outside which this function vanishes.

THEOREM 5 (Titchmarsh). *Let $f_1(t)$ and $f_2(t)$ be two compactly supported integrable functions with the supporting segments $[a, b]$ and $[c, d]$, respectively. Then*

$$\varphi(t) = \int_{-\infty}^{\infty} f_1(t-s) f_2(s)\, ds$$

is a compactly supported function with the supporting segment $[a + c, b + d]$.

The statement that the supporting segment of $\varphi(t)$ belongs to $[a + c, b + d]$ is trivial. The most essential statement of the Titchmarsh convolution theorem is that this segment is exactly $[a + c, b + d]$.

PROOF. The Fourier transforms of the functions $\varphi(t)$, $f_1(t)$, $f_2(t)$ satisfy the relation

$$\Phi(z) = F_1(z) F_2(z).$$

The conjugate diagrams of $F_1(z)$ and $F_2(z)$ are the segments $[ia, ib]$ and $[ic, id]$. According to Theorem 4, the conjugate diagram of the product $\Phi(z)$ is the sum of the diagrams of the factors, i.e., the segment $[ia + ic, ib + id]$. This completes the proof of the theorem.

There are many works containing results of the same type as the Titchmarsh convolution theorem as well as their applications to various problems of harmonic analysis. We mention here Domar [**30**], Ostrovskii and Ulanovskii [**107**], and Borichev [**18**].

In 1938 the journal "Uspekhi Matematicheskikh Nauk", no. 5, published a list of unsolved problems that were suggested by various mathematicians. Problem #17 in this list was the following:

PROBLEM (I. Gelfand). Describe all closed invariant subspaces of the operator of integration:

$$Ix(t) = \int_0^t x(s)\, ds$$

in the space $L^1(0, a)$, $a > 0$.

We shall show that this problem may be solved using the Titchmarsh convolution theorem.

Let E be a closed subspace in $L^1(0,a)$ invariant with respect to the operator I. First, we prove that E is invariant with respect to the right translations. This means that if $x(t) \in E$, then the function

(9) $$x_\tau(t) = \begin{cases} 0 & 0 \le t \le \tau \\ x(t-\tau) & \tau \le t \le a \end{cases}$$

belongs to E for each $\tau \in [0,a]$.

Indeed, since $x \in E$, then

$$(I^n x)(t) \equiv \int_0^t \frac{(t-s)^{n-1}}{(n-1)!} x(s)\, ds \in E .$$

Therefore,

$$\int_0^t P(t-s)x(s)\, ds \in E$$

for all polynomials $P(t)$, and since the set of all polynomials is dense in $C[0,a]$, we have

$$\int_0^t \varphi(t-s)x(s)\, ds = \int_0^t \varphi(s)x(t-s)\, ds \in E$$

for all functions $\varphi \in C[0,a]$.

For a positive number $\varepsilon < \min(\tau, a-\tau)$ we choose a function $\varphi(t) \ge 0$ such that $\varphi(t) = 0$ for $t \in [0, \tau-\varepsilon] \cup [\tau+\varepsilon, a]$, and

$$\int_0^a \varphi(t)\, dt = 1 .$$

The inequality

$$\begin{aligned} \int_0^a \left| \int_0^t \varphi(s)x(t-s)\, ds - x_\tau(t) \right| dt & \\ \le \int_0^\varepsilon |x(t)|\, dt + \max_{s\in[\tau-\varepsilon,\tau+\varepsilon]} & \int_{\tau-\varepsilon}^{\tau+\varepsilon} |x(t-s)|\, dt \\ + \max_{s\in[\tau-\varepsilon,\tau+\varepsilon]} & \int_{\tau+\varepsilon}^{a} |x(t-s) - x(t-\tau)|\, dt \end{aligned}$$

with an arbitrary small $\varepsilon > 0$ shows that, indeed, all functions of the form (9) belong to E.

Let us denote by κ the greatest lower bound of the left endpoints of supporting segments of functions from E, and suppose, first, that $\kappa = 0$.

We denote by $f(t)$ a function of the space $L^\infty(0,a)$ which generates a nontrivial linear functional annihilating the space E. It is evident that the supporting segment $[\alpha, \beta]$ of $f(t)$ is contained in $[0,a]$, and since $\kappa = 0$, there exists a function $x(t) \in E$ with the supporting segment $[\gamma, \delta]$ such that $\gamma \in (0, \beta/2)$. Using the translation property of the space E we have

(10) $$0 = \int_0^a f(t)x_\tau(t)\, dt = \int_\tau^a f(t)x(t-\tau)\, dt$$

for each $\tau \in [0,a]$. If the functions $f(t)$ and $x(t)$ are extended to be zero outside $[0,a]$, then relation (10) can be written in the form

$$(11) \qquad \int_{-\infty}^{\infty} f(t)\,x(t-\tau)\,dt = 0\,, \qquad 0 \le \tau \le a\,.$$

The integral here is the convolution of the functions $f(t)$ and $x(-t)$ with the supporting segments $[\alpha,\beta]$ and $[-\delta,-\gamma]$, respectively. By the Titchmarsh theorem the supporting segment of their convolution is $[\alpha-\delta,\beta-\gamma]$. Since $0<\gamma<\beta/2$, we have $[\alpha-\delta,\beta-\gamma)\cap[0,a]\neq\varnothing$ which contradicts relation (11). Hence $f\equiv 0$, and E coincides with the whole space $L^1(0,a)$.

Now let $\kappa\in(0,a)$. Denote by E_κ the subspace of all functions from $L^1(0,a)$ that vanish on $[0,\kappa]$, and consider the mapping T_κ which transforms the function $x(t)\in E_\kappa$ into the function $y(t)=x(t+\kappa)$ on the segment $[0,a-\kappa]$. It is easy to check that T_κ is an isomorphism between the spaces E_κ and $L^1(0,a-\kappa)$, and that $T_\kappa E$ is a closed subspace in $L^1(0,a-\kappa)$ invariant with respect to the operator I. By the choice of κ there exists a function from $T_\kappa E$ whose supporting segment is arbitrary close to zero. According to the above arguments for $\kappa=0$ we conclude that $T_\kappa E = L^1(0,a-\kappa)$, and hence $E=E_\kappa$.

We have proved that the solution to Gelfand's problem is given by the following theorem.

THEOREM 6. *Every closed subspace $E\subset L^1(0,a)$ invariant with respect to the operator I coincides with one of the spaces E_κ.*

Theorem 6 shows that the operator I is unicellular. This means that the set of its invariant closed subspaces is totally ordered with respect to inclusion. For the operator-theoretic approach to Gelfand's problem and for other related topics, we refer the reader to the monographs Gohberg and Krein [**41**], and Nikol'skii [**104**, Lecture 4].

16.3. Mean periodic functions

The notion of a mean periodic function was introduced by Delsart and was afterwards developed by L. Schwartz.

Let $C(\mathbb{R})$ be the space of all complex functions continuous on the real line with the topology of uniform convergence on compact sets. According to L. Schwartz, a function $f(x)\in C(\mathbb{R})$ is called a mean periodic (m.p.) function if the closure T_f of the linear span of all translations $\{f(x+t)\}_{t\in\mathbb{R}}$ does not coincide with the whole space $C(\mathbb{R})$. This property is equivalent to the existence of some $a>0$ and a nonzero function $\sigma(x)$ of bounded variation such that

$$(12) \qquad \int_0^a f(x+t)\,d\sigma(t) = 0\,, \qquad x\in\mathbb{R}\,.$$

We may assume that the supporting segment of the charge $d\sigma(t)$ coincides with $[0,a]$.

The simplest example of m.p. function is a periodic function. Another example is given by an exponential monomial

$$(13) \qquad f(x) = x^k e^{\lambda x}\,.$$

In this case the corresponding space T_f has dimension k.

Let f be a m.p. function and let E_f be the closure (with respect to the uniform convergence on compact sets) of the linear span of all monomials of the form (13) which are in T_f.

The main problem of the theory of m.p. functions is to prove the equality $E_f = T_f$. This problem is called the *problem of spectral synthesis*[14] and was solved by L. Schwartz back in 1947. Here we present a solution of the problem given by Lyubich and Tkachenko [**87**] using the theorem on the sum of indicators.

If $|f(x)| \leq Ce^{\alpha x}$, then, applying the Laplace transform to relation (12), we have

$$0 = \int_0^\infty e^{-\lambda x}\,dx \int_0^a f(x+t)\,d\sigma(t) = \int_0^a e^{\lambda t}\,d\sigma(t) \int_t^\infty e^{-\lambda x} f(x)\,dx$$
$$= \int_0^a e^{\lambda t}\,d\sigma(t)\Big\{ \int_0^\infty e^{-\lambda x} f(x)\,dx - \int_0^t e^{-\lambda x} f(x)\,dx \Big\}.$$

Hence,

$$\int_0^\infty e^{-\lambda x} f(x)\,dx = \frac{H(\lambda; f)}{A(\lambda)}\,, \tag{14}$$

where

$$A(\lambda) = \int_0^a e^{\lambda t}\,d\sigma(t)$$

and

$$H(\lambda; f) = \int_0^a e^{\lambda t}\,d\sigma(t) \int_0^t e^{-\lambda x} f(x)\,dx\,.$$

The right-hand side of relation (14) is defined independently of any restriction on the growth of the function $f(x)$. It represents a meromorphic function which is called the *Carleman transform* of the function f.

Given a m.p. function f we introduce the formal series

$$f(x) \sim \sum \operatorname{res}_\lambda \Big\{ \frac{H(\lambda; f)}{A(\lambda)} e^{\lambda x} \Big\}\,, \tag{15}$$

with the summation over all zeros of the function $A(\lambda)$. We call this series the generalized Fourier series of the function $f(x)$ and note that each term in the sum has the form $P(x)e^{\lambda x}$, where $P(x)$ is a polynomial. One can easily see that

$$\operatorname{res}_\lambda \frac{H(\lambda; f)}{A(\lambda)} e^{\lambda x} = \operatorname{res}_\lambda \frac{1}{A(\lambda)} \int_0^a e^{\lambda t}\,d\sigma(t) \int_x^t e^{-\lambda s} f(s)\,ds\, e^{\lambda x}\,,$$

and a simple change of variable yields:

$$\operatorname{res}_\lambda \frac{H(\lambda; f)}{A(\lambda)} e^{\lambda x} = \operatorname{res}_\lambda \frac{1}{A(\lambda)} \int_0^a d\sigma(t) \int_x^t f(x+t-s) e^{\lambda s}\,ds\,.$$

Now, by (12) we have

$$\int_0^a d\sigma(t) \int_0^x f(x+t-s) e^{\lambda s}\,ds = \int_0^x e^{\lambda s}\,ds \in t_0^a f(x+t-s)\,d\sigma(t) = 0\,,$$

[14]The problems of spectral synthesis for other classes of functions and with respect to other topologies were investigated by A. Beurling in a series of papers starting from 1945. These results are included in the second volume of his Collected Works.

whence

$$\operatorname{res}_\lambda \frac{H(\lambda; f)}{A(\lambda)} e^{\lambda x} = \operatorname{res}_\lambda \frac{1}{A(\lambda)} \int_0^a d\sigma(t) \int_0^t f(x+t-s) e^{\lambda s}\, ds \,. \tag{16}$$

Therefore, each term of the series (15) can be approximated in $C(\mathbb{R})$ by linear combinations of translations $f(x+t)$ and hence belongs to the subspace T_f.

Now, let $\nu(t)$ be a function of bounded variation on the real axis such that the corresponding linear functional on $C(\mathbb{R})$ annihilates the subspace E_f. Let the support of ν belong to the segment $[b, c]$, and let

$$F(t) = \int_b^c f(t+s)\, d\nu(s) \,. \tag{17}$$

Since

$$\int_0^a F(x+t)\, d\sigma(t) = \int_b^c d\nu(s) \int_0^a f(x+t+s)\, d\sigma(t) = 0 \,, \qquad x \in \mathbb{R} \,, \tag{18}$$

the function $F(t)$ is a m.p. function. Consider the corresponding generalized Fourier series

$$F(x) \sim \sum \operatorname{res}_\lambda \frac{H(\lambda; F)}{A(\lambda)} e^{\lambda x} \,.$$

Using (16) and (17), we have

$$\begin{aligned} \operatorname{res}_\lambda \frac{H(\lambda; F)}{A(\lambda)} e^{\lambda x} &= \operatorname{res}_\lambda \frac{1}{A(\lambda)} \int_0^a d\sigma(t) \int_0^t F(x+t-s) e^{\lambda s}\, ds \\ &= \int_b^c \left\{ \operatorname{res}_\lambda \frac{H(\lambda; f)}{A(\lambda)} e^{\lambda(x+u)} \right\} d\nu(u) \,. \end{aligned}$$

Since the function ν generates a functional which annihilates the subspace E_f, the last integral vanishes for each $\lambda \in \mathbb{C}$, and $H(\lambda; f)/A(\lambda)$ is an entire functions. Let us prove that $F(x) \equiv 0$ for $x \geq 0$. To this end we multiply (18) by $e^{-\lambda x}$ and integrate from 0 to x. We have

$$\begin{aligned} 0 &= \int_0^x e^{-\lambda s}\, ds \int_0^a F(s+t)\, d\sigma(t) \\ &= \int_0^a d\sigma(t) \int_0^x e^{-\lambda s} F(s+t)\, ds = \int_0^a e^{\lambda t}\, d\sigma(t) \int_t^{x+t} F(s) e^{-\lambda s}\, ds \\ &= -H(\lambda; F) + \int_0^a e^{\lambda t}\, d\sigma(t) \int_0^{x+t} F(s) e^{-\lambda s}\, ds \,. \end{aligned}$$

Therefore,

$$\frac{H(\lambda; F)}{A(\lambda)} = \int_0^x e^{-\lambda s} F(s)\, ds + \frac{1}{A(\lambda)} \int_0^a e^{\lambda t}\, d\sigma(t) \int_x^{x+t} F(s) e^{-\lambda s}\, ds \,, \tag{19}$$

and the second term on the right-hand side of (19) is an entire function as are the other entries of this relation. By the theorem on the growth of the quotient of entire functions (Theorem 1, Section 2.4), it is a function of exponential type. Since $A(i\lambda)$ belongs to the class C, we may apply Theorem 4 (on the sum of indicator functions) and conclude that the value of the indicator of this term on the ray $\{\lambda : \arg \lambda = 0\}$ does not exceed $-x$. Hence its indicator diagram belongs to the half-plane $\{\lambda : \operatorname{Re} \lambda \leq -x\}$. If $F(s) \not\equiv 0$ for $s \in [0, x]$, then the indicator diagram of the first term on the right-hand side of (19) is a nontrivial segment in $[-x, 0]$. In this case the indicator diagram of the right-hand side of (19) is nonempty and does not reduce to the point $\lambda = 0$. This is a contradiction, since either $H(\lambda; F) \equiv 0$ and the indicator diagram of the left-hand side of (19) is empty, or $H(\lambda; F) \not\equiv 0$ and by the theorem on the sum of indicators this indicator diagram coincides with the point $\lambda = 0$. This proves that $F(s) \equiv 0$ for $s \in [0, x]$ and, hence,

$$F(0) = \int_b^c f(s)\, d\nu(s) = 0 .$$

The Hahn-Banach theorem implies $T_f = E_f$, which solves the problem of spectral synthesis for the m.p. function $f(x)$.

For other results on m.p. functions dealing with various problems of approximation by exponential functions, quasianalyticity and related problems see Schwartz [**119**], Kahane [**62, 63**], and C. Berenstein and Taylor [**11**].

LECTURE 17

Zeros of Functions of Class C

The distribution of zeros of an entire function of class C is remarkably regular. To study this distribution we shall obtain a relation which may be thought of as a generalization of the Jensen formula.

17.1. The generalized Jensen formula

Let a function $f(z)$ be analytic in the sector

$$\{z : \alpha \le \arg z \le \beta \,, |z| \le r\}$$

and not vanish on its boundary. We apply the argument principle to evaluate the number of zeros of $f(z)$ in this sector. We have

$$\begin{aligned} 2\pi n(r,\alpha,\beta) = \int_0^r \frac{d[\arg f(te^{i\alpha})]}{dt}\,dt - \int_0^r \frac{d[\arg f(te^{i\beta})]}{dt}\,dt \\ + \int_\alpha^\beta \frac{d[\arg f(re^{i\varphi})]}{r\,d\varphi} r\,d\varphi. \end{aligned}$$

By the Cauchy-Riemann conditions this formula can be transformed to

(1)
$$\begin{aligned} 2\pi n(r,\alpha,\beta) = \int_0^r \frac{d\log|f(te^{i\varphi})|_{\varphi=\beta}}{td\varphi}\,dt - \int_0^r \frac{d\log|f(te^{i\varphi})|_{\varphi=\alpha}}{td\varphi}\,dt \\ + r\int_\alpha^\beta \frac{d\log|f(te^{i\varphi})|_{t=r}}{dt}\,d\varphi\,. \end{aligned}$$

Without loss of generality we may suppose that $f(0) = 1$. If

$$J_f^r(\theta) = \int_0^r \frac{\log|f(te^{\theta})|}{t}dt\,,$$

then relation (1) can be written in the form

$$\begin{aligned} 2\pi n(r,\alpha,\beta) = \frac{d}{d\varphi}\left[J_f^r(\varphi)\right]_{\varphi=\beta} - \frac{d}{d\varphi}\left[J_f^r(\varphi)\right]_{\varphi=\alpha} \\ + r\int_\alpha^\beta \frac{d[\log|f(re^{i\varphi})|]_{t=r}}{dr}\,d\varphi\,. \end{aligned}$$

We divide both sides of this equality by $2\pi r$, integrate over r from 0 to r, and set

$$N(r;\alpha,\beta) = \int_0^r \frac{n(t,\alpha,\beta)}{t}\,dt\,.$$

Then we obtain the representation

$$
\begin{aligned}
N(r;\alpha,\beta) &= \frac{1}{2\pi}\frac{d}{d\varphi}\Big[\int_0^r J_f^t(\varphi)\frac{dt}{t}\Big]_{\varphi=\beta} \\
&\quad - \frac{1}{2\pi}\frac{d}{d\varphi}\Big[\int_0^r J_f^t(\varphi)\frac{dt}{t}\Big]_{\varphi=\alpha} + \frac{1}{2\pi}\int_\alpha^\beta \log|f(re^{i\varphi})|\,d\varphi\,.
\end{aligned}
\tag{3}
$$

This is the *generalized Jensen formula.* The usual Jensen formula, Section 2.3, follows if we set $\beta = \alpha + 2\pi$.

17.2. Asymptotic properties of zeros of functions of class C

Let $f(z)$ be an entire function of class C. It follows from Theorem 2, Section 16.1, that its indicator diagram is a segment of the imaginary axis. Multiplying $f(z)$ by $e^{i\lambda z}$ we normalize $f(z)$ in such a way that its indicator diagram is a segment of the imaginary axis which is symmetric with respect to the real axis. In other words, we can assume that

$$h_f(\theta) = \sigma|\sin\theta|\,.$$

The same Theorem 2 also yields

$$\log|f(re^{i\theta})| = h_\varphi(\theta)r + o(r) \tag{4}$$

outside an exceptional set of an arbitrary small view.

For the ray $\{z : \arg z = \theta\}$ not intersecting this exceptional set, we have

$$J_f^r(\theta) = h_f(\theta)r + o(r)\,, \qquad r\to\infty\,,$$

and, finally,

$$\int_0^r J_f^t(\theta)\frac{dt}{t} = h_f(\theta)r + o(r)\,, \qquad r\to\infty\,.$$

LEMMA 1. *Let $n(t)$ be a nondecreasing function for $t > 0$, let $n(t) = 0$ for $0 \le t \le \varepsilon$, and let there exist $\alpha > -2$ such that*

$$\psi(R) = \frac{1}{R^{\alpha+2}}\int_0^R t^\alpha n(t)\,dt$$

approaches the limit d as $R\to\infty$. Then the function $\dfrac{n(R)}{R}$ approaches the limit $(\alpha+2)d$ as $R\to\infty$.

PROOF. Let $k > 1$. Then

$$(kR)^{\alpha+2}\psi(kR) - R^{\alpha+2}\psi(R) = \int_R^{kR} t^\alpha n(t)\,dt\,.$$

Since the function $n(R)$ is nondecreasing, we have[15]

$$
\begin{aligned}
n(R)\frac{(kR)^{\alpha+1}-R^{\alpha+1}}{\alpha+1} &\le R^{\alpha+2}\Big[k^{\alpha+2}\psi(kR) - \psi(R)\Big] \\
&\le n(kR)\frac{(kR)^{\alpha+1}-R^{\alpha+1}}{\alpha+1}\,.
\end{aligned}
$$

[15] We assume that $\alpha \ne -1$. The case $\alpha = -1$ is similar.

Dividing the first inequality by $R^{\alpha+2}$ we obtain

$$\limsup_{R\to\infty} \frac{n(R)}{R} \le \frac{k^{\alpha+2}-1}{k^{\alpha+1}-1}(\alpha+1)d\,,$$

while the second one yields

$$\liminf_{R\to\infty} \frac{n(R)}{R} \ge \frac{1}{k}\frac{k^{\alpha+2}-1}{k^{\alpha+1}-1}(\alpha+1)d\,.$$

Combining the last two inequalities we obtain

$$\lim_{R\to\infty} \frac{n(R)}{R} = (\alpha+2)d\,,$$

which completes the proof of the lemma.

Relation (4) shows that if $f(z)$ belongs to the class C, then there exists the limit

$$\lim_{R\to\infty} \frac{n_f(R)}{R} = 2\sigma\pi\,. \tag{5}$$

Indeed, choosing R such that the circle $\{z : |z| = R\}$ does not intersect exceptional disks, and applying the Jensen formula we have by (4)

$$N_f(R) = \int_0^R \frac{n(t)}{t}\,dt = \frac{2\sigma}{\pi}R + o(R)\,. \tag{6}$$

The set E of all values k such that the circle $\{z : |z| = k\}$ intersects the exceptional disks has a finite logarithmic length (condition (11) of Lecture 15). Taking into account that $N_f(R)$ is a monotonic function we obtain relation (6) for all R. Relation (5) is a consequence of Lemma 1.

The generalized Jensen formula yields an even stronger result. Namely, for $0 < \alpha < \pi$ let us denote by $n_+(r,\alpha)$ and $n_-(r,\alpha)$ the numbers of zeros of the function $f(z)$ in the sectors $\{z : |z| \le r,\ |\arg z| \le \alpha\}$ and $\{z : |z| \le r\ , |\pi - \arg z| \le \alpha\}$, respectively.

THEOREM 1 (Cartwright, Levinson). *Let $f(z)$ be a function of class C, and let $\{a_k\}$, $a_k \ne 0$, be its zero set. Then*

1. $\displaystyle\sum_k \left|\operatorname{Im}\frac{1}{a_k}\right| < \infty\,.$

2. $\displaystyle\lim_{r\to\infty}\frac{n_+(r,\alpha)}{r} = \lim_{r\to\infty}\frac{n_-(r,\alpha)}{r} = \frac{d}{2\pi}\,,$

where d is the width of the indicator diagram of $f(z)$, i.e.,

$$d = h_f\Big(\frac{\pi}{2}\Big) + h_f\Big(-\frac{\pi}{2}\Big)\,.$$

3. *There exists the limit*

$$\lim_{R\to\infty}\sum_{|a_k|<R}\frac{1}{a_k}\,.$$

PROOF. Without loss of generality we may assume that $h_f(\theta) = \frac{d}{2}|\sin\theta|$, $0 \leq \theta \leq 2\pi$. The first statement is a consequence of Theorem 2, Section 14.2, since according to Krein's theorem (Theorem 1 in the previous lecture) the class C coincides with the class of entire functions f such that $\log|f(z)|$ has positive harmonic majorants in the upper and lower half-planes.

One of the consequences of the first statement is that

$$N(r;\alpha,\beta) = o(r)\ , \qquad r \to \infty\ ,$$

if the angle $\{z\ :\ \alpha \leq \arg z \leq \beta\}$ does not contain either the positive or negative rays.

To prove the second statement, we apply the generalized Jensen formula (3). The left-hand side of (3) represents a nondecreasing function of β. Besides, the function

$$\frac{d}{d\varphi}\left[\int_0^r J_f^t(\varphi)\frac{dt}{t}\right]_{\varphi=\beta}$$

has at most a finite number of points of discontinuity. Therefore, the Newton-Leibniz formula is applicable when integrating this function. Integrating both sides of (3) with respect to β from β to $\beta+k$, and with respect to α from α to $\alpha+l$, we obtain

$$\frac{1}{kl}\int_\beta^{\beta+k}\int_\alpha^{\alpha+l} N(r;\varphi,\psi)\,d\varphi\,d\psi = \frac{1}{2\pi}\int_0^r \frac{J_f^t(\beta+k)-J_f^t(\beta)}{tk}\,dt$$
$$-\frac{1}{2\pi}\int_0^r \frac{J_f^t(\alpha+l)-J_f^t(\alpha)}{tl}\,dt + \frac{1}{2\pi lk}\int_\beta^{\beta+k}\int_\alpha^{\alpha+l}\int_\varphi^\psi \log|f(re^{i\theta})|\,d\theta\,d\varphi\,d\psi\ .$$

The numbers α, β, k, l can be chosen such that asymptotic relation (4) holds on the corresponding rays. Besides, without loss of generality we may suppose that this relation holds for all z, $|z| = r$. Then, as $r \to \infty$,

$$\frac{1}{kl}\int_\beta^{\beta+k}\int_\alpha^{\alpha+l} N(r;\varphi,\psi)\,d\varphi\,d\psi = \frac{1}{2\pi}\frac{h_f(\beta+k)-h_f(\beta)}{k}r$$
$$-\frac{1}{2\pi}\frac{h_f(\alpha+l)-h_f(\alpha)}{l}r + \frac{r}{2\pi lk}\int_\beta^{\beta+k}\int_\alpha^{\alpha+l}\int_\varphi^\psi h(\theta)\,d\theta\,d\varphi\,d\psi + o(r)\ .$$

The function $h_f(\varphi) = \frac{d}{2}|\sin\varphi|$ is differentiable for all $\varphi \neq 0,\ \pi$. Therefore, for sufficiently small k and l, we have

$$\text{(7)}\qquad \frac{1}{2\pi}S_f(\alpha,\beta) - \varepsilon + o(1) \leq \frac{N(r,\alpha,\beta)}{r} \leq \frac{1}{2\pi}S_f(\alpha,\beta) + \varepsilon + o(1)\ , \qquad r \to \infty\ ,$$

where

$$S_f(\alpha,\beta) = h_f'(\beta) - h_f'(\alpha) + \int_\alpha^\beta h(\theta)\,d\theta\ .$$

It should also be noted that, for $-\pi < \alpha < \beta < \pi$, we have $S_f(\alpha,\beta) = 0$ if $\alpha\beta > 0$, and $S_f(\alpha,\beta) = d$ if $\alpha\beta < 0$.

Relation (7) yields

$$\lim_{r\to\infty}\frac{N(r,-\alpha,\alpha)}{r} = \frac{d}{2\pi}$$

for $0 < \alpha \leq \pi$ and, according to Lemma 1,

$$\lim_{r\to\infty} \frac{n_+(r,\alpha)}{r} = \frac{d}{2\pi} .$$

Similarly,

$$\lim_{r\to\infty} \frac{n_-(r,\alpha)}{r} = \frac{d}{2\pi} ,$$

which completes the proof of the second statement.

To prove the third statement we apply the Schwarz-Nevanlinna representation (6), Section 2.2. Differentiating both sides of this relation, and setting $z = 0$ we have

$$\frac{f'(0)}{f(0)} = \frac{1}{\pi}\int_0^{2\pi} \frac{\log|f(Re^{i\psi})|}{R} e^{-i\psi}\,d\psi - \sum_{|a_k|<R} \frac{1}{a_k} + \sum_{|a_k|<R} \frac{\bar{a}_k}{R^2} . \tag{8}$$

Let us suppose that the circle $\{z : |z| = R\}$ does not intersect the exceptional set for relation (4). Using this relation we rewrite (8) in the form

$$\sum_{|a_k|<R} \frac{1}{a_k} = -\frac{f'(0)}{f(0)} + \frac{d}{2\pi}\int_0^{2\pi} |\sin\psi| e^{-i\psi}\,d\psi + \sum_{|a_k|<R} \frac{\bar{a}_k}{R^2} + o(1) , \quad R \to \infty . \tag{9}$$

We denote by $I_R\{\bar{a}_k\}$ the sum on the right-hand side of (9), and prove that it approaches some limit as $R \to \infty$. To this end, let us consider separately the zeros from the half-plane $\{z : \operatorname{Re} z \geq 0\}$ and denote them by $a'_k = |a'_k|e^{i\psi'_k}$. To compare the quantities $I_R\{\bar{a}'_k\}$ and $I_R\{|a'_k|\}$, we have

$$\begin{aligned} \left| R^{-2} \sum_{|a'_k|<R} |a'_k| - R^{-2} \sum_{|a'_k|<R} \bar{a}'_k \right| &\leq R^{-2} \sum_{|a'_k|<R} |a'_k|\,|1 - e^{-i\psi'_k}| \\ &\leq \sum_{|a'_k|<R} \frac{|a'_k|^2}{R^2} \left| \frac{1 - e^{-i\psi'_k}}{\sin\psi'_k} \right| \frac{|\sin\psi'_k|}{|a'_k|} . \end{aligned} \tag{10}$$

According to statement 1 the series

$$\sum_k \frac{\sin\psi'_k}{|a'_k|}$$

converges, and hence, the left-hand side of (10) vanishes as $R \to \infty$. For $n_+(t) = n_+(t, \pi/2)$ we have

$$I_R\{|a'_k|\} = \frac{1}{R^2} \sum_{|a'_k|<R} |a'_k| = \frac{1}{R^2}\int_0^R t\,dn_+(t) = \frac{n_+(R)}{R} - \frac{1}{R^2}\int_0^R n_+(t)\,dt ,$$

and, according to statement 2, $n_+(t) = td/2\pi + o(t)$. Therefore,

$$I_R\{|a'_k|\} = \frac{d}{2\pi} - \frac{d}{4\pi} + o(1) = \frac{d}{4\pi} + o(1) , \quad R \to \infty ,$$

whence

$$I_R\{a'_k\} = \frac{d}{4\pi} + o(1) , \quad R \to \infty .$$

If $\{a_k''\}$ is the set of zeros in the half-plane $\{z : \operatorname{Re} z < 0\}$, then in the same way we prove that

$$I_R\{a_k''\} = -\frac{d}{4\pi} + o(1) , \quad R \to \infty .$$

Since the integral in (9) is equal to zero, we obtain

$$\sum_{|a_k|<R} \frac{1}{a_k} = -\frac{f'(0)}{f(0)} + o(1) , \quad R \to \infty ,$$

which yields statement 3.

REMARK 1. Since

$$\Delta = \lim_{R\to\infty} \frac{n(R)}{R} ,$$

we see that $n(R+0) - n(R) = o(R)$, $R \to \infty$. Therefore, if the zeros $\{a_k\}$ are enumerated in increasing orders of $|a_k|$, then the series $\sum_1^\infty a_k^{-1}$ converges, and

$$\sum_1^\infty \frac{1}{a_k} = -\frac{f'(0)}{f(0)} . \tag{11}$$

REMARK 2. By the Hadamard theorem every entire function $f(z)$, $f(0) \neq 0$, of order 1 admits representation of the form

$$f(z) = f(0)e^{cz} \prod_{k=1}^\infty \left(1 - \frac{z}{a_k}\right) e^{z_k/a} .$$

It is easy to check directly that $c = f'(0)/f(0)$. Now we conclude that every entire function $f(z) \in C$ such that $f(0) \neq 0$ and $h_f(\pi/2) = h_f(-\pi/2)$, admits representation

$$f(z) = f(0) \text{ P.V.} \prod \left(1 - \frac{z}{a_k}\right) = f(0) \lim_{R\to\infty} \prod_{|a_k|<R} \left(1 - \frac{z}{a_k}\right) .$$

Here the convergence of the infinite product is conditional, since the a_k are enumerated according to the growth of their absolute values, and it is uniform on each compact set in $\mathbb{C}$.

Further information on zero-distribution for functions of class C can be found in de Branges [**20**], Khabibullin [**67, 68**], and Koosis [**72,** vol. II].

LECTURE 18

Completeness and Minimality of Systems of Exponential Functions in $L^2(a,b)$

We start with reminding the reader that a system $\{h_k\}$ of elements of the Banach space E is called *complete* if the closure of the set of finite linear combinations $\sum a_k h_k$, $a_k \in \mathbb{C}$, coincides with the whole E.

A system of elements $\{h_k\} \in E$ is said to be *minimal* if the relation

$$\Big\| \sum_{k=1}^{m_n} a_k^{(n)} h_k \Big\| \to 0\,, \qquad n \to \infty\,,$$

yields $a_k^{(n)} \to 0$, $n \to \infty$. In other words, for each k, the element h_k does not belong to the closure of the linear span of all other elements.

In this lecture we study the problems of completeness and minimality of exponential systems by methods of the theory entire functions. As in Lecture 3, the reduction to the problem for entire (or, more generally, analytic) functions is performed by using the Hahn-Banach theorem. According to this theorem,

The system $\{h_k\}$ is complete in E if and only if there is no nontrivial functional $f \in E^*$ annihilating all elements $\{h_k\}$, and it is minimal if and only if there exists a biorthogonal system of functionals $\{f_k\} \subset E^*$, which means that $f_k(h_m) = \delta_{km}$.

In many cases where the general form of functionals from E^* is known, the problem of the completeness can be reduced to uniqueness problems for analytic functions and the problems of minimality to the relevant interpolation problem.

For a given sequence $\Lambda = \{\lambda_n\} \subset \mathbb{C}$ of pairwise distinct complex numbers, let us denote by $\mathcal{E}_\Lambda = \{e^{i\lambda t},\ \lambda \in \Lambda\}$ the corresponding system of exponential functions. If this system is not complete in $L^2(-\pi,\pi)$, then there exists a nonzero function $h \in L^2(-\pi,\pi)$ annihilating $\mathcal{E}_\Lambda$. The function

$$(1) \qquad f(\lambda) = \int_{-\pi}^{\pi} h(t) e^{i\lambda t}\, dt\,, \qquad h \in L^2(-\pi,\pi)\,,$$

is an entire function of exponential type $\sigma \le \pi$ such that

$$f \in L^2(-\infty,\infty)\,, \qquad f(\lambda) = 0 \quad \text{for} \quad \lambda \in \Lambda\,.$$

Conversely, if an entire function $f \not\equiv 0$ satisfies these conditions, then by the Paley-Wiener theorem it admits representation (1), and the function h from this representation annihilates the whole system $\mathcal{E}_\Lambda$. Thus, the following theorem is proved.

THEOREM 1. *For a system $\mathcal{E}_\Lambda$ to be incomplete in the space $L^2(-\pi,\pi)$ it is necessary and sufficient that Λ be a subset of the zero set of a function of exponential type $\sigma \le \pi$ which belongs to $L^2(-\infty,\infty)$.*

REMARK 1. Using the linear change of variables it is easy to formulate an incompleteness criterion for a system $\mathcal{E}_\Lambda$ in $L^2(a,b)$ for an arbitrary interval (a,b).

REMARK 2. If some p points of the sequence Λ coincide, then p copies of a point λ_k belong to Λ, and the functions $e^{i\lambda_k t}, te^{i\lambda_k t}, \dots, t^{p-1}e^{i\lambda_k t}$ must be included in the system $\mathcal{E}_\Lambda$. In this case, Theorem 1 states that the multiplicity of λ_k as a zero of f is not less than p. For the sake of simplicity, in what follows, we assume that all points of Λ are pairwise distinct.

An entire function of exponential type which belongs to $L^2(\mathbb{R})$ is an element of class C. Combining Theorem 1 with the theorem on zeros of functions of class C we obtain the following statement:

THEOREM 2. *If a sequence Λ cannot be imbedded into a sequence which satisfies conditions* 1–3 *of Theorem* 1 *from Section* 17.2, *then the system $\mathcal{E}_\Lambda$ is complete in $L^2(-\pi,\pi)$.*

This theorem yields several sufficient conditions of completeness. To formulate one of them let

$$n_+(r,\Lambda) = \operatorname{card}\{\lambda \in \Lambda,\ \operatorname{Re}\lambda \ge 0,\ |\lambda| < r\}\ .$$

If

$$\overline{\delta}(\Lambda) = \lim_{\zeta\downarrow 1}\limsup_{r\to\infty}\frac{n_+(r\zeta,\Lambda) - n_+(r,\Lambda)}{r(\zeta-1)} > 1\ ,$$

then the system $\mathcal{E}_\Lambda$ is complete in $L^2(-\pi,\pi)$.

Indeed, if Λ is a part of the sequence $\mathbb{M}$ of zeros of a function from C and of exponential type π, then the density $\delta_+(\mathbb{M})$ in the right half-plane satisfies the inequality $\delta_+(\mathbb{M}) \ge \overline{\delta}(\Lambda) > 1$, which contradicts condition 2 of Theorem 1 from Section 17.2.

The value $\overline{\delta}(\Lambda)$ was introduced by Pólya. It is called the *maximal density* of the sequence $\Lambda \cap \{\lambda : \operatorname{Re}\Lambda > 0\}$.

Another condition of completeness is the following.

If $|\arg\lambda_k - \pi/2| \le \tau < \pi$ and if $\sum(1+|\lambda_k|)^{-1} = \infty$, then the system $\mathcal{E}_\Lambda$ is complete in $L^2(-l,l)$ for any l.

Indeed, in this case

$$\sum\left|\operatorname{Im}\frac{1}{\lambda_k}\right| \ge \cos\tau\sum\frac{1}{|\lambda_k|} = \infty\ .$$

REMARK. Beurling and Malliavin [**14**] proved that *for any entire function f of class C and any ε there exists an entire function φ of exponential type not greater than ε such that $f(x)\varphi(x) \in L^2(-\infty,\infty)$.*

Therefore, if a sequence Λ is part of the zero set of a function of class C and of exponential type σ, then the system $\mathcal{E}_\Lambda$ is incomplete in the space $L^2(-\sigma-\varepsilon,\sigma+\varepsilon)$ for any $\varepsilon > 0$. When studying the problem of completeness of exponential systems Beurling and Malliavin [**15**] introduced a special density which gives a criterion for a sequence Λ to be part of the zero set of a function of exponential type σ and of class C (see also Krasichkov-Ternovskii [**75**] and Koosis [**72,** vol. II]). In terms of this density Beurling and Malliavin defined the quantity

$$\rho(\Lambda) = \sup\{a : \mathcal{E}_\Lambda \text{ is complete in } L^2(0,a)\},$$

which is called the *completeness radius* of Λ.

PROBLEM 1. If, for some $\mu \notin \Lambda$, the function $e^{i\mu t}$ can be approximated in $L^2(-\pi,\pi)$ by linear combinations of the exponential functions from $\mathcal{E}_\Lambda$, then $\mathcal{E}_\Lambda$ is complete in $L^2(-\pi,\pi)$.

PROBLEM 2. If Λ is the zero set of an entire function $\varphi(z)$ of exponential type such that

$$\liminf_{y\to\pm\infty} |\varphi(iy)|e^{-\pi|y|} > 0\,,$$

then the system $\mathcal{E}_\Lambda$ is complete in $L^2(-\pi,\pi)$.

The following theorem describes a criterion of minimality for an exponential system.

THEOREM 3 (Paley and Wiener). *For a system $\mathcal{E}_\Lambda$ to be minimal in $L^2(-\pi,\pi)$ it is necessary and sufficient that there exist a nonzero entire function $\varphi(\lambda)$ of exponential type not exceeding π vanishing at Λ and such that*

$$\varphi(\lambda)(1+|\lambda|)^{-1} \in L^2(-\infty,\infty)\,. \tag{3}$$

PROOF. SUFFICIENCY. For each $\lambda_k \in \Lambda$ let

$$\varphi_k(\lambda) = \frac{\varphi(\lambda)}{\varphi'(\lambda_k)(\lambda-\lambda_k)}\,.$$

The function $\varphi_k(\lambda)$ is an entire function of exponential type not exceeding π, and by (3) $\varphi_k \in L^2(-\infty,\infty)$. According to Paley-Wiener theorem, there exists a function $h_k \in L^2(-\pi,\pi)$ such that

$$\varphi_k(\lambda) = \int_{-\pi}^{\pi} e^{i\lambda t} h_k(t)\,dt\,.$$

The system of functions $\{h_k\}$ is biorthogonal to $\mathcal{E}_\Lambda$ because

$$\int_{-\pi}^{\pi} e^{\lambda_m t}\varphi_k(t)\,dt = \varphi_k(\lambda_m) = \delta_{km}\,.$$

Therefore, the system $\mathcal{E}_\Lambda$ is minimal.

NECESSITY. Let $\mathcal{E}_\Lambda$ be minimal in $L^2(-\pi,\pi)$. For $\lambda_1 \in \Lambda$, let the function $h_1 \in L^2(-\pi,\pi)$ generate the functional which is orthogonal to $\mathcal{E}_\Lambda \setminus \{e^{i\lambda_1 t}\}$:

$$\int_{-\pi}^{\pi} e^{i\mu t} h_1(t)\,dt = 0\,, \quad \mu \in \Lambda\setminus\{\lambda_1\}\,, \quad h_1 \not\equiv 0\,.$$

The function

$$\varphi(\lambda) = (\lambda-\lambda_1)\int_{-\pi}^{\pi} e^{i\lambda t} h_1(t)\,dt$$

satisfies all the conditions of the theorem. This completes the proof of the theorem.

Now consider the conditions which provide for a system $\mathcal{E}_\Lambda$ to be both complete and minimal in the space $L^2(-\pi,\pi)$.

THEOREM 4. *For a system of functions $\mathcal{E}_\Lambda$ to be both complete and minimal in $L^2(-\pi,\pi)$ it is necessary and sufficient that:*

1. *The infinite product*

$$\varphi(\lambda) = \lim_{R\to\infty} \prod_{|\lambda_k|<R} \left(1 - \frac{z}{\lambda_k}\right)$$

converge uniformly on each compact set in $\mathbb{C}$, $\varphi(\lambda)$ be an entire function of exponential type, and

$$h_\varphi\left(\frac{\pi}{2}\right) + h_\varphi\left(-\frac{\pi}{2}\right) = 2\pi \,.$$

2. $\varphi(\lambda)(1+|\lambda|)^{-1} \in L^2(-\infty,\infty)$.
3. *If $\chi(\lambda)$ is a nonzero entire function of exponential type zero, then*

$$\varphi(\lambda)\chi(\lambda) \notin L^2(-\infty,\infty) \,.$$

NECESSITY. Since the system $\mathcal{E}_\Lambda$ is minimal, there exists an entire function $\theta(\lambda) \not\equiv 0$ which satisfies condition 2 and such that $\theta(\lambda) = 0$, $\lambda \in \Lambda$. If this function has at least one additional zero $\mu \notin \Lambda$, then the function $\psi(\lambda) = (\lambda-\mu)^{-1}\theta(\lambda)$ vanishes on Λ and also belongs to $L^2(-\infty,\infty)$. Its Fourier transform

$$h(t) = \frac{1}{2\pi}\int_{-\infty}^{\infty} \psi(\lambda) e^{-i\lambda t}\, d\lambda \in L^2(-\pi,\pi)$$

generates a nontrivial functional in $L^2(-\pi,\pi)$ which annihilates $\mathcal{E}_\Lambda$. This contradicts the completeness of $\mathcal{E}_\Lambda$ in $L^2(-\pi,\pi)$. Since $\theta(\lambda)$ satisfies condition 2, it belongs to the class C, and hence, admits the representation

$$\theta(\lambda) = c\, e^{ia\lambda} \lim_{R\to\infty} \prod_{|\lambda_k|<R} \left(1 - \frac{\lambda}{\lambda_k}\right), \quad a \in \mathbb{R}\,,$$

where the infinite product converges uniformly on each compact set. For $\varphi(\lambda) = c^{-1}e^{-ia\lambda}\theta(\lambda)$ we have

$$h_\varphi\left(\frac{\pi}{2}\right) + h_\varphi\left(-\frac{\pi}{2}\right) = h_\theta\left(\frac{\pi}{2}\right) + h_\theta\left(-\frac{\pi}{2}\right) \le 2\pi \,.$$

If

$$h_\varphi\left(\frac{\pi}{2}\right) + h_\varphi\left(-\frac{\pi}{2}\right) = 2\sigma < 2\pi \,,$$

then the function

$$\Delta(\lambda) = \varphi(\lambda)\frac{\sin(\pi-\sigma)\lambda}{\lambda}$$

belongs to $L^2(-\infty,\infty)$, vanishes on Λ and is of exponential type π. This contradicts the completeness of $\mathcal{E}_\Lambda$ proving conditions 1 and 2.

Let $\chi(\lambda)$ be an entire function of minimal type with respect to the order 1, and let $\psi(\lambda) = \varphi(\lambda)\chi(\lambda) \in L^2(-\infty,\infty)$. We have

$$h_\psi\left(\frac{\pi}{2}\right) + h_\psi\left(-\frac{\pi}{2}\right) = 2\pi$$

and $\psi(\lambda_k) = 0$, $\lambda_k \in \Lambda$, which again contradicts the completeness of $\mathcal{E}_\Lambda$. Hence, condition 3 is proved.

SUFFICIENCY. Conditions 1 and 2 yield that the system $\mathcal{E}_\Lambda$ is minimal. Suppose it is not complete. Then there exists a nonzero function $h \in L^2(-\pi, \pi)$ such that the function

$$\psi(\lambda) = \int_{-\pi}^{\pi} e^{i\lambda t} h(t)\, dt$$

vanishes on Λ. The function $\psi(\lambda)$ belongs to $L^2(-\infty, \infty)$ and is of exponential type $\sigma \leq \pi$. Let us consider the entire function of exponential type

$$\chi(\lambda) = \frac{\psi(\lambda)}{\varphi(\lambda)} .$$

Since both $\varphi(\lambda)$ and $\psi(\lambda)$ belong to the class C, it follows from the theorem on the sum of indicators that $h_\chi(\theta) = h_\psi(\theta) - h_\varphi(\theta)$. Since $h_\psi(\theta) \leq \sigma|\sin\theta|$ and $h_\varphi(\theta) = \pi|\sin\theta|$, the function $\chi(\lambda)$ is of minimal type with respect to the order 1, and $\psi(\lambda) = \varphi(\lambda)\chi(\lambda) \in L^2(-\infty, \infty)$. We have obtained a contradiction to condition 3, which completes the proof of the theorem.

PROBLEM 3. Let $\theta(\lambda)$ be an entire function of exponential type,

$$h_\theta\Big(\frac{\pi}{2}\Big) + h_\theta\Big(-\frac{\pi}{2}\Big) = 2\pi ,$$

and let

$$\frac{K_1}{(1+|x|)^{1/2}} \leq |\theta(x+ia)| \leq K_2(1+|x|)^{1/2-\varepsilon} , \qquad -\infty < x < \infty ,$$

for some $a \in \mathbb{R}$ and $K_1, K_2 > 0$. If Λ is the zero set of $\theta(\lambda)$, then the system $\mathcal{E}_\Lambda$ is complete and minimal in $L^2(-\pi, \pi)$.

PROBLEM 4. Prove that if $|\lambda_k - k| < q < 1/4$, $k = 0, \pm 1, \pm 2, \ldots$, then the system $\{e^{i\lambda_k t}\}$ is complete and minimal in $L^2(-\pi, \pi)$. This statement is false for $q = 1/4$ (see Theorem 5, Section 23.2).

For further results on the completeness and minimality of exponential systems see Levinson [**84**], Redheffer [**115**], Korevaar [**73**], Koosis [**72,** vol. II] and references therein.

LECTURE 19

Hardy Spaces in the Upper Half-Plane

19.1. Definition and basic properties

The *Hardy class* H^p_+, $p \geq 1$, is the set of functions analytic in the upper half-plane $\mathbb{C}_+$ satisfying the condition

$$\sup_{y>0} \int_{-\infty}^{\infty} |f(x+iy)|^p \, dx < \infty \, ,$$

and endowed with the norm

$$\|f\|_p = \|f\|_{H^p_+} = \Big(\sup_{y>0} \int_{-\infty}^{\infty} |f(x+iy)|^p \, dx \Big)^{1/p} . \tag{1}$$

The class H^p_- is defined similarly for the lower half-plane $\mathbb{C}_-$.

Let $f \in H^p_+$, $z \in \mathbb{C}_+$, and let $K(\rho, z)$ be the disk of radius ρ centered at z. Since $|f(z)|^p$ is a subharmonic function, we have

$$|f(z)|^p \leq \frac{1}{\pi\rho^2} \iint_{K(\rho,z)} |f(\zeta)|^p \, d\sigma_\zeta \, , \tag{2}$$

where $d\sigma$ is the area element. In what follows, we shall frequently use this inequality.

Let us consider the basic properties of the class H^p_+.

1. *The class H^p_+ is a Banach space.* We need only to prove that H^p_+ is complete. Let $\{f_n\}$ be a Cauchy sequence with respect to the norm $\|\cdot\|_p$. By (2) it approaches an analytic function f uniformly on each compact set in $\mathbb{C}$. As $n \to \infty$ in the inequalities

$$\sup_{y>0} \Big(\int_{-\infty}^{\infty} |f_n(x+iy)|^p \, dx \Big)^{1/p} \leq C < \infty \, ,$$

using the Fatou lemma, we obtain $f \in H^p_+$ and, similarly, $\|f - f_n\|_p \to 0$.

2. *The function $f \in H^p_+$ satisfies the estimate*

$$|f(z)| \leq \Big(\frac{2}{\pi y} \Big)^{1/p} \|f\|_p \, , \qquad y = \operatorname{Im} z > 0 \, . \tag{3}$$

To prove this estimate we apply estimate (2) with $\rho = y$ and then replace the disk $K(y, z)$ by the square $\{\zeta = \xi + i\eta : \ |\xi - x| < y, \quad |\eta - y| < y\}$. We obtain

$$|f(z)|^p \leq \frac{1}{\pi y^2} \int_0^{2y} \int_{x-y}^{x+y} |f(xi + i\eta)|^p \, d\xi \, d\eta \leq \frac{2}{\pi y} \|f\|_p^p \, .$$

3. *Let $f \in H^p_+$. Then for each h_1, h_2, $0 < h_1 < h_2$,*

$$f(x+iy) \rightrightarrows 0\,, \qquad |x| \to \infty\,, \quad h_1 \le y \le h_2\,. \tag{4}$$

Indeed, by the definition of H^p_+ we have

$$\int_0^{h_1+h_2} \int_{-\infty}^{\infty} |f(x+iy)|^p \, dx \, dy < \infty\,.$$

Hence,

$$\int_0^{h_1+h_2} \int_{|x|>N} |f(x+iy)|^p \, dx \, dy \to 0\,, \qquad N \to \infty\,.$$

Applying now inequality (2) with $\rho = h_1$ we obtain (4).

4. *Let $f \in H^p_+$. Then*

$$f(z+ih) \to 0\,, \qquad |z| \to \infty\,, \quad \operatorname{Im} z > 0\,, \tag{5}$$

for each $h > 0$. This statement is a direct consequence of statements 2 and 3.

5. *Let $f \in H^p_+$ and $q > p$. Then the function $f_h(x) = f(x+ih)$ belongs to the space $L^q(-\infty, \infty)$ for each $h > 0$.* Indeed, according to (4) the quantity

$$M_h(f) = \sup_x |f(x+ih)|$$

is finite. Therefore

$$\begin{aligned}\int_{-\infty}^{\infty} |f(x+iy)|^q \, dx &= M_h(f)^q \int_{-\infty}^{\infty} \left|\frac{f(x+ih)}{M_h(f)}\right|^q dx \\ &\le M_h(f)^q \int_{-\infty}^{\infty} \left|\frac{f(x+ih)}{M_h(f)}\right|^p dx < \infty\end{aligned}$$

6. *Let $\{\lambda_n\} \subset \mathbb{C}$ be a sequence of points such that $0 < h_1 < \operatorname{Im} \lambda_k < h_2$ and $|\lambda_n - \lambda_k| \ge 2\delta$, $n \ne k$, for some $\delta < h_1$. If $f \in H^p_+$ for some p, $1 \le p < \infty$, then*

$$\sum_n |f(\lambda_n)|^p \le \frac{h_2 + \delta}{\pi \delta^2} \|f\|_p^p\,.$$

PROOF. The properties of the sequence $\{\lambda_n\}$ provide that the disks $K(\lambda_n, \delta)$ lie in $\mathbb{C}_+$ and are pairwise disjoint. Taking into account inequality (2) we obtain

$$\begin{aligned}\sum_n |f(\lambda_n)|^p &\le \frac{1}{\pi\delta^2} \sum_n \iint_{K(\lambda_n,\delta)} |f(\zeta)|^p \, d\sigma_\zeta \\ &\le \frac{1}{\pi\delta^2} \int_0^{h_2+\delta} \int_{-\infty}^{\infty} |f(x+iy)|^p \, dx \, dy \le \frac{h_2+\delta}{\pi\delta^2} \|f\|_p^p\,.\end{aligned}$$

PROBLEM 1. Let $f \in H^p_+$, $p \ge 1$. Then both $|f(z)|^p$ and $\log|f(z)|$ have positive harmonic majorants in $\mathbb{C}_+$.

PROBLEM 2. Let $f \in H^p_+$ and let

$$I_p(y) = \int_{-\infty}^{\infty} |f(x+iy)|^p \, dx\,, \qquad y > 0\,.$$

Prove that $I_p^{1/p}$ is a decreasing convex function, $I_p^{1/p}(+0) = \|f\|_p$ and $I_p^{1/p}(\infty) = 0$.

19.2. Boundary values of functions of H_+^p

We need several auxiliary lemmas.

LEMMA 1. *Let $f \in H_+^p$ and $h > 0$. Then*

$$f(z+ih) = \frac{1}{2i\pi}\int_{-\infty}^{\infty} \frac{f(t+ih)}{t-z}\,dt\,, \qquad \operatorname{Im} z > 0\,, \tag{6}$$

and

$$0 = \frac{1}{2i\pi}\int_{-\infty}^{\infty} \frac{f(t+ih)}{t-\overline{z}}\,dt\,, \qquad \operatorname{Im} z > 0\,. \tag{7}$$

PROOF. Relations (6) and (7) result from integrating the functions

$$\frac{f(\zeta+ih)}{\zeta-z} \quad \text{and} \quad \frac{f(\zeta+ih)}{\zeta-\overline{z}}$$

over the contour which consists of the segment $[-R, R]$ and the half-circle $C_R^+ = \{\zeta : |\zeta| = R, \operatorname{Im}\zeta > 0\}$ followed by passage to the limit as $R \to \infty$. Property 4 from the previous section provides that the integral over C_R^+ tends to zero as $R \to \infty$.

LEMMA 2. *Let $f \in H_+^p$ and let $h > 0$. Then the function $f(z)$ can be represented by the Poisson integral*

$$f(z+ih) = \frac{y}{\pi}\int_{-\infty}^{\infty} \frac{f(t+ih)}{(t-x)^2+y^2}\,dt\,, \qquad z = x+iy\,, \quad y > 0\,. \tag{8}$$

To prove this relation it suffices to use relations (6) and (7) and the identity

$$\frac{1}{t-z} - \frac{1}{t-\overline{z}} = \frac{2iy}{|t-z|^2} = \frac{2iy}{(t-x)^2+y^2}.$$

DEFINITION. A family of functions $k_h(t)$, $h > 0$, $-\infty < t < \infty$, is said to be an *approximate identity* if it possesses the following properties:

1. $k_h(t) > 0,\ -\infty < t < \infty\,,$
2. $\displaystyle\int_{-\infty}^{\infty} k_h(t)\,dt = 1 \quad \text{for all} \quad h > 0\,,$
3. For any $\delta > 0\,,$

$$\int_{|t|>\delta} k_h(t)\,dt \to 0 \quad \text{as} \quad h \to 0\,.$$

The convolution of an approximate identity and an arbitrary function $g \in L^p(-\infty,\infty)$ is defined by the relation

$$(g * k_h)(t) = \int_{-\infty}^{\infty} g(s)k_h(t-s)\,ds = \int_{-\infty}^{\infty} g(t-s)k_h(s)\,ds\,.$$

LEMMA 3. *Let $g \in L^p(-\infty,\infty)$, $p \geq 1$, and let $k_h(t)$ be an approximate identity. Then*

$$\|g * k_h\|_{L^p(-\infty,\infty)} \leq \|g\|_{L^p(-\infty,\infty)}\,.$$

PROOF. Let $p > 1$. Using Minkowski's integral inequality,[16] properties of an approximate identity and Fubini's theorem, we obtain

$$\begin{aligned}\|g * k_h\|^p_{L^p(-\infty,\infty)} &= \int_{-\infty}^{\infty} \left| \int_{-\infty}^{\infty} g(t-s)k_h(s)\,ds\right|^p dt \\ &\leq \int_{-\infty}^{\infty}\int_{\infty}^{\infty} |g(t-s)|^p k_h(s)\,ds\,dt \leq \|g\|^p_{L^p(-\infty,\infty)} \,.\end{aligned}$$

The same estimate is still true for $p = 1$.

LEMMA 4. *Let $g \in L^p(-\infty,\infty)$, $p \geq 1$, and let $k_h(t)$ be an approximate identity. Then*

$$\|g * k_h - g\|_{L^p(-\infty,\infty)} \to 0\,, \qquad h \to 0\,.$$

PROOF. We have

$$\begin{aligned}(g * k_h)(t) - g(t) &= \int_{-\infty}^{\infty} [g(t-s) - g(t)]k_h(s)\,ds \\ &= \int_{|s|<\delta} [g(t-s) - g(t)]k_h(s)\,ds + \int_{|s|\geq\delta} [g(t-s) - g(t)]k_h(s)\,ds\,.\end{aligned}$$

Therefore,

$$(9) \quad \|g * k_h - g\|_{L^p} \leq \sup_{|s|<\delta} \|g(t-s) - g(t)\|_{L^p} + 2\|g\|_{L^p(-\infty,\infty)} \int_{|s|\geq\delta} k_h(s)\,ds\,.$$

Since the translation operator $T_s g(t) = g(t-s)$ is continuous[17] in $L^p(-\infty,\infty)$ for $p \geq 1$, we have $\|g(t-s) - g(t)\|_{L^p(-\infty,\infty)} \to 0$ as $s \to 0$. Therefore, taking δ small enough, we can make the first summand on the right-hand side of (9) arbitrarily small. The second summand vanishes as $h \to 0$. This completes the proof of the lemma.

PROBLEM 3. Prove the following statements.

Let g be a bounded uniformly continuous function on $\mathbb{R}$ and let k_h be an approximate identity. Then $(k_h * g)(t) \rightrightarrows g(t)$, $h \to 0$, $t \in \mathbb{R}$.

Let μ be a finite measure on the real axis. Then the measures

$$(k_h * \mu)(t)dt = \Big(\int_{-\infty}^{\infty} k_h(t-s)d\mu(s)\Big)\,dt$$

approach μ with respect to the weak topology of the space of measures, as $h \to 0$.

[16] It reads

$$\Big\|\int G(t,s)\,d\nu(s)\Big\|_{L^p(\mu)} \leq \int \|G(t,s)\|_{L^p(\mu)}\,d\nu(s)$$

provided that the measures μ and ν are σ-finite, $1 \leq p < \infty$, and the function $(t,s) \mapsto G(t,s)$ is $\mu \times \nu$-measurable. We use it with $G(t,s) = g(t-s)$, $d\mu(t) = dt$, and $d\nu(s) = k_h(s)ds$.

[17] To prove this, it suffices to approximate g by compactly supported continuous functions.

THEOREM 1. *Let $f \in H^p_+$, $p > 1$, and let $f_y(x) = f(x+iy)$. Then there exists a function $f_0 \in L^p(-\infty,\infty)$ such that*

$$\|f_y - f_0\|_{L^p(-\infty,\infty)} \to 0\,, \qquad y \downarrow 0\,, \tag{10}$$

and

$$\|f_0\|_{L^p(-\infty,\infty)} = \|f\|_{H^p_+}\,. \tag{11}$$

PROOF. Since the unit ball in $L^p(-\infty,\infty)$, $p > 1$, is weakly compact, there exists a sequence $h_n \to 0$ such that f_{h_n} converge weakly to a function $f_0 \in L^p(-\infty,\infty)$. Using Lemma 2 with $h = h_n$ and $n \to \infty$, we have

$$f(x+iy) = \frac{y}{\pi}\int_{-\infty}^{\infty} \frac{f_0(t)}{(t-x)^2+y^2}\,dt = (f_0 * P_y)(x)\,, \qquad y > 0\,, \tag{12}$$

where

$$P_y(t) = \frac{y}{\pi(t^2+y^2)}$$

is the Poisson kernel which satisfies the definition of approximate identity. Relation (10) is a consequence of Lemma 4. By Lemma 3 we have

$$\|f_y\|_{L^p(-\infty,\infty)} \le \|f_0\|_{L^p(-\infty,\infty)}\,.$$

On the other hand, combining (10) and Fatou's Lemma we obtain

$$\|f_0\|_{L^p(-\infty,\infty)} \le \liminf_{y\to+0} \|f_y\|_{L^p(-\infty,\infty)}$$

proving equation (11) and Theorem 1.

As stated by Theorem 1, every function $f \in H^p_+$, $p > 1$, has boundary values on the real axis belonging to the space $L^p(-\infty,\infty)$. By virtue of (11) we can consider the space H^p_+ as a subspace of $L^p(-\infty,\infty)$.

REMARK 1. A function $f \in H^p_+$, $p > 1$, may be recovered from its boundary values by means of both the Cauchy integral

$$f(z) = \frac{1}{2i\pi}\int_{-\infty}^{\infty} \frac{f(t)}{t-z}\,dt\,, \qquad \operatorname{Im} z > 0\,, \tag{13}$$

and the Poisson integral

$$f(z) = \frac{y}{\pi}\int_{-\infty}^{\infty} \frac{f(t)}{|t-z|^2}\,dt\,, \qquad \operatorname{Im} z > 0\,. \tag{14}$$

In addition,

$$\frac{1}{2i\pi}\int_{-\infty}^{\infty} \frac{f(t)}{t-\bar{z}} = 0\,, \qquad \operatorname{Im} z > 0\,.$$

It should be mentioned that these relations yield $H^p_+ \cap H^p_- = \{0\}$, $p > 1$.

REMARK 2. Theorem 1 is still true for $p = 1$. This case requires additional considerations since the weak limit of a sequence of functions from the space $L^1(-\infty,\infty)$, generally speaking, may be a measure rather than a function. For details, see Koosis [**71**], Garnett [**37**].

PROBLEM 4. Let $h(t) \in L^p(-\infty,\infty)$ be a nonnegative function. For the existence of a function $f \in H^p_+$ such that $|f(t)| = h(t)$ almost everywhere, it is necessary and sufficient that

$$\int_{-\infty}^{\infty} \frac{\log h(t)}{1+t^2}\, dt > -\infty .$$

19.3. M. Riesz's theorem on conjugate harmonic functions and the general form of linear functionals in H^p_+

Let $u(t) \in L^p(-\infty,\infty)$, $1 < p < \infty$. Consider the harmonic function in $\mathbb{C}_+$

$$u(x+iy) = (u * P_y)(x) = \frac{y}{\pi}\int_{-\infty}^{\infty} \frac{u(t)}{(x-t)^2+y^2}\, dt . \tag{15}$$

Since the Poisson kernel is an approximate identity, we obtain

$$\|u_h - u\|_{L^p(-\infty,\infty)} \to 0 , \qquad h \to 0 , \tag{16}$$

where $u_h(x) = u(x+ih)$, and

$$\int_{-\infty}^{\infty} |u(x+iy)|^p\, dx \le \int_{-\infty}^{\infty} |u(x)|^p\, dx . \tag{17}$$

In exactly the same way as when proving Theorem 1 we find that if a function $u(z)$ is harmonic in $\mathbb{C}_+$ and, for some $p > 1$,

$$\int_{-\infty}^{\infty} |u(x+ih)|^p\, dx \le C < \infty$$

for all $h > 0$, then $u(z)$ has boundary values $u(t) \in L^p(-\infty,\infty)$, and relations (15)–(17) hold.

The conjugate harmonic function $\tilde{u}(z)$ in $\mathbb{C}_+$ is defined by the relation

$$\tilde{u}(z) = \frac{1}{\pi}\int_{-\infty}^{\infty} \frac{(x-t)u(t)}{(x-t)^2+y^2}\, dt , \qquad z = x+iy , \quad y > 0 . \tag{18}$$

THEOREM 2 (M. Riesz). *Let $u(t) \in L^p(-\infty,\infty)$, $p > 1$, and let the function $\tilde{u}(z)$ be defined by (18). Then there exists a constant $K_p < \infty$ which depends on p only, such that*

$$\sup_{h>0} \int_{-\infty}^{\infty} |\tilde{u}(x+ih)|^p\, dx \le K_p \int_{-\infty}^{\infty} |u(x)|^p\, dx . \tag{19}$$

Therefore, the function $\tilde{u}(z)$ possesses boundary values $\tilde{u}(t) \in L^p(-\infty,\infty)$, and

$$\int_{-\infty}^{\infty} |\tilde{u}(t)|^p\, dx \le K_p \int_{-\infty}^{\infty} |u(t)|^p\, dt . \tag{20}$$

PROOF. First, we assume that $p \ge 2$ and that $u(t)$ is a compactly supported function. Relation (18) yields the estimate

$$|\tilde{u}(z)| \le \frac{C_h}{|z|} , \qquad \operatorname{Im} z > h . \tag{21}$$

The function $f(z) = u(z) + i\tilde{u}(z)$ is analytic in $\mathbb{C}_+$ and according to (15) and (21), the function $f(z+ih)$ satisfies the estimate

$$|f(z+ih)| \le \frac{C_h}{|z|}, \qquad \operatorname{Im} z > 0, \tag{22}$$

and so belongs to the space H^p_+ for each $h > 0$.

Let us introduce the function

$$k(\theta) = 1 - \frac{p}{p-1}|\sin\theta|^p .$$

This is a π-periodic function, and

$$k''(\theta) + p^2 k(\theta) = p^2(1 - \sin^{p-2}\theta) \ge 0 .$$

Since $k'(0) = k'(\pi) = 0$, $k(\theta)$ is p-trigonometrically convex function for all $p \ge 2$ (prove it!). Therefore the function

$$G(z) = r^p k(\theta) = |z|^p - \frac{p}{p-1}|y|^p$$

is subharmonic in $\mathbb{C}_+$ (see Problem 2, Section 8.1). It follows from (22) that the integral

$$H(z) = \int_{-\infty}^{\infty} G(f(z+t))\,dt$$

converges. Since $H(z)$ is a subharmonic function, and $H(x+iy)$ is independent of x, we have $H(x+iy) = L(y)$, where $L(y)$ is a convex function of y, $y > 0$ (see Problem 1, Section 7.1).

Estimate (22) yields $L(y) \to 0$ as $y \to \infty$, and since the function $L(y)$ is convex, it must be nonincreasing. Therefore, $L(h) \ge 0$ for each $h > 0$, and hence

$$\frac{p}{p-1}\int_{-\infty}^{\infty} |\tilde{u}_h(x)|^p\,dx \le \int_{-\infty}^{\infty} |f_h(x)|^p\,dx .$$

Applying Minkowski's inequality for the space $L^{p/2}(-\infty,\infty)$ we obtain

$$\begin{aligned}\Big(\int_{-\infty}^{\infty} |f_h(x)|^p\,dx\Big)^{2/p} &= \Big(\int_{-\infty}^{\infty} (|u_h|^2 + |\tilde{u}_h|^2)^{p/2}\,dx\Big)^{2/p} \\ &\le \Big(\int_{-\infty}^{\infty} |\tilde{u}_h^2(x)|^{p/2}\,dx\Big)^{2/p} + \Big(\int_{-\infty}^{\infty} |u_h^2(x)|^{p/2}\,dx\Big)^{2/p} .\end{aligned}$$

Hence,

$$\Big(\frac{p}{p-1}\Big)^{2/p}\Big(\int_{-\infty}^{\infty} |\tilde{u}_h|^p\,dx\Big)^{2/p} \le \Big(\int_{-\infty}^{\infty} |u_h|^p\,dx\Big)^{2/p} + \Big(\int_{-\infty}^{\infty} |\tilde{u}_h|^p\,dx\Big)^{2/p},$$

or

$$\begin{aligned}\int_{-\infty}^{\infty} |\tilde{u}(x+ih)|^p\,dx &\le \Big[\Big(\frac{p}{p-1}\Big)^{2/p} - 1\Big]^{-p/2} \int_{-\infty}^{\infty} |u(x+ih)|^p\,dx \\ &\le K_p \int_{-\infty}^{\infty} |u(x)|^p\,dx .\end{aligned}$$

Until now $u(x)$ was a compactly supported function. If it is an arbitrary function from the space $L^p(-\infty,\infty)$, we find a sequence $\{u_n\} \subset L^p(-\infty,\infty)$ of compactly supported functions such that

$$\|u - u_n\|_{L^p(-\infty,\infty)} \to 0 , \qquad n \to \infty .$$

For any fixed $h > 0$ we have

$$\int_{-\infty}^{\infty} |u_n(x+ih) - u(x+ih)|^p \, dx \le \int_{-\infty}^{\infty} |u_n(x) - u(x)|^p \, dx \to 0 , \qquad n \to \infty ,$$

and also, $\tilde{u}_n(z) \rightrightarrows \tilde{u}(z)$, $h \to \infty$, uniformly on each compact set in $\mathbb{C}_+$. The Fatou lemma yields

$$\int_{-\infty}^{\infty} |\tilde{u}(x+ih)|^p \, dx \le \limsup_{n\to\infty} \int_{-\infty}^{\infty} |\tilde{u}_n(x+ih)|^p \, dx \le K_p \int_{-\infty}^{\infty} |u(x+ih)|^p \, dx$$

proving (19) for $p \ge 2$.

To prove (19) for $1 < p < 2$ we set $q = \dfrac{p}{p-1} > 2$, and let $h > 0$ be fixed. For each $\varepsilon > 0$ choose a compactly supported function $v(t)$ such that

$$\int_{-\infty}^{\infty} |v|^q \, dx = 1 ,$$

and

$$\|\tilde{u}(x+ih)\|_{L^p(-\infty,\infty)} \le \left| \int_{-\infty}^{\infty} \tilde{u}(x+ih) v(x) \, dx \right| + \varepsilon . \tag{23}$$

Combining Fubini's theorem, Hölder's inequality, and estimate (19) with q substituted for p, we obtain

$$\begin{aligned} \left| \int_{-\infty}^{\infty} \tilde{u}(x+ih) v(x) \, dx \right| &= \left| \frac{1}{\pi} \int_{-\infty}^{\infty} \int_{-\infty}^{\infty} \frac{(x-t) u(t) v(x)}{(x-t)^2 + h^2} \, dt \, dx \right| \\ &= \left| \int_{-\infty}^{\infty} u(t) \tilde{v}(t+ih) \, dt \right| \\ &\le \|u\|_{L^p(-\infty,\infty)} \|\tilde{v}_h\|_{L^q(-\infty,\infty)} \le K_p \|u\|_{L^p(-\infty,\infty)} . \end{aligned}$$

Since $\varepsilon > 0$ in (23) can be taken arbitrarily small, we arrive at (19), completing the proof of the theorem.

REMARK 1. The best value of the constant K_p on the right-hand side of (19) was obtained by Pichorides:

$$K_p = \begin{cases} \tan \dfrac{\pi}{2p} & 1 < p \le 2 \\ \cot \dfrac{\pi}{2p} & 2 \le p < \infty . \end{cases}$$

Essén [**35**] proved that this value of K_p may be obtained by modifying the choice of subharmonic function $G(t)$ in the proof of Theorem 2.

REMARK 2. The expression of the conjugate harmonic function may be reduced to the form

$$\tilde{u}(x+iy) = \frac{1}{\pi}\int_{-\infty}^{\infty} \frac{(x-t)u(t)}{(x-t)^2+y^2}\,dt = \frac{1}{\pi}\int_{0}^{\infty} \frac{u(x+s)-u(x-s)}{s}\,\frac{s^2}{s^2+y^2}\,ds\,.$$

For a smooth function $u(t) \in L^p(-\infty,\infty)$, $p > 1$, we have

$$\tilde{u}(x) = \frac{1}{\pi}\,\text{P.V.}\int_{-\infty}^{\infty} \frac{u(t)}{x-t}\,dt\,. \tag{24}$$

The right-hand side of (24) is called the *Hilbert transform* of the function $u(t)$. It is known (see for example Koosis [**71**], Garnett [**37**]) that for each function $u(t) \in L^p(-\infty,\infty)$ its Hilbert transform $\tilde{u}(t)$ exists for almost all t and coincides with the boundary values of $\tilde{u}(x+iy)$. As a result, the Riesz theorem can be formulated as follows.

The Hilbert transform is a bounded operator in $L^p(-\infty,\infty)$, $1 < p < \infty$.

The following corollaries of Riesz's theorem will be used later.

COROLLARY 1. *If* $f \in L^p(-\infty,\infty)$, $1 < p < \infty$, *then the functions*

$$f_{\pm}(z) = \frac{1}{i\pi}\int_{-\infty}^{\infty} \frac{f(t)}{t-z}\,dt\,, \qquad \pm\operatorname{Im} z > 0\,, \tag{25}$$

belong to the spaces $H^p_{\pm}$, *respectively, and*

$$\|f_{\pm}\|_{H^p_{\pm}} \le K_p \|f\|_{L^p(-\infty,\infty)}\,. \tag{26}$$

Here and in what follows, H^p_- stands for the Hardy space in the lower half-plane $\mathbb{C}_- = \{z : \operatorname{Im} z < 0\}$, which is defined similarly to the space H^p_+.

To prove this statement it is sufficient to note that the Poisson kernel $P_y(t) = \frac{y}{\pi}\frac{1}{t^2+x^2}$ and its harmonic conjugate $Q_y(t) = -\frac{1}{\pi}\frac{t}{t^+y^2}$ are the real and imaginary parts of the Cauchy kernel

$$P_y(t-x) + iQ_y(t-x) = \frac{1}{\pi i}\frac{1}{t-z}\,,$$

and then apply estimate (19).

COROLLARY 2. *Each function* $f \in L^p(-\infty,\infty)$, $1 < p < \infty$, *admits the unique representation:*

$$f(t) = f_+(t) - f_-(t)\,, \tag{27}$$

where $f_{\pm}(t)$ *are the boundary values of functions from* $H^p_{\pm}$, *and inequality* (26) *holds.*

PROOF. We set

$$f_+ = \frac{1}{2}(f + i\tilde{f})\,, \qquad f_- = -\frac{1}{2}(f - i\tilde{f})\,,$$

and apply Riesz's theorem. To prove the uniqueness of representation (27) we find that if $f = f_+ - f_- = g_+ - g_-$, then the function $f_+ - g_+ = f_- - g_-$ belongs to $H^p_+ \cap H^p_-$ and, according to Remark 1 to Theorem 1, vanishes identically.

Now we can describe the dual space $(H^p_+)^*$.

THEOREM 3. *Let $1 < p < \infty$. Then the space $(H^p_+)^*$ may be identified with the space H^q_-, $1/q + 1/p = 1$. The functional $E_\psi \in (H^p_+)^*$ which corresponds to a function $\psi \in H^q_-$ has the form*

$$E_\psi(f) = \int_{-\infty}^{\infty} f(t)\psi(t)\,dt\,, \tag{28}$$

and

$$\|E_\psi\|_{(H^p_+)^*} \le \|\psi\|_{H^p_-} \le K_p \|E_\psi\|_{(H^p_+)^*}\,. \tag{29}$$

PROOF. Given a function $\psi \in H^q_-$, equation (28) generates a bounded functional $E_\psi \in (H^p_+)^*$ and the left-hand inequality in (29) holds. Conversely, by the Hahn-Banach theorem each functional $E \in (H^p_+)^*$ admits an extension to a functional on the whole space $L^p(-\infty,\infty)$ with the same norm. Hence

$$E(f) = \int_{-\infty}^{\infty} f(t)\psi(t)\,dt\,, \qquad \psi \in L^q(-\infty,\infty).$$

By Corollary 2 from M. Riesz's theorem we have

$$\psi = \psi_+ - \psi_-\,, \qquad \psi_\pm \in H^q_\pm\,,$$

and

$$\|\psi_\pm\|_{H^p_\pm} \le K_q \|\psi\|_{L^q} = K_q \|E\|_{(H^p_+)^*}\,.$$

Further,

$$\int_{-\infty}^{\infty} f(t)\psi_+(t)\,dt = \lim_{h\to 0} \int_{-\infty}^{\infty} f_h(t)\psi_+(t)\,dt = 0\,,$$

hence, $E = E_\psi$ with $\psi_- = \psi$, proving Theorem 3.

COROLLARY. *If $f \in L^p(-\infty,\infty)$, $p > 1$, then the following three statements are equivalent:*

a) *f is the boundary value of a function from the space H^p_+;*

b) $\displaystyle\int_{-\infty}^{\infty} \frac{f(t)}{t-z}\,dt = 0\,, \qquad \operatorname{Im} z < 0\,;$

c) $\displaystyle\int_{-\infty}^{\infty} f(t)g(t)\,dt = 0$ *for each function* $g \in H^q_+$, $\dfrac{1}{q} + \dfrac{1}{p} = 1.$

For more results on the Hardy spaces and on their applications, see Privalov [**112**], Koosis [**71**] and Garnett [**37**].

19.4. The Paley-Wiener theorem for H^2_+

The space H^2_+ admits the following description:

THEOREM 4 (Paley-Wiener). *For a function $f(z)$ to belong to the space H^2_+ it is necessary and sufficient that f admit the representation*

$$f(z) = \frac{1}{2\pi}\int_0^{\infty} \varphi(t)e^{itz}\,dt\,, \qquad \varphi \in L^2(0,\infty)\,, \tag{30}$$

and in this case

$$\|f\|^2_{H^2_+} = 2\pi \|\varphi\|^2_{L^2(0,\infty)}. \tag{31}$$

PROOF. For $f \in H^2_+$ let us consider the inverse Fourier transform along the line $\operatorname{Im} z = y$:

$$\varphi_y(t) = \int_{-\infty}^{\infty} f(x+iy)e^{-itx}\,dx\,,$$

or

$$\varphi_y(t)e^{ty} = \int_{-\infty+iy}^{\infty+iy} f(z)e^{-itz}\,dz\,.$$

By property 3 of functions from H^2_+, the right-hand side of the latter equation is independent of y. If $\varphi(t) = \varphi_y(t)e^{ty}$, then Parseval's relation yields

$$\sup_{y>0} 2\pi \int_{-\infty}^{\infty} e^{-2ty}|\varphi(t)|^2\,dt = \|f\|^2_{H^2_+} < \infty\,.$$

This implies that $\varphi(t) = 0$ for $t < 0$ and also, that relation (31) holds.

The proof of the inverse statement is straightforward.

It should be mentioned that in the special case $p = 2$ one can easily obtain the results of the previous sections by means of the Paley-Wiener theorem.

PROBLEM 5. Deduce from Theorem 4 the Paley-Wiener theorem on entire functions of exponential type that are square integrable on the real line (Theorem 1, Section 10.1).

LECTURE 20

Interpolation by Entire Functions of Exponential Type

The following is a typical problem in the theory of entire functions:

Let $[\rho, \sigma]$ be the class of entire functions whose growth order does not exceed ρ and whose type with respect to ρ does not exceed σ. Let two sequences $\{\lambda_n\} \in \mathbb{C}$, $\lambda_n \to \infty$, and $\{c_n\} \subset \mathbb{C}$ be given. Find an entire function $f \in [\rho, \sigma]$ such that $f(\lambda_n) = c_n$ for all n.

Usually the numbers λ_n are called *the nodes of interpolation*. Sometimes, instead of restricting the type of $f(z)$, it is required that the indicator of $f(z)$ does not exceed a given ρ-trigonometric function $h(\theta)$ (the class of all such functions is called $[\rho, h(\theta)]$). The most interesting are problems of existence and uniqueness of the solution, as well as its explicit expression. The reconstruction of the solution is based mainly on the classical Lagrange interpolation series and on its generalizations for the case of multiple nodes. For results on interpolation in the classes $[\rho, \sigma]$, $[\rho, h(\theta)]$ and some other classes, see, for example, Levin [82], Leont'ev [**79**], Grishin and Russakovskii [49], Goldberg, Levin, and Ostrovskii [**42,** Chapter 4].

We shall restrict ourselves to the study of interpolation problems in some classes of entire functions of exponential type (EFET) and with integer nodes of interpolation or with nodes close to the integers.

20.1. Spaces L^p_σ and B_σ

Let us denote by L^p_σ, $1 \le p < \infty$, the space of all entire functions of exponential type $\le \sigma$ that belong to the space $L^p(-\infty, \infty)$. It is easy to see that L^p_σ is a closed subspace of $L^p(-\infty, \infty)$. Indeed, the Plancherel-Pólya theorem (see Section 7.4) yields

$$(1) \qquad \int_{-\infty}^{\infty} |f(x+iy)|^p \, dx \le e^{p\sigma|y|} \|f\|^p_{L^p(-\infty,\infty)} .$$

Therefore, we have

$$(2) \qquad \int_{-1}^{1}\int_{-\infty}^{\infty} |f(x+i(y+s))|^p \, dxds \le 2e^{p(1+|y|)} \|f\|^p_{L^p(-\infty,\infty)}$$

for any $y \in \mathbb{R}$, and since $|f|^p$ is a subharmonic function, we obtain

$$(3) \qquad |f(x+iy)|^p \le \frac{2}{\pi} e^{\sigma p(|y|+1)} \|f\|^p_{L^p(-\infty,\infty)} .$$

It follows from this inequality that L^p-convergence of a sequence of functions from L^p_σ implies the uniform convergence in each horizontal strip, and that the limit function is an entire function of exponential type not exceeding σ.

The following properties of the spaces L^p_σ are important.

1. *For each $h \in \mathbb{R}$, the functions $f(z+ih)e^{i\sigma z}$ and $f(z+ih)e^{-i\sigma z}$ belong to the Hardy spaces in the half-planes $\{\operatorname{Im} z > 0\}$ and $\{\operatorname{Im} z < 0\}$, respectively.* This statement is a direct consequence of (1).

2. *For each $-\infty < h_1 < h_2 < \infty$:*

$$f(x+iy) \rightrightarrows 0\,, \qquad |x| \to \infty\,,$$

uniformly in the strip $\{h_1 < y < h_2\}$. In particular, $f(x) \to 0$, as $|x| \to \pm\infty$.

3. *For each function $f \in L^p_\sigma$*

$$|f(iy)| = o(e^{\sigma|y|})\,, \qquad |y| \to \infty\,.$$

This property follows from the previous one combined with the Phragmén-Lindelöf theorem.

4. *Let a sequence $\{\lambda_n\} \subset \mathbb{C}$ be such that for some H, δ*

$$|\operatorname{Im}\lambda_n| \le H < \infty\,, \qquad |\lambda_n - \lambda_m| \ge \delta > 0 \qquad \text{for } n \ne k\,. \tag{4}$$

Then, for each $f \in L^p_\sigma$,

$$\sum_n |f(\lambda_n)|^p \le C\|f\|^p_{L^p(-\infty,\infty)}$$

where the constant C depends on σ, δ and H, but not on f. This follows directly from Property 6, Section 19.1, of functions of the Hardy space.

To simplify notation, we set $L_\sigma = L^1_\sigma$ in what follows. Let us denote[18] by B_σ the space of all entire functions of exponential type not exceeding σ, endowed with the norm

$$\|f\|_{B_\sigma} = \sup_{x\in\mathbb{R}} |f(x)|\,.$$

Using the Phragmén-Lindelöf theorem, it is easy to see that L^p_σ and B_σ are Banach spaces.

20.2. Interpolation with integer nodes

We start with a theorem which is now classical.

THEOREM 1. *Each entire function $f(z) \in L^2_\pi$ admits the representation*

$$f(z) = \sum_{k=-\infty}^{\infty} (-1)^k c_k \frac{\sin \pi z}{\pi(z-k)} \tag{5}$$

where $c_k = f(k)$, $k \in \mathbb{Z}$, and

$$\int_{-\infty}^{\infty} |f(x)|^2\,dx = \sum_{k=-\infty}^{\infty} |f(k)|^2\,. \tag{6}$$

[18] In honor of Serge Bernstein.

The series (5) converges in the L^2_π-norm and uniformly on each compact set in $\mathbb{C}$. Conversely, for each sequence $\{c_k\} \in l^2$ equation (5) defines the function $f(z) \in L^2_\pi$ which solves the interpolation problem $f(k) = c_k$.

PROOF. According to the Paley-Wiener theorem, each function $f \in L^2_\pi$ admits the representation

$$f(z) = \frac{1}{2\pi}\int_{-\pi}^{\pi} \psi(t)e^{-itz}\,dt\,, \qquad \psi \in L^2(-\pi,\pi)\,. \tag{7}$$

The Fourier expansion

$$\psi(t) = \sum_{k=-\infty}^{\infty} c_k e^{ikt} \tag{8}$$

converges in $L^2(-\pi,\pi)$. Substituting it in (7) we obtain relation (5). It is evident that $c_k = f(k)$, $k \in \mathbb{Z}$. In addition, using Parseval's identity we obtain

$$\|f\|_{L^2(-\infty,\infty)} = \frac{1}{\sqrt{2\pi}}\|\psi\|_{L^2(-\pi,\pi)} = \|\{c_k\}\|_{l^2} = \|\{f(k)\}\|_{l^2}\,,$$

and the series on the right-hand side of (5) converges in the $L^2(-\infty,\infty)$-norm.

Conversely, let a sequence $\{c_k\} \in l^2$ be given. Formula (8) defines the function $\psi \in L^2(-\pi,\pi)$ and the function f defined by relation (7) is of exponential type $\sigma \le \pi$ for which representation (5) is valid.

REMARK 1. Another form of Theorem 1 is the following:

The system of functions $\left\{\dfrac{\sin\pi(z-k)}{\pi(z-k)}\right\}_{k=-\infty}^{\infty}$ *forms an orthonormal basis in* L^2_π.

REMARK 2. Theorem 1, as well as some similar results, is of importance in communication theory. According to the Paley-Wiener theorem, each bandlimited signal can be treated as the restriction of some EFET to the real axis. Theorem 1 states that to transmit such a signal it is sufficient to use its samplings only, i.e., its values at some periodic sequence of points. It was V. Kotel'nikov who in the early thirties used this theorem in communication theory. In the forties, C. Shannon applied this theorem to the same area. Now, it is known as the sampling theorem or the Kotel'nikov-Shannon theorem. For more complete exposition of applications of entire functions to the communication theory, optics and some other fields, see Khurgin and Yakovlev [**60**], Higgins [**56**].

20.3. Interpolation in the spaces L^p_π, $1 < p < \infty$, with integer nodes

We begin with a uniqueness theorem for EFET which is close to Carlson's theorem, Section 8.3, and is proved by similar methods.

THEOREM 2. *Let $f(z)$ be an EFET $\sigma \le \pi$, such that $f(n) = 0$, $n \in \mathbb{Z}$, and let*

$$\lim_{|y|\to\infty} f(iy)e^{-\pi|y|} = 0\,. \tag{9}$$

Then $f(z) \equiv 0$.

PROOF. The function $\varphi(z) = f(z)/\sin\pi z$ is an EFET. For any fixed $\delta > 0$ the denominator $\sin\pi z$ admits the lower and upper estimates

$$m_\delta e^{\pi|y|} \le |\sin\pi z| \le M e^{\pi|y|}\,, \qquad y = \operatorname{Im} z\,,$$

where the left estimate is valid for $|z-n| \ge \delta$, $n \in \mathbb{Z}$. Therefore, the indicator diagram of the product $\varphi(z)\sin\pi z$ is the sum of the indicator diagrams of the factors. The indicator diagram of $\sin\pi z$ is the segment $[-i\pi, i\pi]$, and hence $I_f = I_\varphi + [-i\pi, i\pi] \subset \{z : |z| \le \pi\}$. Therefore, $\varphi(z)$ is a function of minimal type with respect to order 1. Condition (9) yields $\varphi(iy) \to 0$, $y \to \pm\infty$, and applying the Phragmén-Lindelöf theorem in the right and left half-planes we find that $\varphi(z)$ is bounded in $\mathbb{C}$. Hence $\varphi(z) \equiv$ const, and since $\varphi(iy) \to 0$, $y \to \pm\infty$, we obtain $\varphi \equiv 0$, which completes the proof of the theorem.

As property 3 from Section 20.1 shows, Theorem 2 can be applied to functions from L^p_σ, $p \ge 1$, leading us to the following statement.

COROLLARY. *Each function from the space L^p_π, $p \ge 1$, is uniquely determined by its values at the integers.*

THEOREM 3 (Plancherel-Pólya). *For any sequence $\{c_k\}_{k=-\infty}^{\infty} \in l^p$, $1 < p < \infty$, the series*

$$f(z) = \sum_{k=-\infty}^{\infty} (-1)^k c_k \frac{\sin\pi z}{\pi(z-k)} \tag{10}$$

converges in the $L^p(-\infty,\infty)$-norm to a functions $f \in L^p_\pi$, which is the unique solution of the interpolation problem $f(k) = c_k$, $k \in \mathbb{Z}$. Conversely, for any function $f \in L^p_\pi$, $1 < p < \infty$, the sequence $\{f(k)\}_{k=-\infty}^{\infty}$ belongs to l^p, and there exist positive constants c and C such that

$$c\Big(\sum_{k=-\infty}^{\infty} |f(k)|^p\Big)^{1/p} \le \|f\|_{L^p(-\infty,\infty)} \le C\Big(\sum_{k=-\infty}^{\infty} |f(k)|^p\Big)^{1/p} \tag{11}$$

for all functions $f \in L^p_\pi$.

COROLLARY. *For $1 < p < \infty$, the mapping*

$$\{c_k\} \mapsto f(z) = \sum_{k=-\infty}^{\infty} (-1)^k c_k \frac{\sin\pi z}{\pi(z-k)}$$

is an isomorphism of the Banach spaces l^p and L^p_π.

PROOF OF THEOREM 3. Let us set

$$\varphi_{n,m}(z) = \sum_{k=n}^{m} (-1)^k \frac{c_k}{z-k}\,, \qquad \Phi_{n,m}(z) = \frac{\sin\pi z}{\pi}\varphi_{n,m}(z)\,.$$

It is evident that the functions $\Phi_{n,m}$ belong to the spaces L^p_π, $1 < p < \infty$. Therefore, for any $h > 0$, we have

$$\begin{aligned} \int_{-\infty}^{\infty} |\Phi_{n,m}(x)|^p\,dx &\le e^{\pi ph}\int_{-\infty}^{\infty} |\Phi_{n,m}(x+ih)|^p\,dx \\ &\le C(h,p)\int_{-\infty}^{\infty} |\varphi_{n,m}(x+ih)|^p\,dx\,. \end{aligned} \tag{12}$$

Thus, to prove the convergence of series (10), it suffices to estimate the norm of functions $\varphi_{m,n}(x+ih)$. Since $\varphi_{m,n}(x+ih) \in H_+^p$, then, using Theorem 3, Section 19.3, which describes the general form of linear functionals in H_+^p, we obtain

$$\|\varphi_{n,m}(x+ih)\|_{H_+^p} = K_p \sup\left\{\left|\int_{-\infty}^{\infty} \varphi_{m,n}(x+ih)\psi(x)\,dx\right| : \psi \in H_-^q,\ \|\psi\|_{H_-^q} \le 1\right\}.$$

To evaluate the integral on the right-hand side we apply the Cauchy formula for residues:

$$\int_{-\infty}^{\infty} \varphi_{n,m}(x+ih)\psi(x)\,dx = -2\pi i \sum_{k=n}^{m} (-1)^k c_k \psi(k-ih)\,.$$

By Hölder's inequality and property 6 of functions from the Hardy spaces (see Section 19.1), we obtain

$$\begin{aligned} &\|\varphi_{n,m}(z+ih)\|_{H_+^p} \\ (13) \qquad &\le K_p\Big(\sum_{k=n}^{m} |c_k|^p\Big)^{1/p} \sup\left\{\Big(\sum_{k=n}^{m} |\psi(k-ih)|^q\Big)^{1/q} : \|\psi\|_{H_-^q} \le 1\right\} \\ &\le K_p\Big(\sum_{k=n}^{m} |c_k|^p\Big)^{1/p}. \end{aligned}$$

Therefore, for sufficiently large N_ε and $m > n > N_\varepsilon$, we have

$$\|\varphi_{n,m}(x+ih)\|_{L^p} \le \varepsilon,$$

and taking (12) into account we find that series (10) converges in the L^p-norm. Since all summands belong to L_π^p and the space L_π^p is complete, the sum of the series (10) belongs to L_π^p as well. The uniqueness of the solution to the interpolation problem follows from the Corollary to Theorem 2.

For $n \to -\infty$, $m \to +\infty$ in relations (12), (13) we obtain

$$\|f\|_{L^p(-\infty,\infty)} \le K_p\Big(\sum_{k=-\infty}^{\infty} |c_k|^p\Big)^{1/p}.$$

The converse inequality follows from property 6, Section 19.1, of functions from L_π^p. This completes the proof of the theorem.

The statement of Theorem 3 is false for L_π and B_π. Indeed, if $\{c_k\} \in l^1$, the series (10) converges uniformly in each horizontal strip and the interpolating function f satisfies (9). Hence, by Theorem 2 it is unique, but now it need not belong to the space L_π. If $\{c_k\}$ is a bounded sequence, the interpolation series (10) may be divergent and also, as the example $f(z) = \sin \pi z$ shows, the solution of the corresponding interpolation problem is not unique.

In the next lecture we shall study the interpolation with integer nodes in the spaces L_σ and B_σ with $\sigma \le \pi$ in more detail.

LECTURE 21

Interpolation by Entire Functions from the Spaces L_π and B_π

In Section 20.3 we studied the interpolation series (10) which determines the behavior of the solution to the interpolation problem for the spaces L^p_π in the case $1 < p < \infty$. The study of the extreme cases $p = 1$ and $p = \infty$ is essentially different. In particular, if $\{c_k\}$ is a bounded sequence, the series (10) may diverge. In the present lecture we give a criterion for solvability of the interpolation problem

$$f(k) = c_k$$

for $\{c_k\}_{k=-\infty}^{\infty} \in l^\infty$ ($\{c_k\}_{k=-\infty}^{\infty} \in l^1$) in the space $B_\sigma = L^\infty_\sigma$ ($L_\sigma = L^1_\sigma$, respectively). We shall also obtain a representation of the solution to this problem via a special interpolation series.

21.1. Interpolation by functions from B_π and L_π

Let a sequence $\{c_k\}_{k=-\infty}^{\infty} \in l^\infty$ be given. We begin with constructing a solution of the interpolation problem

$$f(k) = c_k\ , \qquad k = 0, \pm 1, \pm 2, \dots\ , \tag{1}$$

in the class of EFET, and then we shall study the behavior of this solution on the real axis.

THEOREM 1. *For any bounded sequence $\{c_k\} \in l^\infty$ there exists an entire function $f(z)$ of exponential type $\le \pi$ which solves the interpolation problem* (1) *and is such that*

$$|f(z)|e^{-\pi|y|} = o(|z|)\ , \qquad |z| \to \infty\ . \tag{2}$$

These conditions define the function $f(z)$ up to an additive term of the form $C\sin\pi z$. In addition, the function $f(z)$ admits the representation

$$f(z) = \frac{\sin \pi z}{\pi} \sum_{k=-\infty}^{\infty}{}' (-1)^k c_k \Big[\frac{1}{z-k} + \frac{1}{k}\Big] + C \sin\pi z\ , \tag{3}$$

where the prime in the summation sign means that the second term in the braces is omitted for $k = 0$.

PROOF. First, let us prove that the series

$$g(z) = \frac{\sin \pi z}{\pi} \sum_{k=\infty}^{\infty}{}' (-1)^k c_k \Big[\frac{1}{z-k} + \frac{1}{k}\Big] \tag{4}$$

on the right-hand side of (3) is convergent. Suppose $|c_k| < M$, $k = \pm 0, \pm 1, \ldots$. We have

$$S_{m,n}(z) = \Big| \sum_{|k|=n+1}^{m} (-1)^k c_k \Big(\frac{1}{z-k} + \frac{1}{k}\Big)\Big| \le M|z| \sum_{|k|=n+1}^{m} \frac{1}{|k(z-k)|}$$
$$\le M|z| \Big(\sum_{|k|=n+1}^{m} \frac{1}{k^2}\Big)^{1/2} \Big(\sum_{-\infty}^{\infty} \frac{1}{|z-k|^2}\Big)^{1/2} .$$

If $\delta \in (0, 1/4)$ is fixed and z lies outside the union of the disks $\{z : |z-k| \le \delta\}$, then

$$\sum_{k=-\infty}^{\infty} \frac{1}{|z-k|^2} \le \frac{1}{\delta^2} + 2\sum_{k=1}^{\infty} \frac{1}{k^2} = M_\delta^2 < \infty ,$$

whence

$$|S_{m,n}(z)| \le \varepsilon |z| \quad \text{for} \quad n, m > N_\varepsilon . \tag{5}$$

Combining this estimate and the Maximum Principle we find that the series (4) converges uniformly to a function $g(z)$ on each compact set in the complex plane. Moreover, (5) yields

$$\sum_{k=-\infty}^{\infty}{}' (-1)^k c_k \Big(\frac{1}{z-k} + \frac{1}{k}\Big) = o(|z|) , \qquad z \to \infty ,$$

outside the disks $\{|z-k| \le \delta\}$, $k = 0, \pm 1, \pm 2, \ldots$. Hence, using (4),

$$g(z) e^{-\pi|y|} = o(|z|), \qquad |z| \to \infty . \tag{6}$$

It is evident that the function $g(z)$ is a solution to the interpolation problem (1).

Now let $f(z)$ be an arbitrary entire function which satisfies conditions (1), (2). Since the difference $f(z) - g(z)$ satisfies condition (2) and vanishes at integers, the ratio $\varphi(z) = (f(z) - g(z))/\sin \pi z$ is an entire function with $\varphi(z) = o(|z|)$, $|z| \to \infty$. By Liouville's theorem $\varphi(z)$ is a constant, which proves (3).

REMARK 1. Similar reasoning allows us to replace condition (2) by a weaker condition:

$$|f(iy)| e^{-\pi|y|} = o(|y|) , \qquad |y| \to \infty .$$

PROBLEM 1. Let $f(z)$ admit representation (3) with $\{c_k\} \in l^\infty$. Prove that

$$|f(z)| e^{-\pi|y|} = O(\log |z|) , \qquad |z| \to \infty . \tag{7}$$

PROBLEM 2. Let a sequence $\{c_k\}$ satisfy the condition $|c_k| = O(|k|^{m+s})$, $k \in \mathbb{Z}$, with some integer $m > 0$ and some number $s \in (0, 1)$. Prove that the series

$$\frac{\sin \pi z}{\pi} \sum_{k=-\infty}^{\infty} (-1)^k c_k \Big(\frac{1}{z-k} + \frac{1}{k} + \frac{z}{k^2} + \cdots + \frac{z^m}{k^{m+1}}\Big)$$

converges to an entire function $f(z)$ uniformly on each compact set in $\mathbb{C}$. This function solves the interpolation problem (1), and satisfies the condition

$$|f(z)|\, |z|^{-m-1} e^{-\pi|y|} \to 0, , \qquad |z| \to \infty.$$

Conversely, each entire function $\varphi(z)$ of exponential type $\sigma \leq \pi$ which solves the interpolation problem $f(k) = c_k$ and satisfies the condition

$$|f(iy)|e^{-\pi|y|} = o(|y|^{m+1}), \qquad y \to \pm\infty,$$

admits the representation

$$f(z) = \frac{\sin \pi z}{\pi} \sum_{k=-\infty}^{\infty}{}' (-1)^k c_k \Big[\frac{1}{z-k} + \frac{1}{k} + \cdots + \frac{z^m}{k^{m+1}}\Big] + P(z) \sin \pi z,$$

where $P(z)$ is a polynomial of degree less or equal to m.

There exists a bounded sequence such that the corresponding interpolation problem has no solution in B_σ. To describe conditions that provide the existence of a bounded solution to the interpolation problem, we introduce for each integer τ the functional

(8) $$L_\tau : \{c_k\} \mapsto L_\tau(\{c_k\}) = \sum_{k=-\infty}^{\infty} (d_{k+\tau} - d_k) \frac{k}{k^2+1},$$

where $d_k = (-1)^k c_k$. Since,

$$\frac{k}{k^2+1} - \frac{k+\tau}{(k+\tau)^2+1} = O\Big(\frac{1}{k^2}\Big), \qquad k \to \infty,$$

the series on the right-hand side of (8) is convergent.

THEOREM 2. *Let $\{c_k\}$ be a bounded sequence. In order that there exist an entire function $f(z)$ of exponential type $\leq \pi$, which is bounded on the real axis and solves the interpolation problem* (1), *it is necessary and sufficient that*

(9) $$|L_\tau(\{c_k\})| < M, \qquad \tau = 0, \pm 1, \pm 2, \ldots,$$

for some $M > 0$.

SUFFICIENCY. Let $\{c_k\}$ be a bounded sequence satisfying (9). By Theorem 1 the series (3) converges to $g(z)$ which is an EFET $\sigma \leq \pi$. Therefore, it remains to prove that $g(x)$ is bounded on the real axis.

For an even τ, let us consider the difference

$$\begin{aligned} g(x+\tau+i) - g(x+i) &= \frac{\sin \pi(x+i)}{\pi} \sum_{k=-\infty}^{\infty} d_k \Big(\frac{1}{x+i+\tau-k} - \frac{1}{x+i-k}\Big) \\ &= \frac{\sin \pi(x+i)}{\pi} \sum_{k=-\infty}^{\infty} \frac{d_{k+\tau} - d_k}{x-k+i}. \end{aligned}$$

Since the sequence $\{d_k\}$ is bounded we have

$$\Big| \sum_{k=-\infty}^{\infty} \frac{d_{k+\tau} - d_k}{x-k+i} - \sum_{k=-\infty}^{\infty} \frac{d_{k+\tau} - d_k}{k+i} \Big| \leq C_1$$

for $0 \leq x \leq 2$, where C_1 is independent of $x \in [0,2]$ and τ. Besides,

$$\Big| \sum_{k=-\infty}^{\infty} \frac{d_{k+\tau} - d_k}{k+i} - L_\tau(\{c_k\}) \Big| \leq C_2,$$

where C_2 is also independent of τ. Combining these inequalities with (9), we obtain

$$|g(x+\tau+i) - g(x+i)| \leq M + C_1 + C_2$$

for $0 \leq x \leq 2$, $\tau = 0, \pm 2, \dots$. Now, setting $C_3 = \max_{0\leq x\leq 2} |g(x+i)|$, we have

$$|g(x+i)| \leq C_1 + C_2 + C_3 + M$$

for all $x \in \mathbb{R}$. Therefore, the function $g(z)$ is bounded on a straight line which is parallel to the real axis. By the Phragmén-Lindelöf theorem $g(x)$ is bounded on $\mathbb{R}$ as well.

NECESSITY. Let $f(z)$ be an entire function of exponential type $\sigma \leq \pi$ bounded on the real line and solving the interpolation problem (1). Then by Theorem 1 $f(z)$ admits the representation (3). For an even integer τ we have

$$\begin{aligned} f(\tau+i) - f(i) &= -\frac{\sinh\pi}{\pi i} \sum_{k=-\infty}^{\infty} (-1)^k c_k \Big[\frac{1}{\tau+i-k} - \frac{1}{i-k}\Big] \\ &= -\frac{\sinh\pi}{\pi i} \sum_{k=-\infty}^{\infty} \frac{d_{k+\tau} - d_k}{i-k} \\ &= \frac{\sinh\pi}{\pi i} L_\tau(\{c_k\}) - \frac{\sinh\pi}{\pi i} \sum_{k=-\infty}^{\infty} \frac{(d_{k+\tau} - d_k)(1+ik)}{(k^2+1)(i-k)} . \end{aligned}$$

Since $f(x)$ is bounded on the real axis, so are both $f(x+i)$ and the difference $f(x+i) - f(i)$. Therefore, we have

$$\sup\{|L_\tau(\{c_k\})|\,, \tau = 0, \pm 2, \pm 4, \dots\} < \infty .$$

The estimate for an odd τ follows from the relation

$$L_\tau(\{c_k\}) = L_{\tau-1}(\{c_k\}) + \sum_{k=-\infty}^{\infty} d_{k+\tau-1}\Big[\frac{k-1}{(k-1)^2+1} - \frac{k}{k^2+1}\Big] .$$

This completes the proof of the theorem.

Now let us consider the interpolation problem for a sequence $\{c_k\} \in l^1$. In this case the interpolation series

$$f(z) = \sum_{k=-\infty}^{\infty} (-1)^k c_k \frac{\sin \pi z}{\pi(z-k)} \tag{10}$$

is convergent, but, generally speaking, its sum $f(z)$ does not belong to $L^1(-\infty, \infty)$. To formulate the corresponding summability condition we define, for each integer τ, the functional

$$M_\tau : \{c_k\} \mapsto M_\tau(\{c_k\}) = \sum_{k=-\infty}^{\infty} d_{k+\tau} \frac{k}{k^2+1} . \tag{11}$$

The following statement is similar to Theorem 2.

PROBLEM 3 (Ber). Let $\{c_k\}_{-\infty}^{\infty}$ be a sequence in l^1. In order that there exist a function $f \in L_\pi$ which solves the interpolation problem (1), it is necessary and sufficient that

$$\sum_{\tau=-\infty}^{\infty} |M_\tau(\{c_k\})| < \infty . \tag{12}$$

HINT. To prove the necessity, use the representation (10) of a function $f \in L_\pi$ and the inequality

$$\sum_{\tau=-\infty}^{\infty} |f(\tau + i)| \le \text{const}\, \|f\|_{L_\pi} ;$$

see Section 19.1. To prove sufficiency note that if $f(z)$ is represented by the series (10), then the sequence

$$\Big\{ \int_0^1 |f(x+\tau+i)|\, dx - M_\tau(\{c_k\}) \Big\}_{\tau=-\infty}^{\infty}$$

is summable.

It should also be mentioned that conditions (9) and (12) are equivalent to the boundedness of the *discrete analogs of the Hilbert transform*:

$$\{d_k\}_{-\infty}^{\infty} \mapsto \Big\{ \sum_{k\neq 0} \frac{d_{k+\tau} - d_k}{k} \Big\}_{\tau=-\infty}^{\infty}$$

and

$$\{d_k\}_{-\infty}^{\infty} \mapsto \Big\{ \sum_{k\neq 0} \frac{d_{k+\tau}}{k} \Big\}_{\tau=-\infty}^{\infty}$$

in the spaces l^∞ and l^1, respectively.

PROBLEM 4. Construct a sequence $\{c_k\}_{-\infty}^{\infty} \in l^\infty$ ($\{c_k\}_{-\infty}^{\infty} \in l^1$) such that condition (9) (respectively (12)) fails.

PROBLEM 5. Prove that, for $p = 1$ and $p = \infty$, there exist entire functions of exponential type π such that

$$f(x) = o(|x|) , \qquad x \to \pm\infty , \tag{13}$$

and $\{f(k)\} \in l^p$, $f(x) \notin L^p(-\infty, \infty)$.

PROBLEM 6. Let $p \in [1, \infty]$ and $f(z)$ be an entire function of exponential type π satisfying (13). Then the following statements are equivalent:

1) $f(x) \in L^p(-\infty, \infty)$.

2) There exist $a, b \in \mathbb{R}$ such that $a \neq b$ and

$$\{f(k+ia)\}_{k=-\infty}^{\infty} \in l^p , \qquad \{f(k+ib)\}_{k=-\infty}^{\infty} \in l^p .$$

21.2. Interpolation by functions from L^p_σ with $\sigma < \pi$

As stated in Problem 6 of the previous section, there exist entire functions of exponential type π which satisfy (2), are bounded at the integers, but are unbounded on the real axis.

This is not the case if we require that the type σ of an entire function is less than π. To study such functions we need the following interpolation formula.

THEOREM 3 (Boas, S. N. Bernstein). *Every entire function $f(z)$ of exponential type such that*

$$d = h_f\Big(\frac{\pi}{2}\Big) = h_f\Big(-\frac{\pi}{2}\Big) < \pi$$

and

$$|f(k)| \le M\,, \qquad k = 0, \pm 1, \pm 2, \dots\,,$$

admits the representation

$$f(z) = \frac{\sin \pi z}{\pi \omega} \sum_{k=-\infty}^{\infty} (-1)^k \frac{f(k) \sin \omega (z-k)}{(z-k)^2}\,, \tag{14}$$

where ω is an arbitrary number from the interval $(0, \pi - d)$.

PROOF. Let us fix $z \in \mathbb{C}$ and consider the function of ζ:

$$\varphi(\zeta, z) = f(\zeta) \frac{\sin \omega(\zeta - z)}{\omega(\zeta - z)} - \frac{\sin \pi\zeta}{\pi\omega} \sum_{k=-\infty}^{\infty} (-1)^k \frac{f(k) \sin\omega(z-k)}{(\zeta - k)(z-k)}\,.$$

It is an EFET vanishing at the integers. Therefore, the ratio

$$\psi(\zeta, z) = \frac{\varphi(\zeta, z)}{\sin \pi\zeta} = \frac{f(\zeta)\sin\omega(\zeta - z)}{\omega(\zeta - z)\sin \pi\zeta} - \frac{1}{\pi\omega} \sum_{k=-\infty}^{\infty} (-1)^k \frac{f(k)\sin\omega(z-k)}{(\zeta - k)(z-k)} \tag{15}$$

is an EFET as well. The indicator of the first summand on the right-hand side is equal to $d + \omega - \pi < 0$ at the points $\theta = \pm\pi/2$ and, by the continuity, is negative on some neighborhood of these points. The second summand on the right-hand side of (15) vanishes as $\zeta \to \infty$ along any ray $\{\zeta : \arg\zeta = \theta\}$, for $\theta \ne 0, \pi$. By the Phragmém-Lindelöf theorem we have $\psi(\zeta, z) \equiv 0$. With $\zeta \to z$ we obtain (14).

THEOREM 4 (Cartwright). *Let $f(z)$ be an EFET such that*

$$2d = h_f\Big(\frac{\pi}{2}\Big) + h_f\Big(\frac{\pi}{2}\Big) < 2\pi\,,$$

and let $|f(k)| < M$ for $k = 0, \pm1, \pm2, \dots$. Then

$$|f(x)| \le C(d) M\,, \qquad -\infty < x < \infty\,. \tag{16}$$

PROOF. Without loss of generality we may assume $h_f(\pi/2) = h_f(-\pi/2)$, and, hence, $f(z)$ admits the representation (14). This yields the estimate

$$|f(x \pm i)| \le \frac{e^{\pi+\omega}}{\pi\omega} M \sum_{k=-\infty}^{\infty} \frac{1}{(x-k)^2 + 1} \le M \frac{e^{\pi+\omega}}{\pi\omega} \sum_{k=-\infty}^{\infty} \frac{1}{k^2+1}\,.$$

To obtain (16) it suffices to apply the Phragmén-Lindelöf theorem in the strip $\{z : |\operatorname{Im} z| < 1\}$. The theorem is proven.[19]

Similar reasoning may be applied in the case $p = 1$. Combining these results with the Plancherel-Pólya theorem from the previous lecture, dealing with the case $1 < p < \infty$, we obtain the following result.

THEOREM 5 (Plancherel-Pólya). *Let $f(z)$ be an entire function of exponential type such that*

$$2d_f = h_f\Big(\frac{\pi}{2}\Big) + h_f\Big(-\frac{\pi}{2}\Big) < 2\pi$$

and

$$\{f(k)\}_{k=-\infty}^{\infty} \in l^p$$

for some $p \in [1, \infty]$. Then $f(x) \in L^p(-\infty, \infty)$ and there exist positive numbers c and C not depending on $f(x)$ such that

$$c\|f\|_{L^p(-\infty,\infty)} \le \|\{f(k)\}\|_{l^p} \le C\|f\|_{L^p(-\infty,\infty)}\,.$$

PROOF. We should consider only the case $p = 1$ in which we again use the Boas-Bernstein interpolation formula (14). Let

$$f_{n,m}(z) = \frac{\sin \pi z}{\pi\omega} \sum_{|k|=n+1}^{m} (-1)^k f(k) \frac{\sin \omega(z-k)}{(z-k)^2}\,.$$

Then

$$\int_{-\infty}^{\infty} |f_{m,n}(x \pm i)|\, dx \le \frac{e^{\pi+\omega}}{\pi\omega} \Big(\int_{-\infty}^{\infty} \frac{dx}{x^2+1}\Big) \sum_{|k|=n+1}^{m} |f(k)| = \frac{e^{\pi+\omega}}{\omega} \sum_{|k|=n+1}^{m} |f(k)|\,.$$

The function

$$u_N(z) = \int_{-N}^{N} |f_{n,m}(z+t)|\, dt$$

is a subharmonic bounded function in the strip $\{z : |\operatorname{Im} z| < 1\}$ satisfying the estimate

$$u_N(x \pm i) \le \frac{e^{\pi+\omega}}{\omega} \sum_{|k|=n+1}^{m} |f(k)|\,.$$

By the Phragmèn-Lindelöf theorem this estimate is extended to the whole strip. In particular,

$$\int_{-\infty}^{\infty} |f_{m,n}(x)|\, dx = \limsup_{N\to\infty} u_N(0) \le \frac{e^{\pi+\omega}}{\omega} \sum_{|k|=n+1}^{m} |f(k)|\,.$$

Therefore, the series (14) converges to the function $f(z)$ in the $L^1(-\infty, \infty)$-norm and

$$\|f\|_{L^1(-\infty,\infty)} \le \frac{e^{\pi+\omega}}{\omega} \sum_{k=-\infty}^{\infty} |f(k)|\,.$$

[19]The best value of $C(d)$ is not known yet. In some particular cases ($d = \pi/n$ and $d = \pi - \pi/n$, $n \in \mathbb{N}$) it was found by S. N. Bernstein. Generalizations in various directions of the Cartwright theorem are known. We mention here the papers Agmon [1], Davydova and Logvinenko [**26**], Levin [**81**], and Malliavin [**89**].

The opposite inequality (with another constant) follows from Property 4 of the space L^p_σ (see the previous lecture).

21.3. Interference in a class of entire functions

The following interference phenomenom has been discovered by S. N. Bernstein.

THEOREM 6 (S. N. Bernstein). *Let $f(z)$ be an entire function of exponential type $\sigma \leq \pi$ which satisfies the estimate* (2) *and is such that* $\sup_k |f(k)| < \infty$. *Then*

$$|f(x+1) + f(x)| \leq C \sup_k |f(k)| , \tag{17}$$

where the constant C is independent of f.

PROOF. Expanding $f(z)$ in the interpolation series (3) with $c_k = f(k)$, we obtain

$$f(x+1) + f(x) = \frac{\sin \pi x}{\pi} \sum_{k=-\infty}^{\infty} (-1)^k f(k) \Big(\frac{1}{x-k+1} - \frac{1}{x-k} \Big),$$

which implies (17) with the constant[20]

$$C = \sup_x \sum_{k=-\infty}^{\infty} \left| \frac{\sin \pi x}{(x-k)(x-k+1)} \right| .$$

This proves the theorem.

The similar effect takes place with respect to $L^1(-\infty, \infty)$-norm.

PROBLEM 8 (Ber). Let $f(z)$ be an entire function of exponential type $\sigma \leq \pi$, which satisfies condition (13) and is such that $\sum_{k=-\infty}^{\infty} |f(k)| < \infty$. Then

$$\int_{-\infty}^{\infty} |f(x+1) + f(x)| \, dx \leq C \sum_{k=-\infty}^{\infty} |f(k)| .$$

An operator which transforms each entire function $f(z)$ of exponential type $\sigma \leq \pi$, such that condition (13) is satisfied and $\{f(k)\} \in l^\infty$ (or $\{f(k)\} \in l^1$) into a function from the space B_π (respectively L_π), is called the interference operator. Such operators (not necessarily of the difference form) were studied in Boas [**17**], Akhiezer [**4,** pp. 207–211], Levin and Din Than Hoa [**83**], Ber [**10**]. In particular, the general form of an interference operator commuting with translation by an integer can be described.

[20] The best constant $C = 8/\pi$ was found by S. N. Bernstein.

LECTURE 22

Sine-Type Functions

22.1. Interpolation with nodes at the zeros of a sine-type function

Theorems on interpolation with integer nodes can be extended to more general sets of nodes. Starting with an evident observation that the integers are the zeros of the function $\sin \pi z$, we shall consider the interpolation problem with the nodes at zero sets of a more general class of functions which we now define.

DEFINITION. An entire function $F(z)$ is called a *sine-type function* with the width of indicator diagram 2σ if

a)

$$h_F\left(\frac{\pi}{2}\right) = h_F\left(-\frac{\pi}{2}\right) = \sigma\,; \tag{1}$$

b) all zeros λ_k of $F(z)$ are simple and satisfy the separation condition

$$\inf_{k\neq n} |\lambda_k - \lambda_n| = 2\delta > 0\,; \tag{2}$$

c)

$$\sup_k\{|\operatorname{Im}\lambda_k|\} = H < \infty\,; \tag{3}$$

d) there exist constants h, c, C, such that

$$0 < c < |F(x+ih)| < C < \infty\,, \qquad -\infty < x < \infty\,. \tag{4}$$

The zeros of a sine-type function lie in a horizontal strip. We enumerate them in increasing order of their real parts; i.e., $\operatorname{Re}\lambda_k \le \operatorname{Re}\lambda_{k+1}$, $k = 0, \pm 1, \pm 2, \dots$, and assume these relations to be fulfilled unless otherwise stated. Before investigating interpolation problems we prove some auxiliary statements.

LEMMA 1. *For any $\eta > 0$ there exists $m_\eta > 0$ such that*

$$|F(z)| > m_\eta e^{\sigma|\operatorname{Im} z|} \tag{5}$$

if $\operatorname{dist}(z, \{\lambda_k\}) > \eta$.

PROOF. Let $H_1 > \max\{H, |h|\}$. First, we shall prove (5) under the additional restriction $|\operatorname{Im} z| < H_1$. Relation (4) together with the Phragmén-Lindelöf theorem implies that $F(z)$ is bounded in the strip $\Pi = \{z : |\operatorname{Im} z| \le 2H_1\}$. Therefore, $\{F(z+\tau) : -\infty < \tau < \infty\}$ is a normal family of functions. If relation (5) fails in the strip $\Pi_1 = \{z : |\operatorname{Im} z| < H_1\}$, then there exists a sequence $\{z_j\} \subset \Pi_1$, $z_j = x_j + iy_j$, such that $F(z_j) \to 0$, $j \to \infty$, and $\operatorname{dist}(z_j, \{\lambda_k\}) > \eta$, $j = 1, 2 \dots$. Introducing, if needed, a subsequence of $\{z_j\}$ we can assume that the functions

$F(z+x_j)$ converge to a function $F_0(z)$ uniformly on each compact set in Π, and that there exists the limit $\lim_{j\to\infty} y_j = y_0 \in [-H_1, H_1]$. From the left-hand side inequality in (4) we have $|F_0(x+ih)| \geq c$, $-\infty < x < \infty$, and, hence, $F_0(z)$ is not identically zero. On the other hand,

$$F_0(iy_0) = \lim_{j\to\infty} F_j(iy_j) = \lim_{j\to\infty} F(x_j + iy_j) = 0 .$$

It follows from Hurwitz's theorem that there exists a number j_η such that, for $j > j_\eta$, all the functions $F_j(z) = F(z+x_j)$ have at least one zero in the disk $\{z : |z - iy_0| < \eta/2\}$, and hence, the function $F(z)$ has a zero in the disk $\{z : |z - z_j| < \eta\}$ contradicting the conditions $\operatorname{dist}(z_j, \{\lambda_k\}) > \eta$.

To obtain estimate (5) in the half-plane $\{z : \operatorname{Im} z > H_1\}$ we may use the representation

$$\log|F(x+iy)| = \frac{y-H_1}{\pi} \int_{-\infty}^{\infty} \frac{\log|F(t+iH_1)|}{|t-z+iH_1|^2}\, dt + (y-H_1) h_F\left(\frac{\pi}{2}\right) ,$$

and the estimate $|F(t+iH_1)| \geq m_\eta e^{\sigma H_1}$, which has just been proved. The estimate in the half-plane $\{z : \operatorname{Im} z < -H_1\}$ is derived in the similar way. This completes the proof of the lemma.

REMARK. By the Phragmén-Lindelöf theorem the inequality which is opposite to (5) with another constant instead of m_η, holds as well. Therefore, we have

$$(6) \qquad 0 < c < |F(z)| e^{-\sigma|\operatorname{Im} z|} < C < \infty , \qquad \operatorname{dist}(z, \{\lambda_k\}) > \eta ,$$

with appropriate constants c and C. This inequality is equivalent to the definition of the sine-type functions with simple zeros satisfying (2).

LEMMA 2. *For every sine-type function F there exist constants N_1, N_2 such that*

$$(7) \qquad 0 < N_1 < |F'(\lambda_k)| < N_2 < \infty , \qquad k = 0, \pm 1, \pm 2, \dots .$$

PROOF. Let δ be the number in the separation condition (2) from the definition of sine-type function. According to the previous lemma, we have

$$0 < m_\delta \leq \min_{|z-\lambda_k|=\delta} \left| \frac{F(z)}{z-\lambda_k} \right| \leq \max_{|z-\lambda_k|=\delta} \left| \frac{F(z)}{z-\lambda_k} \right| \leq M_\delta < \infty ,$$

with some m_δ, M_δ. Inequalities (7) now follow from the Minimum and Maximum Principles for analytic functions.

The following property of sine-type functions will be used in Lecture 23.

LEMMA 3. *The zeros of a sine-type function satisfy the relation*

$$(8) \qquad \sup_k \{|\lambda_k - \lambda_{k+1}|\} < \infty .$$

PROOF. Let the contrary be true. Since all zeros λ_k lie in the strip $\{z : |\operatorname{Im} z| \le H\}$, there exists a sequence $\{x_k\}$ such that the vertical strip $\{z : |\operatorname{Re} z - x_k| \le k\}$ is free from zeros of the function $F(z)$. Taking a subsequence of $\{\operatorname{Re} x_k\}$, if needed, we can assume (as in the proof of Lemma 1) that the sequence of functions $F(z + x_k)$ converges to a nonzero function $F_0(z)$ uniformly on each compact set in $\mathbb{C}$. According to the choice of the points $\{x_k\}$ this function has no zero in $\mathbb{C}$, and hence, $F_0(z) = e^{f(z)}$ for some entire function $f(z)$. In view of relation (6) we have

$$(9) \qquad 0 < ce^{\sigma|\operatorname{Im} z|} < |F_0(z)| < Ce^{\sigma|\operatorname{Im} z|} < \infty\,, \qquad z \in \mathbb{C}\,.$$

Therefore, $F_0(z)$ is an EFET and $f(z) = \alpha z + \beta$ for some $\alpha\,, \beta \in \mathbb{C}$. In this case, the indicator diagram of the function $F_0(z)$ coincides with the point $\overline{\alpha}$. On the other hand, according to (9), it must be the segment $[-i\sigma, i\sigma]$. This contradiction proves the lemma.

A parametric description of the class of sine-type functions with real zeros in terms of special conformal mappings was obtained recently by Eremenko and Sodin [**33**].

The following statement shows that the zero sets of sine-type functions have the same interpolation properties as the set of integers.

THEOREM 1. *Let $F(z)$ be a sine-type function with indicator diagram of width 2σ, and let $\{\lambda_k\}_{k=-\infty}^{\infty}$ be its zero set. Then the mapping*

$$(10) \qquad \{c_k\}_{k=-\infty}^{\infty} \mapsto f(z) = \sum_{k=-\infty}^{\infty} c_k \frac{F(z)}{F'(\lambda_k)(z - \lambda_k)}$$

is an isomorphism between l^p and L^p_σ for each $p \in (1, \infty)$. The series on the right-hand side of (10) *converges in the $L^p(-\infty, \infty)$-norm. The inverse mapping is defined by the relation*

$$(11) \qquad f \mapsto \{f(\lambda_k)\}_{k=-\infty}^{\infty}\,.$$

PROOF. Let $f \in L^p_\sigma$. Property 4 of the space L^p_σ (see Section 20.1) gives

$$\sum_{k=-\infty}^{\infty} |f(\lambda_k)|^p \le C\|f\|^p_{L^p(-\infty,\infty)}\,,$$

and hence (11) defines a continuous mapping from L^p_σ into l^p.

Now, we prove that for an arbitrary sequence $\{c_k\}_{-\infty}^{\infty} \in l^p$ the series (10) converges in the space L^p_σ and

$$\|f\|_{L^p(-\infty,\infty)} \le \text{const}\ \|\{c_k\}\|_{l^p}\,.$$

Let, for the sake of simplicity, all zeros $\{\lambda_k\}$ belong to a horizontal strip $\{z : 0 < \eta < \operatorname{Im} z < H < \infty\}$. Otherwise, we could use the same reasoning for the function $F(z + 2iH)$. From (6) we obtain $0 < c < |F(x)| < C < \infty$, $-\infty < x < \infty$. To prove the L^p-convergence of the series

$$(12) \qquad \varphi(z) = \sum_{k=-\infty}^{\infty} \frac{c_k}{F'(\lambda_k)(z - \lambda_k)}$$

we note that each partial sum

$$\varphi_{m,n}(z) = \sum_{k=m}^{n} \frac{c_k}{F'(\lambda_k)(z-\lambda_k)}$$

belongs to the Hardy space H^p_-. Much as in the proof of the Plancherel-Pólya theorem (Theorem 3, Lecture 20), its L^p-norm can be found using the relation

$$\|\varphi_{m,n}\|_{H^p_-} \leq \text{const} \cdot \sup\left\{\left|\int_{-\infty}^{\infty} \varphi_{m,n}(x)\psi(x)\,dx\right| : \psi \in H^q_+,\ \|\psi\|_{H^q_+} \leq 1\right\},$$

with $q = \dfrac{p}{p-1}$. Applying the residue theorem to evaluate the integral, we obtain

$$\|\varphi_{m,n}\|_{H^p_-} \leq \text{const}\sup\left\{\left|\sum_{k=m}^{n} \frac{c_k}{F'(\lambda_k)}\psi(\lambda_k)\right| : \psi \in H^q_+, \|\psi\|_{H^q_+} \leq 1\right\}.$$

Now, the Hölder inequality together with relations (7) and property 6 of Hardy spaces (see Section 19.1) yields

$$\begin{aligned}
&\|\varphi_{m,n}\|_{H^p_-} \\
&\quad \leq \text{const}\sup\left\{\Big(\sum_{k=m}^{n}|c_k|^p\Big)^{1/p}\Big(\sum_{k=m}^{n}|\psi(\lambda_k)|^q\Big)^{1/q} : \psi \in H^q_+, \|\psi\|_{H^q_+} \leq 1\right\} \\
&\quad \leq \text{const}\Big(\sum_{k=m}^{n}|c_k|^p\Big)^{1/p}
\end{aligned}$$

proving the convergence of the series. Passing to the limit as $m \to -\infty$, $n \to \infty$, we obtain the bound for the norm of the series. Theorem is proved.

Dzhrbashyan and Rafaelyan [**29**] considered the class EFET satisfying the estimate

$$0 < c < (1+|z|)^{\kappa}|F(z)|e^{-\sigma|\operatorname{Im} z|} < C < \infty, \qquad \operatorname{dist}(z, \{\lambda_k\}) > \eta$$

for some κ. In the case $|\kappa| < 1/2$ they obtained results similar to Theorem 1.

22.2. Functions whose zeros are close to the integers

In this section we study the following problem. Given a bounded sequence $\{d_k\}_{k=-\infty}^{\infty}$ of complex numbers, consider the function

$$f(z) = \lim_{R\to\infty} \prod_{|k+d_k|\leq R} \left(1 - \frac{z}{k+d_k}\right) \tag{13}$$

which vanishes at the points $k+d_k$. (If $k+d_k = 0$ for some k, we replace the corresponding factor in (13) by z.) For $d_k = 0$, $k = 0, \pm1, \pm2, \dots$, we have $f(z) = \dfrac{1}{\pi}\sin\pi z$. In the general case $f(z)$ may be treated as a perturbation of $\dfrac{1}{\pi}\sin\pi z$. What conditions on $\{d_k\}$ ensure $f(z)$ to be a sine-type function?

Let $a_0 = \sup_k\{|\operatorname{Im} d_k|\}$ and let the branch of $\arg(k + d_k - z)$ in the half-plane $\{z : \operatorname{Im} z > a_0\}$ be such that $\arg(k + d_k - z) \in (-\pi, 0)$. Assume also that $\arg(k + d_k) \in (-3\pi/2, \pi/2]$. By this choice of arguments we have

$$\arg\left(1 - \frac{z}{k + d_k}\right) = \arg(k+d_k - z) - \arg(k+d_k) \to 0\,, \qquad k \to \pm\infty\,, \quad \operatorname{Im} z > a_0\,.$$

Let us now define the function

$$\theta(z) = \log\{f(z)e^{i\pi \operatorname{Re} z}\}\,, \qquad \operatorname{Im} z > a_0\,. \tag{14}$$

Its behavior along a horizontal line may be studied using the functionals $\{L_\tau\}$ constructed in Section 21.1:

$$L_\tau : \{c_k\} \mapsto \sum_{k=-\infty}^{\infty} (d_{k+\tau} - d_k)\frac{k}{k^2+1}\,, \qquad c_k = (-1)^k d_k\,, \quad \tau \in \mathbb{Z}\,. \tag{15}$$

THEOREM 2. *The function $\theta(x + ia)$ is bounded with respect to x for all $a > a_0$ if and only if*

$$|L_\tau(\{(-1)^k d_k\})| < M\,, \qquad \tau = 0, \pm1, \pm2, \dots\,, \tag{16}$$

for some $M > 0$.

PROOF. We have

$$\begin{aligned}
\theta(x+\tau+ia) - \theta(x+ia) &= \log\left\{\frac{f(x+\tau+ia)}{f(x+ia)}e^{i\pi\tau}\right\} \\
&= \lim_{N\to\infty}\left\{\log\prod_{-N}^{N}\frac{k+d_k-\tau-x-ia}{k+d_k-x-ia} + i\pi\tau\right\} \\
&= \lim_{N\to\infty}\log\prod_{-N}^{N-\tau}\frac{k+d_{k+\tau}-x-ia}{k+d_k-x-ia} \\
&\quad + \lim_{N\to\infty}\left\{\log\prod_{l=1}^{\tau}\frac{-N-l+d_{-N-l+\tau}-x-ia}{N-\tau+l+d_{N-\tau+l}-x-ia} + i\pi\tau\right\}.
\end{aligned}$$

Each factor in the last product converges to $e^{-i\pi}$ as $N \to \infty$. Therefore, the product of these factors converges to $-i\pi\tau$ and the last term vanishes as $N \to \infty$. Hence,

$$\theta(x+\tau+ia) - \theta(x+ia) = \lim_{N\to\infty}\sum_{-N}^{N}\log\left(1 + \frac{d_{k+\tau}-d_k}{k+d_k-x-ia}\right). \tag{17}$$

The function $\theta(x + ia)$ is bounded for $-\infty < x < \infty$ if and only if the differences $\theta(x + \tau + ia) - \theta(x + ia)$ are bounded for $0 \le x \le 1$ uniformly with respect to $\tau = 0, \pm1, \pm2, \dots$. With a being sufficiently large, we can assume that

$$|(d_{k+\tau} - d_k)/(k + d_k - x - ia)| < \frac{1}{2}, \quad 0 \le x \le 1.$$

Since $|\log(1+\alpha)-\alpha|<|\alpha|^2$ for $|\alpha|<1/2$, we conclude that the uniform boundedness of $\theta(x+\tau+ia)-\theta(x+ia)$ with respect to τ holds if and only if the sum

$$\sum_{k=-\infty}^{\infty}\frac{d_{k+\tau}-d_k}{k+d_k-x-ia} \tag{18}$$

is bounded in x and τ. This in turn is equivalent to (16), which completes the proof of the theorem.

The above theorem combined with Theorem 2, Section 21.1, and with the definition of a sine-type function yields the following statement.

COROLLARY. *Let $\{d_k\}$ be a bounded sequence and let the function $f(z)$ be defined by* (13). *For $f(z)$ to be a sine-type function it is necessary and sufficient that*

1) $\inf_{k\neq l}|k+d_k-l-d_l|>0$;

2) *there exist a function $g\in B_\pi$ which solves the interpolation problem $g(k)=(-1)^k d_k$, $k=0,\pm1,\pm2,\dots$.*

For other applications of the functionals L_τ for describing zeros of some classes of entire functions of exponential type, see Levin [**82,** Appendix VI], Kheifits [**69**].

LECTURE 23

Riesz Bases Formed by Exponential Functions in $L^2(-\pi,\pi)$

In Lecture 18 we studied the completeness and minimality of exponential systems in the space $L^2(-\pi,\pi)$. Here, we turn to a more delicate property of an exponential system to be a Riesz base in $L^2(-\pi,\pi)$; this requires more advanced notions from functional analysis.

23.1. Definition and properties of Riesz bases

First, we remind the reader the definition of a Riesz base in a Hilbert space.

DEFINITION A. A system of elements $\{x_n\}$ in a Hilbert space H is called a *Riesz base* if every element $x \in H$ can be represented in the form $x = \sum c_n(x)x_n$ and there exist positive numbers c and C such that

$$c\|x\|^2 \le \sum |c_n(x)|^2 \le C\|x\|^2, \qquad x \in H.$$

Each Riesz base is a complete and minimal system in H but not vice versa. For a detailed exposition of the properties of Riesz bases, see, for example, Gohberg and Krein [**40**].

Using the notion of Riesz base, we may formulate Theorem 1 of the previous lecture as follows.

THEOREM 1. *Let $F(z)$ be a sine-type function, let the width of its indicator diagram be 2π, and let $\{\lambda_n\}$ be its zero set. Then the system of functions*

$$\left\{\frac{F(z)}{F'(\lambda_n)(z-\lambda_n)}\right\}_{n=-\infty}^{\infty} \tag{1}$$

is a Riesz base in L^2_π.

Let $\{\varphi_n\}$ be an orthonormal base in H. It follows from the definition that *a system $\{x_n\} \subset H$ is a Riesz base if and only if the operator*

$$B : \sum c_n\varphi_n \mapsto \sum c_n x_n$$

is bounded and invertible. In other words, a system of elements of a Hilbert space is a Riesz base if and only if it is the image of an orthonormal base under the action of an invertible bounded operator. Since its biorthogonal system is the image of the same orthonormal base under the action of the inverse of the dual operator, the latter system also forms a Riesz base.

Let $\chi_n(t) \in L^2(-\pi,\pi)$ be the functions representing the functions

$$F_n(z) = \frac{F(z)}{F'(\lambda_n)(z-\lambda_n)}$$

according to Paley-Wiener theorem. This means that

$$F_n(z) = \int_{-\pi}^{\pi} e^{itz}\chi_n(t)\,dt\,. \tag{2}$$

Then

$$\int_{-\pi}^{\pi} e^{it\lambda_k}\chi_n(t)dt = F_n(\lambda_k) = \delta_{kn},$$

which means that $\{\chi_n(t)\}_{n=-\infty}^{\infty}$ is biorthogonal in $L^2(-\pi,\pi)$ to the system $\{e^{i\lambda_n t}\}_{n=-\infty}^{\infty}$. Since the Fourier transform is an isomorphism between the spaces L^2_π and $L^2(-\pi,\pi)$ we conclude that the system $\{\chi_n(t)\}_{n=-\infty}^{\infty}$ is a Riesz base in $L^2(-\pi,\pi)$. We have arrived at the following theorem:

THEOREM 2. *Let $\{\lambda_n\}_{n=-\infty}^{\infty}$ be the sequence of zeros of a sine-type function with indicator diagram of width 2π. Then the system $\{e^{i\lambda_n t}\}_{n=-\infty}^{\infty}$ is a Riesz base in $L^2(-\pi,\pi)$.*

REMARK. Examining the asymptotic relations

$$||e^{i\lambda_n t}||^2 = \int_{-\pi}^{\pi} |e^{i\lambda_n t}|^2 dt \asymp \frac{1}{1+|\operatorname{Im}\lambda_n|},$$

$$||e^{i\lambda_n t} - e^{i\lambda_k t}||^2 = \int_{-\pi}^{\pi} |e^{i\lambda_n t}(1-e^{i(\lambda_k-\lambda_n)t})|^2 dt \asymp |\lambda_n - \lambda_k|\,,$$

valid for $|n|, |k| \to \infty$, we readily see that conditions (2) and (3) from Lecture 22 are necessary for the system $\{e^{i\lambda_n t}\}_{n=-\infty}^{\infty}$ to form a Riesz base in $L^2(-\pi,\pi)$.

Let us now consider the *problem of stability of exponential bases*, which goes back to Paley and Wiener [**109,** Chapter VII]. First, we shall prove the following auxiliary statement.

LEMMA 1. *Let $\{x_k\}$ be a Riesz base in a Hilbert space H. Then there exists $\alpha > 0$ such that each system $\{y_k\} \subset H$ satisfying*

$$\|\sum c_k(x_k - y_k)\|^2 \le \alpha^2 \sum |c_k|^2 \tag{3}$$

for any finite set $\{c_k\} \subset \mathbb{C}$, is also a Riesz base in H.

PROOF. Since $\{x_k\}$ is a Riesz base, there exists an invertible operator $B: H \to H$ and a complete orthonormal system $\{\varphi_k\}$ such that $x_k = B\varphi_k$. We define the operator $C: H \to H$ setting $C\varphi_k = y_k - x_k$. It follows from (3) that $\|C\| \le \alpha$. The operator $A = B + C$ transforms φ_k into y_k and is invertible if $\alpha < \|B^{-1}\|^{-1}$. For all such α the system $\{y_k\}$ is a Riesz base in H.

THEOREM 3 (Golovin). *Let the system $\{e^{i\lambda_k t}\}_{k=-\infty}^{\infty}$ be a Riesz base in $L^2(-\pi,\pi)$. Then there exists $\varepsilon > 0$ such that the system $\{e^{i\mu_k t}\}_{k=-\infty}^{\infty}$ is a Riesz base in $L^2(-\pi,\pi)$ for each sequence $\{\mu_k\}_{k=-\infty}^{\infty}$ satisfying*

$$|\lambda_k - \mu_k| \le \varepsilon\,, \qquad k = 0, \pm 1, \pm 2, \dots\,. \tag{4}$$

PROOF. Since the set $\{e^{i\lambda_k t}\}_{k=-\infty}^{\infty}$ is a Riesz base in $L^2(-\pi,\pi)$, the sequence $\{\lambda_k\}_{k=-\infty}^{\infty}$ satisfies conditions (2), (3) from Section 22.1 with some $\delta > 0$. If $\{\mu_k\}_{k=-\infty}^{\infty}$ satisfies (4), then for any sequence $\{c_k\}_{k=-\infty}^{\infty}$ with a finite number of nonzero elements, using the Paley-Wiener theorem, we obtain

$$\begin{aligned}
&\Big\| \sum_{k=-\infty}^{\infty} c_k(e^{i\lambda_k t} - e^{i\mu_k t})\Big\|_{L^2(-\pi,\pi)} \\
&= \sup\Big\{\Big|\int_{-\pi}^{\pi} \varphi(t) \sum_{k=-\infty}^{\infty} c_k(e^{i\lambda_k t} - e^{i\mu_k t})\,dt\Big| : \|\varphi\|_{L^2(-\pi,\pi)} = 1\Big\} \qquad (5)\\
&= \sup\Big\{\Big|\sum_{k=-\infty}^{\infty} c_k\big(f(\lambda_k) - f(\mu_k)\big)\Big| : \|f\|_{L^2_\pi} = 1\Big\}.
\end{aligned}$$

For each $k \in \mathbb{Z}$ we have

$$|f(\lambda_k) - f(\mu_k)| \le \varepsilon \max_{\zeta\in[\lambda_k,\mu_k]} |f'(\zeta)| = \varepsilon|f'(\zeta_k)| \tag{6}$$

where ζ_k is a point of the segment $[\lambda_k, \mu_k]$ depending on f. If $\varepsilon < \delta/2$, then $|\zeta_k - \zeta_j| \ge \delta$ for $j \ne k$. Since $f \in L^2_\pi$ and $\|f\|_{L^2_\pi} \le 1$, the Paley-Wiener theorem implies $f' \in L^2_\pi$ and $\|f'\|_{L^2_\pi} \le \pi$. Then by property 4 of the space L^p_σ (see Section 20.1) we have

$$\sum_{k=-\infty}^{\infty} |f'(\zeta_k)|^2 \le C^2\,, \tag{7}$$

where C depends on H and δ only. Coming back to (5), and taking into account (6) and (7), we obtain

$$\begin{aligned}
&\Big\| \sum_{k=-\infty}^{\infty} c_k(e^{i\lambda_k t} - e^{i\mu_k t})\Big\| \\
&\le \varepsilon\Big(\sum_{k=-\infty}^{\infty} |c_k|^2\Big)^{1/2} \sup\Big\{\Big(\sum_{k=-\infty}^{\infty} |f'(\zeta_k)|^2\Big)^{1/2} : \|f\|_{L^2_\pi} \le 1\Big\} \\
&\le \varepsilon C\Big(\sum_{k=-\infty}^{\infty} |c_k|^2\Big)^{1/2}.
\end{aligned}$$

To complete the proof it is sufficient to apply Lemma 1.

Now we introduce another equivalent definition of a Riesz base.

DEFINITION B. A system $\{x_n\}$ in a Hilbert space H is called a Riesz base if it is minimal and there exist numbers c and C such that

$$c\|x\|^2 \le \sum |(x, x_n)|^2 \le C\|x\|^2\,, \qquad x \in H\,. \tag{8}$$

It follows immediately from (8) that the system $\{x_n\}$ is complete in H. For a proof of the equivalence of the two definitions we refer the readers to Gohberg and Krein [**40**] or to Nikol'skii [**114**, Lecture VI].

Let $H = L^2(-\pi,\pi)$ and let $\{x_n\} = \{e^{i\lambda_n t}\}_{n=-\infty}^{\infty}$. By the Paley-Wiener theorem condition (8) may be written in the form

$$c\int_{-\infty}^{\infty} |f(\zeta)|^2\,d\zeta \le \sum_{n=-\infty}^{\infty} |f(\lambda_n)|^2 \le C\int_{-\infty}^{\infty} |f(\zeta)|^2\,d\zeta\,, \qquad f\in L^2_\pi\,. \tag{9}$$

Combining (9) with the condition of minimality of an exponential system given in Lecture 18, we obtain the following statement.

LEMMA 2. *For a system of exponential functions $\{e^{i\lambda_n t}\}_{n=-\infty}^{\infty}$ to form a Riesz base in $L^2(-\pi,\pi)$ it is necessary and sufficient that (9) hold and there exist an entire function $\varphi(\lambda)$ of exponential type π which vanishes on the set $\{\lambda_n\}_{n=-\infty}^{\infty}$ and such that $\varphi(\lambda)(1+|\lambda|)^{-1}\in L^2(-\infty,\infty)$.*

23.2. The 1/4-theorem

The following problem was formulated by Paley and Wiener [**109**, Chapter VII]. *Describe all numbers $d>0$ such that every functional system $\{e^{i\mu_n t}\}_{n=-\infty}^{\infty}$ with a sequence $\{\mu_n\}_{n=-\infty}^{\infty}$ satisfying the inequalities*

$$|\mu_n - n| < d\,, \qquad n = 0,\pm1,\pm2,\dots$$

is a Riesz base in $L^2(-\pi,\pi)$.

Paley and Wiener proved that for $d<1/\pi^2$ the system $\{e^{i\mu_n t}\}_{n=-\infty}^{\infty}$ is a Riesz base. A bit later, Ingham showed that for $d=1/4$ the property to be a Riesz base may fail. We will present Ingham's example at the end of this section. Several authors improved the value of the Paley-Wiener constant d before 1964 when the complete answer was obtained by M. I. Kadets who proved that any $d<1/4$ fits. Here, we prove a more general theorem unifying both Theorem 2 and the Kadets theorem.

THEOREM 4 (V. Katsnelson). *Let a real sequence $\{\lambda_n\}_{n=-\infty}^{\infty}$ be the zero set of a sine-type function $F(z)$ with the width of indicator diagram equal to 2π. Let, for some $d\in(0,1/4)$, a sequence of real numbers $\{\delta_n\}_{n=-\infty}^{\infty}$ satisfy the conditions*

$$|\delta_n| < d\rho_n\,, \qquad n=0,\pm1,\pm2,\dots\,,$$

where $\rho_n = \inf_{k\ne n}|\lambda_k-\lambda_n|$. Then the system of exponential functions

$$\{e^{i(\lambda_n+\delta_n)t}\}_{n=-\infty}^{\infty}$$

forms a Riesz base in $L^2(-\pi,\pi)$.

We split the proof into several steps,[21] assuming during the proof that $F(0)=1$.

LEMMA 3. *Let $G(\lambda)$ be a sine-type function, let 2σ be the width of its indicator diagram, and let $\{\lambda_n\}_{n=-\infty}^{\infty}$ be its zero set. Then*

$$\sum_{n=-\infty}^{\infty}\frac{g(\lambda_n)}{G'(\lambda_n)} = 0 \qquad \text{for all } g\in L_\sigma\,. \tag{10}$$

[21] We follow the original paper Katsnelson [**65**] where the case of complex λ_n is treated.

PROOF. Since G is a sine-type function and $g \in L_\sigma$, we have

$$\sum_{n=-\infty}^{\infty} \frac{|g(\lambda_n)|}{|G'(\lambda_n)|} \le C\|g\|_{L_\sigma}$$

and hence the left-hand side of (10) defines a linear functional on L_σ. Therefore, it suffices to verify (10) for functions g from a dense set in L_σ, for example for all g's satisfying $g(z) = o(|z|^{-1})e^{\pi|\operatorname{Im} z|}$, $z \to \infty$. Let g belong to this set. According to the lower estimate (5) for sine-type functions proved in Lemma 1, Section 22.1 we have

$$\left|\frac{g(z)}{G(z)}\right| = o(|z|^{-1}), \qquad z \to \infty, \qquad \operatorname{dist}(z, \lambda_n) \ge \eta > 0. \tag{11}$$

Hence the sum of the residues of the meromorphic function $g(z)/G(z)$ vanishes, which proves (10).

Let us assume now that the points $\{\lambda_n\}_{n=-\infty}^{\infty}$ are enumerated in such a way that $\lambda_n < \lambda_{n+1}$ and define the sequence $\{\widehat{\lambda}_n\}_{n=-\infty}^{\infty}$:

$$\widehat{\lambda}_n = \frac{1}{2}(\lambda_n + \lambda_{n+1}), \qquad n = 0, \pm 1, \dots .$$

Let

$$\rho_n = \inf_{j\ne n} |\lambda_n - \lambda_j| = \min(\lambda_n - \lambda_{n-1}, \lambda_{n+1} - \lambda_n),$$

and let $\{\delta_n\}_{n=-\infty}^{\infty}$ be a real sequence which satisfies $|\delta_n| \le d\rho_n$ with some $d \in (0, 1/4)$. For $\theta \in [0,1]$ we set $\lambda_n'(\theta) = \lambda_n + \theta\delta_n$, and $\lambda_n''(\theta) = \widehat{\lambda}_n - \theta\delta_n$. If $\inf_{n\ne k} |\lambda_n - \lambda_k| = 2\delta > 0$, then

$$|\lambda_n'(\theta) - \lambda_n''(\theta)| = \left|\frac{\lambda_n - \lambda_{n+1}}{2} + 2\theta\delta_n\right| \ge \frac{1}{2}\rho_n - 2\delta_n \ge (1 - 4d)\delta > 0, \tag{12}$$

and each set $\Lambda_\theta = \{\lambda_n'(\theta)\}_{n=-\infty}^{\infty} \cup \{\lambda_n''(\theta)\}_{n=-\infty}^{\infty}$ is separated (this is the place where the condition $d \in (0, 1/4)$ works!). Moreover, we have in addition that $\inf_n \rho_n > 0$, and there exists a number $\delta > 0$ such that

$$\inf_{n,k} |\lambda_n'(\theta) - \lambda_k''(\theta)| > \delta$$

for all $\theta \in [0,1]$. Hence the set $\Lambda = \bigcup_{\theta\in[0,1]} \Lambda_\theta$ does not cover the entire real line, and without loss of generality we can assume that $\lambda_n'(\theta) \ne 0, \lambda_n''(\theta) \ne 0$ for all $n \in \mathbb{Z}$ and $\theta \in [0,1]$: otherwise we replace $\{\lambda_n\}_{n=-\infty}^{\infty}$ by a set $\{\lambda_n + a\}_{n=-\infty}^{\infty}$ with a suitable $a \in \mathbb{R}$, and note that the systems $\{e^{i\lambda_n t}\}_{n=-\infty}^{\infty}$ and $\{e^{i(\lambda_n + a)t}\}_{n=-\infty}^{\infty}$ are or are not the Riesz bases in $L^2(0, \pi)$ simultaneously.

LEMMA 4. *The product*

$$\widehat{F}(z) = \text{P.V.} \prod (1 - z/\widehat{\lambda}_n) \tag{13}$$

converges uniformly on each compact set in $\mathbb{C}$ and defines a function $\widehat{F}$ such that

$$0 < c \le |\widehat{F}(t + iy)e^{-\pi|y|}| \le C < \infty, \quad |y| > 1, \tag{14}$$

with some positive numbers c and C independent of τ.

PROOF. To prove that the product in (13) converges and that (14) holds we note that

$$\frac{\left(1-\frac{z}{\lambda_n}\right)\left(1-\frac{z}{\lambda_{n+1}}\right)}{\left(1-\frac{z}{\widehat{\lambda}_n}\right)^2}=\frac{(\lambda_n-z)(\lambda_{n+1}-z)}{(\widehat{\lambda}_n-z)^2}\,\frac{\widehat{\lambda}_n^2}{\lambda_n\lambda_{n+1}}.$$

Thus

$$\frac{\left(1-\frac{z}{\lambda_n}\right)\left(1-\frac{z}{\lambda_{n+1}}\right)}{\left(1-\frac{z}{\widehat{\lambda}_n}\right)^2}=\left\{1-\frac{(\lambda_{n+1}-\lambda_n)^2}{4(\widehat{\lambda}_n-z)^2}\right\}\left\{1-\frac{(\lambda_{n+1}-\lambda_n)^2}{4\widehat{\lambda}_n^2}\right\}^{-1}.$$

According to Lemma 3, Section 22.1, we have

$$\sup_n|\lambda_{n+1}-\lambda_n|<\infty\,, \tag{15}$$

while the separation condition (2), Section 22.1 yields that there exists a number N such that each interval $I_k=[k,k+1]$ contains at most N points of the set $\{\lambda_n\}_{n=-\infty}^{\infty}$. Therefore, there is no more than $N+2$ points of the set $\{\widehat{\lambda}_n\}_{n=-\infty}^{\infty}$ inside I_k. If $\operatorname{Re} z=x, \operatorname{Im} z=\tau, |\tau|>1$, then

$$\begin{aligned}\sup_x\sum_{n=-\infty}^{\infty}\frac{1}{|\widehat{\lambda}_n-z|^2}&\le\sup_x\sum_{k=-\infty}^{\infty}\sum_{\widehat{\lambda}_n\in I_k}\frac{1}{|\widehat{\lambda}_n-x|^2+\tau^2}\\&\le 2(N+2)\sum_{k=-\infty}^{\infty}\frac{1}{k^2+\tau^2}<\infty\,.\end{aligned}\tag{16}$$

Using (15) and (16), we find that the product

$$\prod_{n=-\infty}^{\infty}\frac{\left(1-\frac{z}{\lambda_n}\right)\left(1-\frac{z}{\lambda_{n+1}}\right)}{\left(1-\frac{z}{\widehat{\lambda}_n}\right)^2}$$

converges and defines an analytic function bounded in $\{z:|\operatorname{Im} z|\ge\tau\}$ for $|\tau|>1$. Moreover, for all sufficiently large $|\tau|$ the reciprocal of it is also analytic and bounded in $\{z:|\operatorname{Im} z|\ge\tau\}$.

Since $F(z)$ is a sine-type function, it belongs to the class C and hence has the representation

$$F(z)=\text{P.V.}\prod\left(1-\frac{z}{\lambda_n}\right).$$

Thus the product in (13) converges as well. Moreover, inequalities (6), Section 22.1 are valid, and applying (15) and (16) again, we obtain (14).

LEMMA 5. *Let the sequences $\{\lambda_n\}_{n=-\infty}^{\infty}$ and $\{\widehat{\lambda}_n\}_{n=-\infty}^{\infty}$ be as in the previous lemma, and let $\{\delta_n\}_{n=-\infty}^{\infty}$ be a real sequence which satisfies $|\delta_n|<d\inf_{k\neq n}|\lambda_n-\lambda_k|$ with some $d\in(0,1/4)$. Then, for each $\theta\in[0,1]$, the infinite product*

$$G_\theta(z)=\text{P.V.}\prod\left(1-\frac{z}{\lambda_n+\theta\delta_n}\right)\left(1-\frac{z}{\widehat{\lambda}_n-\theta\delta_n}\right)\tag{17}$$

converges uniformly on each compact set in $\mathbb{C}$ and defines an entire function of exponential type 2π such that

$$0 < c \le |G_\theta(x+iy)e^{-2\pi|y|}| \le C < \infty\,, \quad |y| > 1, \tag{18}$$

where c and C are positive numbers independent of θ, x and y.

Skipping the proof, which is similar to that of the previous lemma, we only note that it suffices to consider the relation

$$\frac{G_\theta(z)}{F(z)\widehat{F}(z)} = \prod_{n=-\infty}^{\infty} \frac{\left(1 - \dfrac{z}{\lambda_n + \theta\delta_n}\right)\left(1 - \dfrac{z}{\widehat{\lambda}_n - \theta\delta_n}\right)}{\left(1 - \dfrac{z}{\lambda_n}\right)\left(1 - \dfrac{z}{\widehat{\lambda}_n}\right)}$$

$$= \prod_{n=-\infty}^{\infty}\left(1 + \frac{(\lambda_{n+1}-\lambda_n)\theta\delta_n - 2\theta^2\delta_n^2}{2(\lambda_n - z)(\widehat{\lambda}_n - z)}\right)\cdot \prod_{n=-\infty}^{\infty}\left(1 + \frac{(\lambda_{n+1}-\lambda_n)\theta\delta_n - 2\theta^2\delta_n^2}{2\lambda_n\widehat{\lambda}_n}\right)^{-1}.$$

LEMMA 6. *Let $F(z)$ be a sine-type function with the width of indicator diagram equal to 2π and real zeros $\{\lambda_n\}_{n=-\infty}^{\infty}$. Let $\rho_n = \inf_{j\neq n}|\lambda_n - \lambda_j|$ and let $\{\delta_n\}_{n=-\infty}^{\infty}$ be a real sequence which satisfies*

$$|\delta_n| < d\rho_n \tag{19}$$

for some $d \in (0, 1/4)$. Then there exist positive numbers c and C such that for all $f \in L^2_\pi$ and $\theta \in [0,1]$ the following inequalities hold:

$$c\sum_{n=-\infty}^{\infty} |f(\lambda_n + \theta\delta_n)|^2 \le \int_{-\infty}^{\infty} |f(t)|^2\,dt \le C\sum_{n=-\infty}^{\infty} |f(\lambda_n + \theta\delta_n)|^2\,. \tag{20}$$

PROOF. It follows from (12) that, for each $\theta \in [0,1]$, the set Λ_θ is separated, which jointly with Lemma 5 proves that the function G_θ defined by (17) is a sine-type function with the width of indicator diagram 4π. Using (12) and (18) we prove, in addition to the statement of Theorem 1, Section 22.1 that there exist positive numbers c_1 and C_1 not depending on θ such that

$$c_1\int_{-\infty}^{\infty}|f(t)|^2\,dt \le \sum_{n=-\infty}^{\infty}|f(\lambda_n'(\theta))|^2 + \sum_{n=-\infty}^{\infty}|f(\lambda_n''(\theta))|^2 \le C_1\int_{-\infty}^{\infty}|f(t)|^2\,dt$$

for all $f \in L^2_\pi$. Therefore, to prove relation (20) it suffices to check that

$$c_2\sum_{n=-\infty}^{\infty}|f(\lambda_n''(\theta))|^2 \le \sum_{n=-\infty}^{\infty}|f(\lambda_n'(\theta))|^2 \le C_2\sum_{n=-\infty}^{\infty}|f(\lambda_n''(\theta))|^2, \quad f\in L^2_\pi \tag{21}$$

with c_2 and C_2 independent of θ and $f \in L^2_\pi$. Applying Lemma 3 to the functions $G_\theta(z)$ and $f(z)\overline{f(\overline{z})} \in L^2_\pi$, we obtain

$$\sum_{n=-\infty}^{\infty}\frac{|f(\lambda_n'(\theta))|^2}{G_\theta'(\lambda_n'(\theta))} = -\sum_{n=-\infty}^{\infty}\frac{|f(\lambda_n''(\theta))|^2}{G_\theta'(\lambda_n''(\theta))}\,, \qquad f \in L^2_\pi\,. \tag{22}$$

Since G_θ is a real function and the numbers $\lambda_n'(\theta)$ and $\lambda_n''(\theta)$ are interlacing, the signs of all numbers $G_\theta'(\lambda_n'(\theta))$ coincide and are opposite to those of $G_\theta'(\lambda_n''(\theta))$. Again using (12) we find that there exist positive numbers c_3 and C_3 such that

$$0 < c_3 \le |G_\theta'(\lambda_n'(\theta))| \le C_3 < \infty, \qquad 0 < c_3 \le |G_\theta'(\lambda_n''(\theta))| \le C_3 < \infty.$$

In view of (22) we have

$$c_3C_3^{-1}\sum_{n=-\infty}^{\infty}|f(\lambda_n'(\theta))|^2\le C_3c_3^{-1}\sum_{n=-\infty}^{\infty}|f(\lambda_n''(\theta))|^2$$

and (21) implies (20), proving Lemma 6.

To prove Theorem 4 we note that, according to Lemmas 2 and 6, the system $\{e^{i(\lambda_n+\delta_n)}\}_{n=-\infty}^{\infty}$ forms a Riesz basis in $L^2(-\pi,\pi)$ if the infinite product

$$F_1(z)=\text{P.V.}\prod\Big(1-\frac{z}{\lambda_n+\delta_n}\Big) \tag{23}$$

is convergent and if

$$\int_{-\infty}^{\infty}\frac{|F_1(t)|^2}{1+t^2}\,dt<\infty\,. \tag{24}$$

An equivalent form of the latter relation is

$$f_1(z)\stackrel{\text{def}}{=}\frac{F_1(z)}{z-\lambda_0-\delta_0}\in L^2_\pi\,.$$

To prove (24) let us consider the family of functions

$$F_\theta(z)=\text{P.V.}\prod_{n=-\infty}^{\infty}\Big(1-\frac{z}{\lambda_n+\theta\delta_n}\Big)\,,\qquad f_\theta(z)=\frac{F_\theta(z)}{z-\lambda_0-\theta\delta_0}\,,\qquad \theta\in[0,1]\,.$$

Since, for each $z\ne\lambda_n$,

$$\Big(1+\frac{\theta\delta_n}{\lambda_n-z}\Big)\Big(1+\frac{\theta\delta_n}{\lambda_n}\Big)^{-1}=1+O(\lambda_n^{-2})\,,\qquad n\to\infty\,,$$

the ratio

$$\frac{F_\theta(z)}{F_0(z)}=\prod_{n=-\infty}^{\infty}\frac{1-\dfrac{z}{\lambda_n+\delta_n\theta}}{1-\dfrac{z}{\lambda_n}}=\prod_{n=-\infty}^{\infty}\frac{1+\dfrac{\theta\delta_n}{\lambda_n-z}}{1+\dfrac{\theta\delta_n}{\lambda_n}}$$

converges for all $\theta\in[0,1]$ on every compact set of $\mathbb{C}$ not containing points of the set $\{\lambda\}_{n=-\infty}^{\infty}$, which proves that $F_\theta(z)$ is an entire function of exponential type. If $f_\theta\in L^2_\pi$, then, taking into account that $f_\theta(\lambda_n+\theta\delta_n)=0$ for $n\ne0$, the relation (20) yields

$$\int_{-\infty}^{\infty}|f_\theta(t)|^2\,dt\le C|f_\theta(\lambda_0+\theta\delta_0)|^2=C|F_\theta'(\lambda_0+\theta\delta_0)|^2\,. \tag{25}$$

For $k\ne0$ we have $|(\lambda_0+\theta\delta_0)-(\lambda_k+\theta\delta_k)|\ge2^{-1}\inf_{k\ne n}|\lambda_k-\lambda_n|=\delta>0$, and the right-hand side of (25) is bounded for $\theta\in[0,1]$. Since the family of entire functions $\{f_\theta(z)\}$ is continuous in θ with respect to convergence on compact sets of $\mathbb{C}$, estimate (25) and the Fatou lemma imply that the set $S=\{\theta:\theta\in[0,1];f_\theta\in L^2_\pi\}$ is closed. On the other hand, this set coincides with the set of all θ such that the set $\{e^{it(\lambda_n+\theta\delta_n)}\}_{k=-\infty}^{\infty}$ is a Riesz base in $L^2(-\pi,\pi)$. According to Theorem 3, this set is open in $[0,1]$. Since $0\in S$, we have $S\ne\varnothing$ and hence $S=[0,1]$. This proves (24) and Theorem 4.

The following result shows that the condition $d \in (0, 1/4)$ in Theorem 4 is the best possible.

THEOREM 5 (Ingham). *If $\lambda_n = n - \frac{1}{4}\operatorname{sgn} n$, $n \neq 0$, $\lambda_0 = 0$, then the system of exponential functions $\{e^{i\lambda_n t}\}_{n=-\infty}^{\infty}$ does not form a Riesz base in $L^2(-\pi, \pi)$.*

PROOF. We shall prove that the system $\{e^{i\lambda_n t}\}_{n=-\infty}^{\infty}$ is not minimal in the space $L^2(-\pi, \pi)$. To this end, we introduce the function

$$F(z) = z \prod_{n=1}^{\infty} \left(1 - \frac{z^2}{(n-1/4)^2}\right)$$

which vanishes exactly on the set $\{\lambda_n\}$. As soon as we prove that

$$|F(x+i)|^2 \asymp 1 + |x| \tag{26}$$

we find that $z^{-1}F(z) \notin L^2_\pi$, implying the system $\{e^{i\lambda_n t}\}$ is not minimal.

To prove (26), let us consider the function

$$\varphi(x) = \frac{F(x+i)}{\sin \pi(x+i)} = \frac{1}{\pi} \prod_{n=1}^{\infty} \frac{1 + \dfrac{x+i}{n-1/4}}{1 + \dfrac{x+i}{n}} \cdot \prod_{n=1}^{\infty} \frac{1 - \dfrac{x+i}{n-1/4}}{1 - \dfrac{x+i}{n}} = \Pi_1(x)\Pi_2(x) . \tag{27}$$

For the first product we have

$$\begin{aligned}
\Pi_1(x) &= \prod_{n=1}^{\infty} \frac{1 + \dfrac{x+i-1/4}{n}}{1 + \dfrac{x+i}{n}} \, \frac{n}{n-1/4} \\
&= \prod_{n=1}^{\infty} \frac{\left(1 + \dfrac{x+i-1/4}{n}\right) \exp\left(-\dfrac{x+i-1/4}{n}\right)}{\left(1 + \dfrac{x+i}{n}\right) \exp\left(-\dfrac{x+i}{n}\right)} \left(1 + \frac{1/4}{n-1/4}\right) \exp\left(-\frac{1/4}{n}\right) \\
&= C \frac{\Gamma(x+i)}{\Gamma(x-1/4+i)},
\end{aligned}$$

where Γ is the Euler gamma function and C is a nonzero constant. The Stirling formula gives

$$\Pi_1(x) \asymp 1 + |x|^{1/4} , \qquad x \to +\infty . \tag{28}$$

For the second product a similar representation is valid

$$\Pi_2(x) = \prod_{n=1}^{\infty} \frac{1 - \dfrac{x+i}{n-1/4}}{1 - \dfrac{x+i}{n}} = C \frac{\Gamma(-x-i)}{\Gamma(-x-1/4-i)}$$

with some nonzero constant C. Using the identity $\Gamma(z)\Gamma(1-z) = \dfrac{\pi}{\sin \pi z}$, and, again, the Stirling formula we prove

$$\Pi_2(x) \asymp 1 + |x|^{1/4} , \qquad x \to +\infty . \tag{29}$$

Combining estimates (28) and (29) with (27), we obtain (26) as $x \to +\infty$ and, since F is odd, as $x \to -\infty$. Theorem 4 is proven.

PROBLEM 1. Construct a sequence $\{\lambda_n\}_{n=-\infty}^{\infty}$ such that $|\lambda_n - n| < 1/4$, $n = 0, \pm 1, \pm 2, \dots$, and the system $\{e^{i\lambda_n t}\}_{n=-\infty}^{\infty}$ does not form a Riesz base in $L^2(-\pi,\pi)$.

PROBLEM 2. Let $\lambda_0 = 0$, $\lambda_n = n + \frac{1}{4}\operatorname{sgn} n$, $n \neq 0$, and let

$$F(z) = z \prod_{n=1}^{\infty} \left(1 - \frac{z^2}{\lambda_n^2}\right).$$

Prove that the function $F(z)$ satisfies the estimate

$$F(x+i) \asymp (1+|x|)^{-1/2}, \qquad x \to \pm\infty. \tag{30}$$

Relation (30) yields that the system $\{e^{i\lambda_n t}\}$ generated by the sequence $\{\lambda_n\}$ is complete and minimal in $L^2(-\pi,\pi)$ (see Problem 3 from Lecture 18). Prove that the system $\{e^{i\lambda_n t}\}$ does not form a Riesz base in $L^2(-\pi,\pi)$.

To conclude this lecture we mention without proof the criterion for a system of exponentials to form a Riesz base in $L^2(-\pi,\pi)$. This criterion has been discovered by Pavlov, see Hruščev, Nikol'skii and Pavlov **[59]** which also presents the history of the subject.

Now let the sequence Λ belong to the half-plane $\{\lambda : \operatorname{Im}\lambda > h\}$. We assume that $h > 0$, which involves no loss of generality since the translation $\Lambda \longrightarrow \Lambda + w$, $w \in \mathbb{C}$ does not alter the property of the sequence Λ to generate a Riesz base of exponentials. The exponentials $e^{i\lambda_n t}$ no longer satisfy the conditions

$$\|e^{i\lambda_n t}\|_{L^2(-\pi,\pi)} \asymp 1, \quad \lambda_n \in \Lambda,$$

which are necessary for the Riesz base property to hold, so we need to introduce an additional normalization.

Denote

$$\mathcal{E}_\Lambda = \{(1+|\operatorname{Im}\lambda_n|)^{1/2} e^{-\pi|\operatorname{Im}\lambda_n|} e^{i\lambda_n t};\ \lambda_n \in \Lambda\}.$$

The following statement holds:

For the system $\mathcal{E}_\Lambda$ to be a Riesz base in $L^2(-\pi,\pi)$ it is necessary and sufficient that

a) Λ satisfy Carleson's condition

$$\inf_n \prod_{m \neq n} \left|\frac{\lambda_m - \lambda_n}{\lambda_m - \bar{\lambda}_n}\right| > 0;$$

b) the product

$$F(z) = \text{P.V.} \prod \left(1 - \frac{z}{\lambda_n}\right)$$

converge uniformly on each compact set in $\mathbb{C}$ and the Hilbert transform be bounded in the space

$$L^2_w(-\infty,\infty) = \left\{f : \int_{-\infty}^{\infty} |f(x)|^2 w(x)\,dx < \infty\right\}, \qquad w(x) = |F(x)|^2.$$

In the case when Λ is located in a horizontal strip, the Carleson condition is equivalent to the separation condition $\inf_{k\neq n}|\lambda_k - \lambda_n| > 0$. Thus both conditions a) and b) hold if $\{\lambda_n\}$ is the set of zeros of a sine-type function.

Criteria for the Hilbert transform to be bounded in $L^2_w(-\infty,\infty)$ may be formulated in terms of the weight $w(x)$ (see, for example, Garnett [**37**]). One such criterion is the A_2-condition of Muckenhoupt:

$$\sup_I \left\{ \frac{1}{|I|} \int_I w\,dx \; \frac{1}{|I|} \int_I w^{-1}\,dx \right\} < \infty ,$$

where supremum is taken over all intervals $I \subset \mathbb{R}$.

More recent developments can be found in Minkin [**99**] and Seip [**120**].

APPENDIX

Completeness of the Eigenfunction System of a Quadratic Operator Pencil

When separating variables in partial differential equations one can meet problems of completeness and minimality for functional systems of more complicated structure than exponentials. In some cases such systems are formed by solutions of ordinary differential equations with polynomial dependence on a spectral parameter. Here we consider an example of such a system generated by the boundary value problem

$$f''(x) - 2\lambda \cos\theta f'(x) + \lambda^2 f(x) = 0 , \qquad f(0) = f(\pi) = 0 .$$

Such an equation admits nonzero solutions only if $\lambda = n/\sin\theta$, $n = \pm 1, \pm 2, \dots$, and the entire system of solutions has the form

$$\mathcal{K}_a = \{f_n(x)\}_{n=-\infty,\, n\neq 0}^{\infty} , \qquad f_n(x) = e^{anx} \sin nx ,$$

with $a = \cot\theta$.

If $\cos\theta = 0$, the system $\mathcal{K}_a$ contains both functions $\sin nx$ and $-\sin nx, n \neq 0$. In a sense, the system under consideration is "overcomplete", and we can expect something more than its completeness in the space $L^2(0,\pi)$. More specifically, we shall investigate the following problem:

Given a functional system $f_n(x)$, is it possible to represent an arbitrary vector $(\varphi_0, \varphi_1) \in L^2(0,\pi) \oplus L^2(0,\pi)$ as the limit of finite linear combinations of the form

$$\Big(\sum_{n\neq 0} c_n f_n(x), \sum_{n\neq 0} n c_n f_n(x)\Big) .$$

If such an approximation is possible, then, following M. Keldysh, it is said that the system $f_n(x)$ is *twofold complete* in the space $L^2(0,\pi)$.

A.1. Twofold completeness of the system $\mathcal{K}_a$

THEOREM 1. *The system $\mathcal{K}_a$ is twofold complete in $L^2(0,\pi)$ for any $a \in \mathbb{R}$.*

PROOF. Let a vector $(f,g) \in L^2(0,\pi) \oplus L^2(0,\pi)$ annihilate all the vectors of the form $(e^{ant}\sin nt, ne^{ant}\sin nt)$, $n = \pm 1, \pm 2, \dots$. This means that

$$\int_0^\pi f(t) e^{ant} \sin nt\, dt + n \int_0^\pi g(t) e^{ant} \sin nt\, dt = 0 . \tag{1}$$

Let us set

$$F(t) = \int_t^\pi f(t)\, dt .$$

Integrating by parts in the first integral of (1) we obtain

$$\int_0^\pi \Big[(a+i)F(t)+g(t)\Big]e^{(a+i)nt}\,dt - \int_0^\pi \Big[(a-i)F(t)+g(t)\Big]e^{(a-i)nt}\,dt = 0\,, \quad n=\pm1,\pm2,\dots\,.$$

The EFET

$$(2)\quad \Phi(\lambda) = \int_0^\pi \Big[(a+i)F(t)+g(t)\Big]e^{(a+i)\lambda t}\,dt - \int_0^\pi \Big[(a-i)F(t)+g(t)\Big]e^{(a-i)\lambda t}\,dt$$

vanishes at $\lambda \in \mathbb{Z}\setminus\{0\}$, and its indicator diagram I_Φ lies inside the triangle T_a with vertices at the points 0, $(a\pm i)\pi$. It is easily seen that $\Phi(i\eta)e^{-\pi|\eta|} \to 0$, $\eta \to \pm\infty$. Therefore the function

$$\Psi(\lambda) = \frac{\lambda\Phi(\lambda)}{\sin \pi\lambda}$$

is an EFET, and

$$\Psi(i\eta) = o(|\eta|)\,, \qquad \eta \to \pm\infty\,.$$

Outside the disks $\{\lambda : |\lambda - n| \le \delta\}$ an estimate $|\sin\pi\lambda| \ge m_\delta \exp\pi|\operatorname{Im}\lambda|$ is fulfilled, and if $\Phi \not\equiv 0$, then $I_\Phi = I_\Psi + [-i\pi, i\pi]$. The only set I_Ψ that agrees with the condition that $I_\Phi \subseteq T_a$ is $I_\Psi = \{a\}$. Therefore, the indicator diagram of the function $\Psi_1(\lambda) = e^{-a\pi\lambda}\Psi(\lambda)$ is the point $I_{\Psi_1} = \{0\}$. Since $\Psi_1(i\eta) = o(|\eta|)$, $\eta \to \pm\infty$, the Phragmén-Lindelöf theorem yields $\Psi_1(z) \equiv$ const. Returning to the function Φ we obtain

$$\Phi(\lambda) = ce^{a\pi\lambda}\frac{\sin\pi\lambda}{\lambda} = \frac{c}{2i}\Big[(a+i)\int_0^\pi e^{(a+i)t\lambda}\,dt - (a-i)\int_0^\pi e^{(a-i)\lambda t}\,dt\Big]$$

including the case $\Phi \equiv 0$ which corresponds to $c = 0$. Substituting this expression into (2), we obtain

$$\int_0^\pi \Big[(a+i)(F(t)-c_1)+g(t)\Big]e^{(a+i)\lambda t}\,dt - \int_0^\pi \Big[(a-i)(F(t)-c_1)+g(t)\Big]e^{(a-i)\lambda t}\,dt = 0$$

and, applying the Laplace transform,

$$\int_0^\pi \frac{(a+i)(F(t)-c_1)+g(t)}{(a+i)t-z}\,dt - \int_0^\pi \frac{(a-i)(F(t)-c_1)+g(t)}{(a-i)t-z}\,dt = 0\,, \quad \operatorname{Re} z > \max\{0, a\pi\}\,.$$

This relation is analytically continued to the set $\mathbb{C}\setminus\{[0,(a+i)\pi] \cup [0,(a-i)\pi]\}$. The Sokhotskii-Plemelj formula yields

$$(a+i)(F(t)-c_1)+g(t) = (a-i)(F(t)-c_1)+g(t) = 0\,, \qquad t\in[0,\pi]\,.$$

Therefore, $F(t) \equiv c_1$, and $f(t) = -F_1'(t) \equiv 0$, which proves the twofold completeness of the system $\mathcal{K}_a$.

PROBLEM 1. Prove that the system $\mathcal{K}_a$ is twofold minimal in $L^2(0,\pi)$. In other words no vector $(e^{ant}\sin nt, ne^{ant}\sin nt)$ belongs to the closure of linear span of all other vectors of the same form.

A.2. Completeness of the system $\mathcal{K}_a^+$

To avoid "overcompleteness" of the system $\mathcal{K}_a$ in the case $\cos\theta = 0$, it is sufficient to introduce the subsystem $\{\sin nx\}_{n=1}^{\infty}$ which is complete and minimal in $L^2(0,\pi)$. For an arbitrary θ, the corresponding subsystem has the form

$$\mathcal{K}_a^+ = \{f_n(x)\}_{n=1}^{\infty} , \qquad f_n(x) = e^{anx}\sin nx ,$$

and we arrive at the problem of the completeness of the system $\mathcal{K}_a^+$.

THEOREM 2. [22] *The system $\mathcal{K}_a^+$ is complete in the space $L^2(0,\pi)$.*

PROOF. Let a function $\psi \in L^2(0,\pi)$ annihilate $\mathcal{K}_a^+$. This means that

$$(3)\qquad \begin{aligned} \int_0^{\pi} \psi(t)e^{ant}\sin \pi t\,dt &= \frac{1}{2i}\int_0^{\pi} \psi(t)e^{(a+i)nt}\,dt \\ &- \frac{1}{2i}\int_0^{\pi} \psi(t)e^{(a-i)nt}\,dt = 0 , \qquad n = 1,2,\dots . \end{aligned}$$

Introduce the auxiliary function

$$\Psi(t) = \int_t^{\pi} \psi(\tau)\,d\tau .$$

Integrating by parts each integral on the right-hand side of (3) we obtain

$$(4)\qquad \int_0^{\pi} \Psi(t)e^{(a+i)(n-1)t}\,de^{(a+i)t} - \int_0^{\pi} \Psi(t)e^{(a-i)(n-1)t}\,de^{(a-i)t} = 0 .$$

Setting $\zeta = e^{(a+i)t}$ in the first integral and $\zeta = e^{(a-i)t}$ in the second, we must integrate over the curves $\gamma^+ = \{\zeta = e^{(a+i)t} : t \in [0,\pi]\}$, and $\gamma^- = \{\zeta = e^{(a-i)t} : t \in [0,\pi]\}$, respectively (see Figure 6). Let the function $\widehat{\Psi}(\zeta)$ be defined on the closed curve $\gamma = \gamma^+ \cup \gamma^-$ as follows:

$$\widehat{\psi}\big(e^{(a\pm i)t}\big) = \Psi(t) , \qquad t \in [0,\pi] .$$

Relation (4) takes the form:

$$(5)\qquad \int_{\gamma} \widehat{\Psi}(\zeta)\zeta^{n-1}\,d\zeta = 0 , \quad n = 1,2,\dots .$$

Since $\psi \in L^2(0,\pi)$, the function $\widehat{\Psi}$ satisfies the Hölder condition with the exponent $1/2$. Let G be the domain bounded by the curve γ. Consider the functions

$$F_+(\zeta) = \frac{1}{2\pi i}\int_{\gamma} \frac{\widehat{\Psi}(\xi)}{\xi - \zeta}\,d\xi , \qquad \zeta \in G ,$$

$$F_-(\zeta) = \frac{1}{2\pi i}\int_{\gamma} \frac{\widehat{\Psi}(\xi)}{\xi - \zeta}\,d\xi , \qquad \zeta \in \mathbb{C}\backslash (G \cup \gamma) .$$

Since $\widehat{\Psi}$ satisfies a Hölder condition on γ, the functions $F_\pm(\zeta)$ are continuous on the closures of their domains, and the Sokhotskii-Plemelj formula yields

$$(6)\qquad \widehat{\Psi}(\zeta) = F_+(\zeta) - F_-(\zeta) , \qquad \zeta \in \gamma ,$$

[22]This theorem was proved in various ways by Dzhavadov and Gasymov, V. P. Gurarii, Kostyuchenko and Shkalikov, and the author. Here, we follow essentially Gurarii's proof.

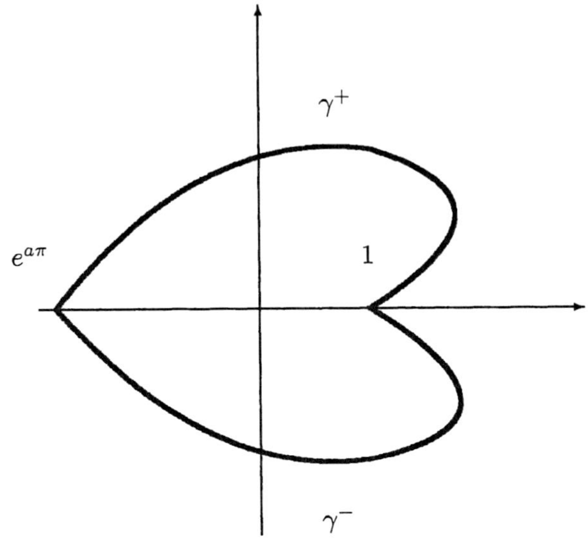

FIGURE 6

where $F_\pm(\zeta)$ are the boundary values of the corresponding functions. At infinity the function $F_-(\zeta)$ admits the representation

$$F_-(\zeta) = \sum_{k=0}^{\infty} \frac{1}{\zeta^{k+1}} \frac{1}{2\pi i} \int_\gamma \widehat{\Psi}(\xi)\xi^k \, d\xi \,,$$

and relation (5) yields $F_- \equiv 0$. By (6) we conclude that

$$\widehat{\Psi}(\zeta) = F_+(\zeta) \,, \qquad \zeta \in \gamma \,.$$

Thus the function $\widehat{\Psi}(\zeta)$ admits a continuation to a function analytic in G and continuous in the closure of G. Without loss of generality we may assume that the function ψ which annihilates the system $\mathcal{K}_a^+$ is real-valued. Then the functions $\Psi(t)$, $t \in [0,\pi]$, and $\widehat{\Psi}(\zeta)$, $\zeta \in \gamma$, are real as well. In particular, the set $\widehat{\Psi}(\gamma) = \{\widehat{\Psi}(\zeta) : \zeta \in \gamma\}$ is a bounded subset in $\mathbb{R}$. If $\widehat{\Psi}(\zeta)$ is a nonconstant function, the set $\widehat{\Psi}(G) = \{\widehat{\Psi}(\zeta) : \zeta \in G\}$ is a domain, and its boundary is a part of $\widehat{\Psi}(\gamma)$. Therefore, $\widehat{\Psi}(G)$ is an unbounded set. This is a contradiction, since Ψ is a continuous function in the closure of G. Thus, $\widehat{\Psi}(\zeta) \equiv \text{const}$, $\Psi(t) \equiv \text{const}$ and, finally, $\psi(t) = -\Psi'(t) \equiv 0$. We have proved, therefore, that the only function from $L^2(0,\pi)$ annihilating $\mathcal{K}_a^+$ is identically equal to zero. This means that the system $\mathcal{K}_a^+$ is complete in $L^2(0,\pi)$.

PROBLEM. Prove that $\mathcal{K}_a^+$ is complete in all spaces $L^2(0,l)$, $0 < l < 2\pi$.

It may be proved that the system $\mathcal{K}_a^+$ is minimal in $L^2(0,\pi)$ (see Shkalikov [**104**]).

For the completeness and minimality problems for eigensystems of other operator pencils, see Gohberg and Krein [**40**], Markus [**94**], Lyubarskii [**86**].

Part III. Some Additional Problems of the Theory of Entire Functions

In this part we present some special topics of the theory of entire functions.

First of all we derive the Carleman formula for functions meromorphic in a sector, which is an effective tool for applications of entire functions. This formula is applied to describe exact conditions for nontriviality of some classes of infinitely differentiable functions which, in particular, originate as the test spaces in the theory of distributions and related partial differential equations.

The same formula is used to prove a theorem due to Matsaev which states that some lower bounds for an entire function may imply that the function is of finite order. The statements of this type have been successfully applied to a study of spectral properties of nonselfadjoint compact operators in Hilbert spaces.

In Lecture 27 we discuss a class P of entire functions without zeros in a half-plane and prove an analogue of the Hermite-Biehler theorem for it. One of the most important properties of a function $f \in P$ is that the zeros of its real and imaginary parts are real and interlacing. This is a very typical situation encountered in the theory of ordinary selfadjoint differential operators.

We conclude the monograph by exposing the famous inequality discovered by S. N. Bernstein on the derivative of EFET, which is of great importance for the theory of approximation on the real line.

LECTURE 24

The Formulas of Carleman and R. Nevanlinna and their Applications

In this lecture we shall derive integral formulas relating the modulus and zeros of functions holomorphic in a half-plane. These formulas, similar to those of Jensen and Poisson-Jensen, have important applications in the theory of entire functions.

24.1. The Carleman formula

Let $f(z)$ be a meromorphic function in a closed sector $\overline{S} = \{z : \rho \leq |z| \leq R, \operatorname{Im} z \geq 0\}$ whose zeros and poles do not lie on the boundary ∂S. We exclude disks of sufficiently small radius $\varepsilon > 0$ with centers at the zeros a_n and the poles b_n of the function $f(z)$ from the domain S and apply the second Green's formula to the functions $u(z) = \ln |f(z)|$ and $v(z) = -\operatorname{Im}(\frac{1}{z} + \frac{z}{R^2})$ in the resulting domain S_ε. We obtain

$$\int_{\partial S_\varepsilon} \left(u\frac{\partial v}{\partial n} - v\frac{\partial u}{\partial n}\right) ds = 0,$$

where $\partial/\partial n$ is the differentiation along the inner normal and ds is the element of arc length.

The function $v(z)$ is harmonic in S; the equations

$$v = 0\,, \qquad \frac{\partial v}{\partial n} = \frac{2}{R^2}\sin\varphi\,, \quad z = Re^{i\varphi}$$

hold on the half-circle $\{|z| = R, \operatorname{Im} z \geq 0\}$, while the equations

$$v = 0\,, \qquad \frac{\partial v}{\partial n} = \frac{1}{t^2} - \frac{1}{R^2}$$

hold on the intervals $\{z = \pm t : \rho < t < R\}$. As $\varepsilon \to 0$, the integrals over the excluded disks give a total contribution equal to

$$2\pi \sum_{a_n \in S} \operatorname{Im}\left(\frac{1}{a_n} + \frac{a_n}{R^2}\right) - 2\pi \sum_{b_n \in S} \operatorname{Im}\left(\frac{1}{b_n} + \frac{b_n}{R^2}\right).$$

Thus,

$$\begin{aligned}
0 &= \lim_{\varepsilon\to 0}\int_{\partial S_\varepsilon} \left(u\frac{\partial v}{\partial n} - v\frac{\partial u}{\partial n}\right) ds \\
&= \frac{2}{R^2}\int_0^\pi \log|f(Re^{i\varphi})|\sin\varphi\, d\varphi + 2\pi\int_\rho^R \left(\frac{1}{t^2} - \frac{1}{R^2}\right)\log|f(t)f(-t)|\, dt \\
&\quad + 2\pi \sum_{a_n\in S} \operatorname{Im}\left(\frac{1}{a_n} + \frac{a_n}{R^2}\right) - 2\pi \sum_{b_n \in S}\operatorname{Im}\left(\frac{1}{b_n} + \frac{b_n}{R^2}\right) + A_f(\rho, R)\,,
\end{aligned}$$

where the remainder term $A_f(\rho, R)$ has the form

$$A_f(\rho,R) = -\frac{1}{2\pi}\int_0^\pi \Big[\Big(\frac{1}{\rho^2}+\frac{1}{R^2}\Big)\log|f(\rho e^{i\varphi})| + \Big(\frac{1}{\rho}-\frac{\rho}{R^2}\Big)\frac{\partial}{\partial\rho}\log|f(\rho e^{i\varphi})|\Big]\rho\sin\varphi\,d\varphi\,. \tag{1}$$

Taking into consideration that

$$\operatorname{Im}\Big(\frac{1}{c}+\frac{c}{R^2}\Big) = -\Big(\frac{1}{|c|}-\frac{|c|}{R^2}\Big)\sin(\arg c)\,,$$

we obtain Carleman's formula

$$\begin{aligned}\sum_{\rho<|a_n|<R}\Big(\frac{1}{|a_n|}-\frac{|a_n|}{R^2}\Big)\sin\alpha_n - \sum_{\rho<|b_n|<R}\Big(\frac{1}{|b_n|}-\frac{|b_n|}{R^2}\Big)\sin\beta_n \\ = \frac{1}{2\pi}\int_\rho^R\Big(\frac{1}{t^2}-\frac{1}{R^2}\Big)\log|f(t)f(-t)|\,dt \\ + \frac{1}{\pi R}\int_0^\pi \log|f(Re^{i\varphi})|\sin\varphi\,d\varphi + A_f(\rho,R)\,,\end{aligned} \tag{2}$$

where $a_n = |a_n|e^{i\alpha_n}$, $b_n = |b_n|e^{i\beta_n}$ are zeros and poles of the function $f(z)$, and the remainder term $A_f(\rho, R)$ is expressed by formula (1).

We have derived Carleman's formula under the additional restriction that there are neither zeros nor poles of the function $f(z)$ on the boundary of the semi-annulus S. However, because of continuous dependence on ρ and R of all terms, Carleman's formula remains valid for functions with zeros and poles on the half-circles $\{|z| = \rho, \operatorname{Im} z \geq 0\}$ and $\{|z| = R, \operatorname{Im} z \geq 0\}$. If $f(z)$ has zeros and poles on the real axis one should first exclude small semi-disks around these zeros and poles and then pass to the limit. Moreover, one may only require the function $f(z)$ to be meromorphic in the open sector S and continuous up to its boundary.

If $f(z)$ is meromorphic in the upper half-plane and continuous up the real axis, then due to expression (1)

$$A_f(\rho,R) = O(1), \qquad R\to\infty\,.$$

REMARK 1. Let the function $f(z)$ be holomorphic in a neighborhood of the point $z = 0$, and let $f(0) = 1$, i.e., $f(z) = 1+cz+O(z^2)$, $z \to 0$. Then $\log|f(\rho e^{i\varphi})| = \rho\operatorname{Re}(ce^{i\varphi}) + O(\rho^2)$, $\rho \to 0$. Substituting this expression into (1), we obtain

$$\lim_{\rho\to 0} A_f(\rho,R) = -\frac{1}{\pi}\int_0^\pi \operatorname{Re}(ce^{i\varphi})\sin\varphi\,d\varphi = \frac{\operatorname{Im} c}{2} = \frac{1}{2}\operatorname{Im} f'(0)\,.$$

REMARK 2. The Nevanlinna characteristics for the upper half-plane may be related to functions meromorphic in the upper half-plane (see Goldberg and Ostrovskii [**43**]). In particular, the counting functions for zeros and poles in the upper half-plane have the form

$$c_f(t,0) = \sum_{|a_n|\leq t}\sin\alpha_n\,, \qquad c_f(t,\infty) = \sum_{|b_n|\leq t}\sin\beta_n\,,$$

$$c_f(t) = c_f(t,0) - c_f(t,\infty)\,.$$

If $f(0) \neq 0, \infty$, then

$$\sum_{|a_n|<R} \left(\frac{1}{|a_n|} - \frac{|a_n|}{R^2}\right) \sin \alpha_n - \sum_{|b_n|<R} \left(\frac{1}{|b_n|} - \frac{|b_n|}{R^2}\right) \sin \beta_n$$
$$= \int_0^R \left(\frac{1}{t} - \frac{t}{R^2}\right) dc_f(t) = \int_0^R \left(\frac{1}{t^2} + \frac{1}{R^2}\right) c_f(t)\, dt\,.$$

We state a frequently applicable uniqueness theorem that follows from Carleman's formula.

THEOREM 1 (F. and R. Nevanlinna). *Let a positive function $\Lambda(r)$ be such that*

$$\int^\infty \frac{\Lambda(r)}{r^2}\, dr = \infty\,.$$

Let $f(z)$ be a memomorphic function in the closed upper half-plane which, for sufficiently large r and constants λ, μ, and ν satisfies the following three estimates:

1. $\log |f(r)f(-r)| \leq 2\pi\lambda(1 + o(1))\Lambda(r)$,
2. $\displaystyle \frac{1}{\pi r} \int_0^\pi \log |f(re^{i\varphi})| \sin \varphi \, d\varphi \leq \mu(1 + o(1)) \int_1^r \frac{\Lambda(t)}{t^2}\, dt,$
3. $c_f(r) \geq \nu(1 + o(1))\Lambda(r).$

If $\nu > \mu + \lambda$, then $f(z) \equiv 0$.

For example, if $f(z)$ is a holomorphic function of exponential type in $\mathbb{C}_+$, then, assuming $\Lambda(r) = r$, $\mu = 0$, we obtain a generalization of the Carlson theorem from Section 8.3:

If

$$\limsup_{r\to\infty} \frac{\log |f(r)f(-r)|}{2\pi r} < \liminf_{r\to\infty} \frac{c_f(r, 0)}{r}\,,$$

then $f(z) \equiv 0$.

PROBLEM 1 (N. I. Akhiezer). Let $f(z)$ be an entire function of exponential type and let $a_n = |a_n|e^{i\alpha_n}$ be the sequence of its nonzero roots. Derive that each of the three inequalities:

(a) $\displaystyle \sum_{n=1}^\infty \frac{|\sin \alpha_n|}{|a_n|} < \infty;$

(b) $\displaystyle \sup_{R>1} \int_1^R \frac{\log |f(t)f(-t)|}{t^2}\, dt < \infty;$

(c) $\displaystyle \sup_{R>1} \left| \int_1^R \frac{\log |f(t)f(-t)|}{t^2}\, dt \right| < \infty$

implies the other two.

24.2. The Phragmén-Lindelöf principle as formulated by F. and R. Nevanlinna

The argument given above provides a simple way of proving the Phragmén-Lindelöf principle in the elegant formulation of F. and R. Nevanlinna.

Let $u(z)$ be a function harmonic in the closed upper half-plane. Then (after substituting $u(z)$ for $\log|f(z)|$) the above-given proof of Carleman's formula (2) is valid and leads to the formula

$$\frac{1}{\pi R}\int_0^\pi u(Re^{i\varphi})\sin\varphi\,d\varphi+\frac{1}{2\pi}\int_\rho^R\Big(\frac{1}{t^2}-\frac{1}{R^2}\Big)[u(t)+u(-t)]\,dt+A_u(\rho,R)=0\,. \tag{3}$$

If $u(0)=0$, then the integral

$$\int_0 \frac{u(t)+u(-t)}{t^2}\,dt<\infty \tag{4}$$

converges, and $A_u(\rho,R)$ approaches a constant as $\rho\to 0$. Passing in (3) to the limit as $\rho\to 0$, we obtain

$$\frac{1}{\pi R}\int_0^\pi u(Re^{i\varphi})\sin\varphi\,d\varphi+\frac{1}{2\pi}\int_0^R\Big(\frac{1}{t^2}-\frac{1}{R^2}\Big)[u(t)+u(-t)]\,dt=\text{const}\,. \tag{5}$$

Now let us demonstrate how condition (4) may be removed. From the explicit form (1) of the remainder term $A_u(\rho,R)$ it follows that

$$\lim_{\rho\to 0}[A_u(\rho,R)-A_u(\rho,r)]=0$$

for fixed values $r<R$. Now let us subtract from equation (3) a similar equation with r instead of R and pass to the limit as $\rho\to 0$. We obtain

$$\begin{aligned}&\frac{1}{\pi R}\int_0^\pi u(Re^{i\varphi})\sin\varphi\,d\varphi-\frac{1}{\pi r}\int_0^\pi u(re^{i\varphi})\sin\varphi\,d\varphi\\&+\frac{1}{2\pi}\int_r^R\frac{u(t)+u(-t)}{t^2}\,dt-\frac{1}{2\pi R^2}\int_0^R[u(t)+u(-t)]dt\\&+\frac{1}{2\pi r^2}\int_0^r[u(t)+u(-t)]\,dt=0\,.\end{aligned} \tag{6}$$

Set

$$m(r)=\frac{1}{\pi}\int_0^\pi u(re^{i\varphi})\sin\varphi\,d\varphi\,,$$
$$U(r)=2\int_0^r[u(t)+u(-t)]\,dt\,.$$

Then, for $r<R$,

$$\begin{aligned}\int_r^R\frac{U(t)}{t^3}dt&=\frac{1}{2}\frac{U(r)}{r^2}-\frac{1}{2}\frac{U(R)}{R^2}+\frac{1}{2}\int_r^R\frac{dU(t)}{t^2}\\&=\frac{1}{r^2}\int_0^r[u(t)+u(-t)]\,dt-\frac{1}{R^2}\int_0^R[u(t)+u(-t)]\,dt\\&+\int_r^R\frac{u(t)+u(-t)}{t^2}\,dt\,.\end{aligned}$$

As a result, equation (6) is written in the form

$$\frac{m(R)}{R} - \frac{m(r)}{r} + \frac{1}{2\pi}\int_r^R \frac{U(t)}{t^3}\,dt = 0\,, \tag{7}$$

where condition (4) now is not used. The same relation holds for $u(z)$ harmonic in the upper half-plane and continuous up to the real axis.

Now, let $u(z)$ be a subharmonic function in $\overline{\mathbb{C}}_+$. Having replaced the function $u(z)$ with its least harmonic majorant[23] in an arbitrary half-disk $\{z : \operatorname{Im} z \geq 0,\ |z| \leq R\}$, we obtain, instead of identity (7), the inequality

$$\frac{m(r)}{r} \leq \frac{m(R)}{R} + \frac{1}{2\pi}\int_r^R \frac{U(t)}{t^3}\,dt\,,$$

or, choosing an arbitrary $\rho < r$, the inequality

$$\frac{m(r)}{r} + \frac{1}{2\pi}\int_\rho^r \frac{U(t)}{t^3}\,dt \leq \frac{m(R)}{R} + \frac{1}{2\pi}\int_\rho^R \frac{U(t)}{t^3}\,dt\,.$$

Thus, the quantities $m(r)$ and $U(r)$ are related by the following theorem:

THEOREM 2 (Ahlfors [**3**]). *Let $u(z)$ be a subharmonic function in $\overline{\mathbb{C}}_+$, then the sum*

$$\frac{m(r)}{r} + \frac{1}{2\pi}\int_\rho^r \frac{U(t)}{t^3}\,dt$$

is a nondecreasing function of r. If $u(z)$ is harmonic in $\overline{\mathbb{C}}_+$, then this sum is independent of r.

Suppose the subharmonic function $u(z)$ has nonpositive values on the real axis. Then the integral

$$\int_\rho^r \frac{U(t)}{t^3}\,dt$$

does not increase with r, hence the ratio $m(r)/r$ does not decrease with r. Since the function $u^+(z)$ is also subharmonic, the ratio $m_+(r)/r$, with

$$m_+(r) = \frac{1}{\pi}\int_0^\pi u^+(re^{i\varphi})\sin\varphi\,d\varphi\,,$$

does not decrease with r. Thus we have proved the Phragmén-Lindelöf principle as formulated F. and R. Nevanlinna.

THEOREM 3. *If a subharmonic function in $\overline{\mathbb{C}}_+$ has nonpositive values on the real axis, then the functions $m(r)/r$ and $m_+(r)/r$ are nondecreasing. In particular, the condition*

$$\liminf_{r\to\infty} \frac{m_+(r)}{r} = 0$$

implies $u^+(z) \equiv 0$, i.e., the function $u(z)$ is nonpositive everywhere in $\mathbb{C}_+$.

A stronger theorem was obtained in Ahlfors [**2**].

[23] I.e., harmonic function with the same boundary values as those of the function $u(z)$.

24.3. R. Nevanlinna's formula for a half-disk

We remind the reader that if a domain Ω has a piecewise-smooth boundary $\partial\Omega$, then a positive function $G_z(\zeta)$, $\zeta \in \Omega$, which is equal to zero at every point $\zeta \in \partial\Omega$, is harmonic in $\Omega\backslash\{z\}$ and has the asymptotic expression

$$G_z(\zeta) = \log\frac{1}{|\zeta - z|} + O(1)\,, \qquad \zeta \to z\,, \quad z \in \Omega\,,$$

in a neighborhood of the point $\zeta = z$, is called the Green function of the domain $\Omega \subset \mathbb{C}$. If $u(z)$ is harmonic in the domain Ω and continuous up to its boundary, then its values inside the domain Ω can be recovered with the aid of the Poisson kernel

$$(8) \qquad P_z(\zeta) = \frac{1}{2\pi}\frac{\partial G_z}{\partial n}\,, \qquad \zeta \in \partial\Omega\,, \qquad z \in \Omega\,,$$

(differentiation is taken along the inner normal), by the formula

$$u(z) = \int_{\partial\Omega} u(\zeta)P_z(\zeta)\,ds\,.$$

If Ω is a simply connected domain and $g_z(\zeta)$ is a conformal mapping of Ω onto the unit disk $\mathbb{D}$ such that $g_z(z) = 0$, then

$$(9) \qquad G_z(\zeta) = \log\frac{1}{|g_z(\zeta)|}\,,$$

and hence

$$(10) \qquad P_z(\zeta)\,ds = \frac{1}{2\pi i}[\log g_z(\zeta)]'\,d\zeta\,.$$

By using formulas (9) and (10) we shall find the expression for the Green function and for the Poisson kernel in the half-disk $\mathbb{D}_R^+ = \{\zeta : |\zeta| < R, \operatorname{Im}\zeta > 0\}$. The function

$$\varphi_z(\zeta) = \frac{R(\zeta - z)}{R^2 - \zeta\overline{z}}$$

maps $D_R = \{\zeta : |\zeta| < R\}$ conformally onto the unit disk $\mathbb{D}$, $\varphi_z(z) = 0$. That is why the function

$$(11) \qquad g_z(\zeta) = \frac{\varphi_z(\zeta)}{\varphi_{\overline{z}}(\zeta)} = \frac{\zeta - z}{\zeta - \overline{z}}\cdot\frac{R^2 - \zeta z}{R^2 - \zeta\overline{z}}$$

maps the semi-disk $\mathbb{D}_R^+$ conformally onto $\mathbb{D}$, and $g_z(z) = 0$. Substituting (11) into (9), we obtain the expression for the Green function

$$(12) \qquad G_z(\zeta) = \log\left|\frac{\zeta - \overline{z}}{\zeta - z}\cdot\frac{R^2 - \zeta\overline{z}}{R^2 - \zeta z}\right|\,.$$

Further,

(13)

$$
\begin{aligned}
P_z(\zeta)\,ds &= \frac{1}{2\pi i}[\log g_z(\zeta)]'\,d\zeta \\
&= \frac{1}{2\pi i}\Big[\frac{1}{\zeta - z} - \frac{1}{\zeta - \bar{z}} + \frac{\bar{z}}{R^2 - \zeta\bar{z}} - \frac{z}{R^2 - \zeta z}\Big]\,d\zeta \\
&= \begin{cases} \dfrac{R^2 - |z|^2}{2\pi}\Big(\dfrac{1}{|Re^{i\theta} - z|^2} - \dfrac{1}{|Re^{i\theta} - \bar{z}|^2}\Big)\,d\theta\,, & \zeta = Re^{i\theta},\ 0 < \theta < \pi\,, \\ \dfrac{\operatorname{Im} z}{\pi}\Big(\dfrac{1}{|t - z|^2} - \dfrac{R^2}{|R^2 - tz|^2}\Big)\,dt\,, & \zeta = t,\ -R < t < R\,. \end{cases}
\end{aligned}
$$

Let $u(z)$ be a function harmonic in $\mathbb{D}_R^+$ with continuous boundary values. Then

$$
\begin{aligned}
u(z) = {} & \frac{R^2 - |z|^2}{2\pi}\int_0^\pi \Big(\frac{1}{|Re^{i\theta} - z|^2} - \frac{1}{|Re^{i\theta} - \bar{z}|^2}\Big)u(Re^{i\theta})\,d\theta \\
& + \frac{\operatorname{Im} z}{\pi}\int_{-R}^{R}\Big(\frac{1}{|t - z|^2} - \frac{R^2}{|R^2 - tz|^2}\Big)u(t)\,dt\,.
\end{aligned}
\tag{14}
$$

Now, let $f(z)$ be a meromorphic function in the half-disk $\overline{\mathbb{D}}_R^+$, $\{a_n\}$ be its zeros and $\{b_n\}$ its poles. Using expression (12) for the Green function of the domain $\mathbb{D}_R^+$, we deduce that the function

$$
\begin{aligned}
u(z) = \log|f(z)| & - \sum_{a_n \in \mathbb{D}_R^+} \log\left|\frac{z - a_n}{z - \bar{a}_n}\cdot\frac{R^2 - a_n z}{R^2 - \bar{a}_n z}\right| \\
& + \sum_{b_n \in \mathbb{D}_R^+} \log\left|\frac{z - b_n}{z - \bar{b}_n}\cdot\frac{R^2 - b_n z}{R^2 - \bar{b}_n z}\right|
\end{aligned}
$$

which is harmonic in $\mathbb{D}_R^+$, has the same boundary values as the function $\log|f(z)|$. Applying representation (14) to $u(z)$, we obtain the R. Nevanlinna formula for a half-disk

(15)

$$
\begin{aligned}
\log|f(z)| = {} & \frac{R^2 - |z|^2}{2\pi}\int_0^\pi \Big(\frac{1}{|Re^{i\theta} - z|^2} - \frac{1}{|Re^{i\theta} - \bar{z}|^2}\Big)\log|f(Re^{i\theta})|\,d\theta \\
& + \frac{\operatorname{Im} z}{\pi}\int_{-R}^{R}\Big(\frac{1}{|t - z|^2} - \frac{R^2}{|R^2 - tz|^2}\Big)\log|f(t)|dt \\
& + \sum_{a_n \in \mathbb{D}_R^+} \log\left|\frac{z - a_n}{z - \bar{a}_n}\cdot\frac{R^2 - z a_n}{R^2 - \bar{a}_n z}\right| - \sum_{b_n \in \mathbb{D}_R^+} \log\left|\frac{z - b_n}{z - \bar{b}_n}\cdot\frac{R^2 - b_n z}{R^2 - \bar{b}_n z}\right|.
\end{aligned}
$$

PROBLEM 2. Deduce the Carleman formula for subharmonic functions.

HINT. Use the R. Nevanlinna representation (15) which can be transferred directly to subharmonic functions.

LECTURE 25

Uniqueness Problems for Fourier Transforms and for Infinitely Differentiable Functions

25.1. Uniqueness theorem for Fourier transforms

The Fourier transform

$$\tilde{f}(\lambda) = \int_{-\infty}^{\infty} f(x)e^{i\lambda x}\, dx \tag{1}$$

of a nontrivial function $f \in L^2(\mathbb{R})$ with compact support belongs to the class C, and as proved in Section 14.2, cannot decrease along the real axis so fast that the integral

$$\int_{-\infty}^{\infty} \frac{\log^- |\tilde{f}(t)|}{1+t^2}\, dt$$

diverges. The relation arising in this situation between the functions f and $\tilde{f}$ is of rather general nature and can be formulated as the statement that the two functions cannot decrease simultaneously too fast at infinity. This phenomenon is called the Uncertainty Principle; it has been encountered in Section 8.3. For a detailed description of results related to this principle, see Havin and Jöricke [**61**] and Nazarov [**101**].

Following Beurling and Dzhrbashyan, we shall prove theorems on relation between the rates of decrease of f and $\tilde{f}$ along the real axis, and afterwards we shall apply these theorems to describe conditions for some classes of infinitely differentiable functions to be nontrivial.

In what follows, without a special mention, we denote by $p(t)$ a positive continuous function on the half-axis $[0, \infty)$ satisfying the conditions:

a) $p(0) = p'(0) = 0$;
b) $\lim_{t\to\infty} p(t)t^{-1} = +\infty$.

For every fixed $x \geq 0$, the quantity

$$p^*(x) = \sup_{t\geq 0}\{xt - p(t)\} \tag{2}$$

is finite. It is called the *Legendre transform*, or the *Young dual function* for p. Being the upper envelope of a system of linear functions, $p^*(x)$ is a convex function, and it follows from (2) that $p^*(x)/x$ increases with x. In addition, (2) implies

$$xt \leq p^*(x) + p(t)\,, \qquad x, t \geq 0\,. \tag{3}$$

Therefore,

$$\sup_{x\geq 0}\{xt - p^*(x)\} \leq p(t)\,. \tag{4}$$

If $p(t)$ is a convex function, then, for every $t \geq 0$, there exists a value $x = x(t)$ such that (3), and hence (4), is an equality. This implies that if $p(t)$ is a convex function, then $(p^*)^* = p$.

THEOREM 1 (Dzhrbashyan). *Let $f(t)\exp p(|t|) \in L^1(\mathbb{R})$. If $\tilde{f}$ is the Fourier transform* (1) *of f, and if*

$$\begin{aligned}\liminf_{R\to\infty}\Big\{\frac{1}{R}\int_0^\pi p^*(R\sin\theta)\sin\theta\,d\theta \\ + \frac{1}{R^2}\int_1^R x\Big(\int_1^x \frac{\log|\tilde{f}(u)\tilde{f}(-u)|}{u^2}\,du\Big)dx\Big\} = -\infty\,,\end{aligned} \tag{5}$$

then f and $\tilde{f}$ are equal to zero a.e. on $\mathbb{R}$.

PROOF. Since $p(t) \geq 0$, we have $f \in L^1(\mathbb{R})$, and $\tilde{f}$ is a continuous function on the whole real axis. Moreover, by (3), $\tilde{f}$ is an entire function satisfying the estimate

$$|\tilde{f}(x+iy)| \leq \int_{-\infty}^{\infty} |f(t)|e^{|ty|}\,dt \leq \int_{-\infty}^{\infty} f(t)e^{p(t)}\,dt\cdot e^{p^*(y)}\,. \tag{6}$$

If f is not equal to zero a.e., then applying the Carleman formula from Section 24.1 and integrating by parts we obtain

$$\begin{aligned}&\frac{1}{\pi R}\int_0^\pi \log|\tilde{f}(R\sin\theta)|\sin\theta\,d\theta \\ &\qquad + \frac{1}{\pi R^2}\int_1^R x\,dx\int_1^x \frac{\log|\tilde{f}(u)\tilde{f}(-u)|du}{u^2} + O(1) \\ &\qquad = \frac{1}{\pi R}\int_0^\pi \log|\tilde{f}(R\sin\theta)|\sin\theta\,d\theta \\ &\qquad + \frac{1}{2\pi}\int_1^R \Big(\frac{1}{u^2} - \frac{1}{R^2}\Big)\log|\tilde{f}(u)\tilde{f}(-u)|\,du + O(1) \geq 0\,.\end{aligned}$$

Hence, the function

$$\frac{1}{R}\int_0^\pi p^*(R\sin\theta)\sin\theta\,d\theta + \frac{1}{R^2}\int_1^R x\,dx\int_1^x \frac{\log|\tilde{f}(u)\tilde{f}(-u)|}{u^2}\,du$$

is bounded from below as $R\to\infty$, contradicting condition (5). We conclude that $\tilde{f}(\lambda)\equiv 0$, and $f(t) = 0$ a.e. on $\mathbb{R}$.

REMARK. If, instead of $f(t)\exp p(|t|) \in L^1(\mathbb{R})$, we assume that $f \in L^1(\mathbb{R})$ and $f(t)\exp p(t) \in L^1(-\infty, 0)$, then Theorem 1 remains valid. Indeed, these assumptions imply that $\tilde{f}(\lambda)$ is a function analytic in the half-plane $\{\lambda : \operatorname{Im}\lambda > 0\}$ and is continuous in its closure. Therefore, relation (6) is valid for $y \geq 0$. The previous arguments prove that $f = 0$ a.e. on $\mathbb{R}$.

COROLLARY 1. *If $f(t) \exp p(|t|) \in L^1(\mathbb{R})$ and, instead of* (5), *the condition*

$$\liminf_{R\to\infty} \left\{ \frac{p^*(R)}{R} + \frac{2}{\pi R^2} \int_1^R x \Big(\int_1^x \frac{\log|\tilde f(u)\tilde f(-u)|}{u^2}\, du \Big)\, dx \right\} = -\infty \tag{7}$$

holds in the hypothesis of Theorem 1, *then f and $\tilde f$ are equal to zero a.e. on* $\mathbb{R}$.

Indeed, since $p^*(x)/x$ is an increasing function, we have that

$$\frac{1}{R} \int_0^{\pi} p^*(R\sin\theta)\sin\theta\, d\theta < \frac{p^*(R)}{R} \int_0^{\pi} \sin^2\theta\, d\theta = \frac{\pi p^*(R)}{2R}\,,$$

and (7) implies (5).

COROLLARY 2. *If $f(t)\exp p(|t|) \in L^1(\mathbb{R})$, and instead of* (5), *we assume that $\tilde f(x) = O(e^{-\alpha(|x|)})$ as $|x| \to \infty$, and*

$$\liminf_{R\to\infty} \left\{ \frac{p^*(R)}{R} - \frac{4}{\pi R^2} \int_1^R x \Big(\int_1^x \frac{\alpha(u)}{u^2}\, du \Big)\, dx \right\} = -\infty\,, \tag{8}$$

then $f = \tilde f = 0$ a.e. on $\mathbb{R}$.

If we set $p(t) = At^p$ and $\alpha(t) = Bt^{p'}$, where $1/p + 1/p' < 1$, then since $p^*(t) = A^* t^q$ for $1/p + 1/q = 1$ we have $q < p'$ and (8) is satisfied. It leads us to the uniqueness theorem of Morgan from Section 8.3.

Here is another variation on the same theme.

THEOREM 2 (Beurling). *Let $f \in L^1(\mathbb{R})$ and let*

$$\iint_{\mathbb{R}^2} |f(x)\tilde f(y)| e^{|xy|}\, dx dy < \infty\,.$$

Then f equals zero a.e. on $\mathbb{R}$.

PROOF.[24] Let

$$M(y) = \int_{-\infty}^{\infty} |f(x)| e^{|xy|}\, dx\,,$$

$$N(x) = \int_{-\infty}^{\infty} |\tilde f(y)| e^{|xy|}\, dy\,.$$

It is evident that $M(y)$ and $N(x)$ are increasing functions of $|y|$ and $|x|$, respectively, and,

$$\int_{-\infty}^{\infty} |\tilde f(y)| M(y)\, dy = \int_{-\infty}^{\infty} |f(x)| N(x)\, dx = \iint_{\mathbb{R}^2} |f(x)\tilde f(y)| e^{|xy|}\, dx dy < \infty\,. \tag{9}$$

Let us assume, first, that the function $f(x)$ has compact support. If $f \not\equiv 0$, then the function $M(y)$ grows exponentially. Hence, $|\tilde f(y)|$ tends to zero, at least, exponentially, and f has an analytic continuation into a horizontal strip. Since $f(x)$ has bounded support, $f \equiv 0$. If the function $\tilde f$ has a compact support, then we may use the same arguments.

[24]This theorem was proved by A. Beurling in the early forties, but the proof was published by Hörmander [**58**] only recently. Here, we follow the latter paper.

Thus we conclude that none of the functions f and $\tilde{f}$ has a compact support and hence M and N grow faster than an arbitrary exponential function. By the inversion formula the functions f and $\tilde{f}$ are entire functions admitting the estimates

$$\begin{aligned} |f(x+iy)| &\le N(y) , \\ |\tilde{f}(x+iy)| &\le M(y) . \end{aligned} \tag{10}$$

Let

$$F(z) = \int_0^z f(s)f(is)\,ds .$$

The first estimate in (10) together with (9) implies that

$$\int_{-\infty}^{\infty} |f(x)|\,|f(\pm ix)|\,dx \le \int_{-\infty}^{\infty} |f(x)|N(x)\,dx < \infty , \tag{11}$$

and hence the entire function $F(z)$ is bounded on the coordinate axes. Further, the estimate (10) gives the inequality

$$M_F(r) \le rN^2(r) .$$

If

$$\liminf_{|x|\to\infty} N(x)e^{-cx^2/2} = 0$$

for each $c > 0$, then the version of Phragmén-Lindelöf theorem given in Section 24.1 and applied to the function $F(z^2)$ implies that $F(z)$ is a bounded function in $\mathbb{C}$, and hence $F(z) \equiv 0$. Since the same argument is valid for the function $\tilde{f}$, we conclude that if $f \not\equiv 0$, then there exists $c > 0$ such that

$$N(x) \ge e^{cx^2/2} , \qquad M(y) \ge e^{cy^2/2}$$

for large $|x|$ and $|y|$. Since

$$f(z) = (2\pi)^{-1} \int_{-\infty}^{\infty} \tilde{f}(y)M(y)\{e^{-izy}/M(y)\}\,dy ,$$

we conclude from (9) that

$$|f(z)| \le K \sup_y e^{|y|\,|\operatorname{Im} z| - cy^2/2} = Ke^{|\operatorname{Im} z|^2/2c} ,$$

proving that either f is of order at most two or it equals zero identically.

Since the majorant $N(r)$ is increasing, it follows from (10) and (9) that for every $\alpha \in [0, 2\pi]$,

$$\int_0^{\infty} |f(r)f(re^{i\alpha})|\,dr \le \int_0^{\infty} |f(r)|N(r)\,dr = K' < \infty .$$

Let us show that

$$\int_0^{\infty} |f(re^{i\alpha/2})|^2\,dr < K' , \qquad 0 < \alpha < \frac{\pi}{2} \tag{12}$$

with the same constant K' for all $\alpha, 0 < \alpha < \pi/2$. Indeed, if φ is a smooth compactly supported function on the real axis, $|\varphi| \le 1$, then the analytic function

$$\Phi(z) = z \int_0^{\infty} f(tz)\overline{f(t\overline{z}e^{i\alpha})}\varphi(t)\,dt$$

is bounded by a constant K' on the boundary of the angle $\{z : 0 < \arg z < \alpha\}$. Since the order of $\Phi(z)$ does not exceed 2, by virtue of the Phragmén-Lindelöf theorem the same estimate holds inside the angle, proving (12). As $\alpha \to \pi/2$ we conclude that

$$\int_0^\infty |f(re^{i\pi/4})|^2\,dr \le K' .$$

In the same way the estimates

$$\int_0^\infty |f(\pm re^{\pm i\pi/4})|^2\,dr < \infty$$

are proved. Hence,

$$|F(\pm re^{\pm i\pi/4})|$$

are bounded functions, and by the Phragmén-Lindelöf theorem, $F(z)$ is constant. Therefore, $f(z)f(iz) \equiv 0$, proving Theorem 2.

COROLLARY. *If $p(x)$ and $q(x)$ are nonnegative functions on the real axis, if*

$$(13)\qquad |xy| \le p(x) + q(y) ,$$

and if

$$\int_{-\infty}^\infty |f(x)|e^{p(x)}\,dx < \infty , \qquad \int_{-\infty}^\infty |\tilde f(y)|e^{q(y)}\,dy < \infty ,$$

then f vanishes a.e. on $\mathbb{R}$.

25.2. Construction of entire functions decaying on the real axis

The following problem due to Gelfand and Shilov [**39**, Chapter IV] is of interest in the theory of distributions.

Let S_l^m be the class of all entire functions $f(z)$ satisfying the condition

$$(14)\qquad |f(z)| \le C\exp\{l(A|y|) - m(B|z|)\} , \qquad z = x + iy ,$$

where $l(x)$ and $m(x)$ are positive increasing functions, and A, B and C are numbers depending, perhaps, on f. What are conditions for l and m such that there exists an entire function $f(z) \in S_l^m$ not identically equal to zero ?

Let us remark that the proof of Theorem 4, Section 8.3, due to Shilov, was reduced to the proof of triviality of the class S_l^m for $m(x) = x^{1/\alpha}$, $l(x) = x^{1/(1-\beta)}$ with $\alpha + \beta < 1$ (see inequality (7) of Lecture 8).

Carleman's formula yields a condition of triviality of the classes S_l^m.

THEOREM 3 (Dzhrbashyan). *Let positive functions $l(x)$ and $m(x)$ be given such that $l(x)/x$ and $m(x)$ monotonically increase as x increases, and let*

$$(15)\qquad \liminf_{x\to\infty}\left[\frac{l(\theta x)}{x} - \int_1^x \frac{m(u)}{u^2}\,du\right] = -\infty$$

for every $\theta > 0$. Then the class S_l^m is trivial.

PROOF. Let an entire function $f(z)$ be not identically zero. Introducing the function $f(z/B)$ and using the Carleman formula we have

$$O(1) \leq \frac{1}{\pi R}\int_0^{\pi}\Big[l\Big(\frac{A}{B}R\sin\theta\Big) - m(R)\Big]\sin\theta\, d\theta$$
$$-\frac{1}{\pi}\int_1^R\Big(\frac{1}{u^2}-\frac{1}{R^2}\Big)m(u)\,du\,, \quad R\to\infty\,.$$

If we set $\theta = A/B$ and take into account that $l(x)/x$ and $m(x)$ are monotonic, we obtain

$$O(1) \leq \frac{2l(\theta R)}{R} - \frac{2m(R)}{R} - \int_1^R \frac{m(u)}{u^2}\,du$$
$$+\frac{1}{R^2}\int_1^R m(u)\,du < \frac{l(2\theta R)}{R} - \int_1^R \frac{m(u)}{u^2}\,du$$

in contradiction to (15). This proves the theorem.

Under additional assumptions on the behavior of the function $m(x)$ Babenko [**9**] obtained a converse to Theorem 2.

THEOREM 4 (Babenko). *Let functions $l(x) > 0$ and $m(x) > 0$ be such that $l(x)/x$ and $xm'(x)$ monotonically increase as x increases, let*

$$\lim_{x\to\infty} xm'(x) = +\infty\,,$$

and let, for sufficiently large k,

$$x^{-k}\int_1^x \frac{m(u)}{u^2}\,du \searrow 0 \tag{16}$$

as $x\to\infty$. If the inequality

$$\liminf_{x\to\infty}\Big[\frac{l(\theta x)}{x} - \int_1^x \frac{m(u)}{u^2}\,du\Big] > -\infty \tag{17}$$

holds at least for one value $\theta > 0$, then the class S_l^m is nontrivial.

PROOF. Let p be a positive integer and let

$$h(z) = 1 - \Big(1 - \frac{\sin^2 z}{z^2}\Big)^p\,.$$

It is evident that $h(z)$ satisfies the following conditions:

$$h(0) = 1\,, \qquad h'(0) = h''(0) = \cdots = h^{(2p-1)}(0) = 0\,, \tag{18}$$

$$0 \leq h(x) < 1 \quad \text{for} \quad x \neq 0\,. \tag{19}$$

Then the function

$$H(z) = \Big(\frac{\sin z}{z}\Big)^2 h(z)$$

satisfies the inequality

$$H(x) \leq \frac{1}{|x|}\,, \qquad x\in\mathbb{R}\,. \tag{20}$$

It is clear that the function $H(z)$ also meets conditions (18) and (19). We fix a number p such that the inequality $2p > k+1$ is valid with k as in (16), and denote by σ the exponential type of $H(z)$: $\sigma = 2+2p$. By virtue of the Phragmén-Lindelöf theorem, (19) and (20),

$$\log|h(z)| \le \sigma|y|\,, \qquad z = x+iy \in \mathbb{C}\,, \tag{21}$$

$$\log|H(z)| \le \sigma|y| - \log|z|\,, \qquad z \in \mathbb{C}\,. \tag{22}$$

Let us choose a number $\delta > 0$ small enough to satisfy the inequality

$$4\sigma\delta(1 + (\log 2)^{-1}) < \frac{1}{4} \tag{23}$$

and such that the relations

$$\log|H(z)| \le 0\,, \tag{24}$$

$$\log|H(z)| \le -\log|z| \tag{25}$$

are valid in the sector

$$G_\delta = \{z : |\arg z| \le \delta \quad \text{or} \quad |\arg z - \pi| \le \delta; \quad |z| \le 2\}\,.$$

In what follows, without loss of generality, we assume $m(x) = 0$ for $x \le 1$. Let us set $n(x) = [xm'(x)]$, where $[t]$ is the integral part of t, and

$$m_*(x) = \int_0^\infty \frac{n(t)}{t}\,dt\,.$$

Then $m_*(x) \le m(x) \le m_*(x) + \log x$. We denote by $\rho_1 \le \rho_2 \le \cdots \le \rho_k \le \cdots$ the jump points of the function $n(x)$. Since the function $m(x)$ is monotonic and condition (16) is valid,

$$\frac{m(x)}{x^{k+1}} \le 2x^{-k}\int_x^{2x} \frac{m(u)}{u^2}\,du \to 0\,, \qquad x \to \infty\,.$$

Integrating by parts three times and using condition (16) with the inequality $2p > k+1$, we obtain

$$\begin{aligned}
|z|^{2p} \sum_{k=n(|z|)+1}^{\infty} \frac{1}{\rho_k^{2p}} &= |z|^{2p}\int_{|z|}^\infty \frac{dn(u)}{u^{2p}} \le 2p|z|^{2p}\int_{|z|}^\infty \frac{n(u)}{u^{2p+1}}\,du \\
&= 2p|z|^{2p}\int_{|z|}^\infty \frac{dm_*(u)}{u^{2p}} \le (2p)^2|z|^{2p}\int_{|z|}^\infty \frac{m_*(u)}{u^{2p+1}}\,du \\
&\le (2p)^3|z|^{2p}\int_{|z|}^\infty \frac{dt}{t^{2p}}\int_0^t \frac{m_*(u)}{u^2}\,du \\
&\le (2p)^3|z|^{2p}\int_{|z|}^\infty \frac{dt}{t^{2p}}\left(\frac{t}{|z|}\right)^k\int_0^{|z|} \frac{m(u)}{u^2}\,du \\
&\le C_1|z|\int_0^{|z|} \frac{m(u)}{u^2}\,du\,.
\end{aligned} \tag{26}$$

Here and in what follows each C_j is a positive number not depending on z.

It follows from (18) (for $H(z)$) and (26) that the infinite product

$$\prod_{k=1}^{\infty} H^2(z/\rho_k) \tag{27}$$

converges uniformly on each compact set to an entire function $f(z)$. We shall show that $f(z)$ belongs to the class S_l^m. First of all,

$$\begin{aligned}\log|f(z)| &= 2\int_0^\infty \log\Big|H\Big(\frac{z}{t}\Big)\Big|\,dn(t)\\ &= 2\Big(\int_0^{|z|}+\int_{|z|}^\infty\Big)\log\Big|H\Big(\frac{z}{t}\Big)\Big|\,dn(t) = J_1+J_2\,.\end{aligned} \tag{28}$$

To estimate the first integral, let us use inequality (22):

$$\begin{aligned}J_1 &= 2\int_0^{|z|}\log\Big|H\Big(\frac{z}{t}\Big)\Big|\,dn(t) \le 2\sigma|y|\int_0^{|z|}\frac{dn(t)}{t} - 2\int_0^{|z|}\frac{n(t)}{t}\,dt\\ &= 2\sigma|y|\Big[\frac{n(|z|)}{|z|}+\int_0^{|z|}\frac{n(t)}{t^2}\,dt\Big] - 2m_*(|z|)\,.\end{aligned}$$

With account taken of the inequalities

$$n(|z|)\le m_*(e|z|)\,,\qquad m_*(|z|)\le 2|z|\int_0^{2|z|}\frac{m(u)}{u^2}\,du\,,$$

we obtain

$$J_1 \le C_2|y|\int_0^{2e|z|}\frac{m(u)}{u^2}\,du - 2m_*(|z|)\,. \tag{29}$$

To estimate the second integral on the right-hand side of (28), we use first the inequality $\log|H(z)|\le C_3|z|^{2p}$, $|z|\le 1$, which follows from conditions (18), and then estimate (26):

$$J_2 \le C_3|z|^{2p}\int_{|z|}^\infty\frac{dn(t)}{t^{2p}} \le C_4|z|\int_0^{|z|}\frac{m(u)}{u^2}\,du\,. \tag{30}$$

Substituting (29) and (30) into (28) we arrive at the inequality

$$\log|f(z)| \le C_5|y|\int_0^{C_5|y|}\frac{m(u)}{u^2}\,du - m(|z|)\,, \tag{31}$$

valid for all z such that $\delta\le\arg z\le\pi-\delta$ or $-\pi-\delta\le\arg z\le-\delta$ where δ is a number chosen in accord with (23).

Let us now estimate the function $f(z)$ inside the angle

$$\{z:\ |\arg z|\le\delta\}\cup\{z:\ |\arg z-\pi|\le\delta\}\,.$$

We have

$$\log|f(z)| = 2\Big(\int_0^{|z|/2}+\int_{|z|/2}^{|z|}+\int_{|z|}^\infty\Big)\log\Big|H\Big(\frac{z}{t}\Big)\Big|\,dn(t)\,.$$

By virtue of inequality (24), when estimating the function $\log|f(z)|$ from above, the third integral may be disregarded. To estimate the first and the second integrals, we apply inequalities (22) and (25), respectively. Since $\sigma > 1$, we obtain, after integrating by parts,

$$
\begin{aligned}
\log|f(z)| &\le 2\sigma \int_0^{|z|/2} \Big(\frac{|y|}{t} - \log\frac{|z|}{t}\Big)\, dn(t) - 2\int_{|z|/2}^{|z|} \log\frac{|z|}{t}\, dn(t) \\
&\le 2\sigma|y|\Big[\frac{n(|z|/2)}{|z|/2} + \int_0^{|z|/2} \frac{n(t)}{t^2}\, dt\Big] - 2\int_0^{|z|} \frac{n(t)}{t}\, dt \qquad (32)\\
&= 2\sigma|y|\Big[\frac{n(|z|/2)}{|z|/2} + \frac{m_*(|z|/2)}{|z|/2} + \int_0^{|z|/2} \frac{m_*(u)}{u^2}\, du\Big] - 2m_*(|z|)\,.
\end{aligned}
$$

Now

$$n(|z|/2) \le m_*(|z|)(\log 2)^{-1}\,,$$

and

$$\int_0^{|z|/2} \frac{m_*(u)}{u^2}\, du \le \int_0^{4\sigma|y|} \frac{m_*(u)}{u^2}\, du + \frac{m_*(|z|/2)}{4\sigma|y|}\,.$$

If we substitute these bounds in (32) and take into account that now $|y/z| \le \delta$, we obtain that

$$
\begin{aligned}
\log|f(z)| &\le 4\sigma\delta(1 + (\log 2)^{-1})m_*(|z|) + \frac{1}{2}m_*(|z|) \\
&\quad + 2\sigma|y| \int_0^{4\sigma|y|} \frac{m_*(u)}{u^2}\, du - 2m_*(|z|) \\
&= 2\sigma|y| \int_0^{4\sigma|y|} \frac{m_*(u)}{u^2}\, du + \Big[4\sigma\delta(1 + (\log 2)^{-1}) - \frac{3}{2}\Big] m_*(|z|)\,.
\end{aligned}
$$

By virtue of (23) we have

$$\log|f(z)| \le 2\sigma|y| \int_0^{4\sigma|y|} \frac{m(u)}{u^2}\, du - m(|z|)\,. \qquad (33)$$

Estimates (31) and (33) imply that, everywhere in the complex plane, the following inequality holds:

$$\log|f(z)| \le C_6|y| \int_0^{C_6|y|} \frac{m(u)}{u^2}\, du - m(|z|)\,.$$

According to condition (17),

$$|f(z)| \le Ce^{l(C_6\theta|y|) - m(|z|)}\,,$$

implying that $f(z)$ is a nontrivial element of the class S_l^m. Theorem 4 is proved.

Some results close to Theorem 4 are contained in papers of Mandelbrojt [**92**], Katznelson and Mandelbrojt [**64**]. See also Redheffer [**115**, pages 26–29].

25.3. Uniqueness problem of Gelfand and Shilov for infinitely differentiable functions

Gelfand and Shilov posed the problem of the existence of a nonzero infinitely differentiable function $f(t)$ on the real axis which satisfies the inequalities

$$|t^k f^{(n)}(t)| \le C_f A^k B^n m_n l_k, \qquad n,k = 0,1,2\dots. \tag{34}$$

If $l_k = A^{-k}$ for all $k \ge 0$, then it follows from (34) that $f(t) \equiv 0$ for $|t| > 1$ and the problem is reduced to the Hadamard problem of nonquasianalyticity of the class C_{m_n} (cf. Lecture 14). If $l_k = k^{\alpha k}$, $m_n = n^{\beta n}$, $\alpha + \beta < 1$, then, as was proved by Shilov (see Lecture 8), condition (34) implies $f(t) \equiv 0$.

In what follows, we denote by $C(l_k, m_n)$ the class of all infinitely differentiable functions on the whole real axis, satisfying (34) with some positive numbers A, B, and C, perhaps depending on f. Classes of this type and their multidimensional analogs arise in the theory of distributions and in the theory of partial differential equations (see Gelfand and Shilov [**39**], Palamodov [**108**]). We shall assume that the sequences $\{l_k\}$ and $\{m_k\}$ from the definition of $C(l_k, m_k)$ satisfy the following condition:

$$\lim_{n\to\infty} \frac{m_n}{C^n} = \lim_{k\to\infty} \frac{l_k}{C^k} = +\infty \tag{35}$$

for each $C > 0$.

We need the following definition.

DEFINITION. A sequence of positive numbers $\{m_n\}$ is called *logarithmically convex* if, for all k and n,

$$\log m_{k+n} \le \log m_k + \log m_n .$$

Let us, once more, recall the definition of A. Ostrowski's function (see Section 14.3) for a sequence $\{m_n\}$:

$$M(x) = \sup_{n\ge 0} \frac{x^n}{m_n} .$$

By virtue of (35) this function is finite for each $x > 0$. Let the function $\varphi(u)$ on $(0,\infty)$ be continuous and linear on each interval $[n, n+1]$ and such that $\varphi(n) = \log m_n$. Then the function

$$\log M(e^t) = \sup_{n\ge 0}(nt - \log m_n)$$

coincides with the Legendre transform of $\varphi(u)$. In particular, if $\{m_n\}$ is a logarithmically convex sequence, then the function $\varphi(u)$ is convex, and hence $\varphi(u) = (\log M(e^t))^*$. Therefore, in this case the initial sequence $\{m_n\}$ can be recovered from its Ostrowski function in the following way:

$$m_n = \sup_{x>0} \frac{x^n}{M(x)} . \tag{36}$$

Our aim now is to give necessary and sufficient conditions for $C(l_k, m_n)$ to be nontrivial if $\{l_k\}$ and $\{m_n\}$ are logarithmically convex sequences.[25] To this end,

[25] In our exposition we follow Dzhrbashyan [**27**] and Babenko [**9**]. Some other nontriviality conditions for the same classes but under different restrictions on $\{l_k\}$ and $\{m_n\}$ are given in Babenko [**8**], Mandelbrojt [**92, 93**].

we need a description of the class $\tilde{C}(l_k, m_n)$ formed by the Fourier transforms of functions from $C(l_k, m_n)$.

THEOREM 5. *Let $\{l_k\}$ and $\{m_n\}$ be logarithmically convex sequences, and let $L(x)$ and $M(x)$ be their Ostrowski functions. Then*

$$\tilde{C}(l_k, m_n) \subset S_{l_*}^m \subset \tilde{C}(l_k, m_{n+2}) ,$$

where $m(x) = \log M(x)$, $l^*(x)$ is the Legendre transform of the function $l(x) = \log L(x)$, and the class S_l^m is defined by (14).

PROOF. Let us first prove the inclusion $\tilde{C}(l_k, m_n) \subset S_{l^*}^m$. If f is a nontrivial function of class $C(l_k, m_n)$, then it follows from (34) that

$$|f^{(n)}(t)| \le C\frac{1+t^2}{L(t/A)} \cdot \frac{B^n m_n}{1+t^2} = \frac{C}{L_1(t)} \cdot \frac{B^n m_n}{1+t^2} ,$$

where $L_1(t) = L(t/A)/(1+t^2)$. To estimate the Fourier transform $\tilde{f}(z)$, we integrate by parts and obtain

$$\tilde{f}(z) = \frac{(-1)^n}{(iz)^n} \int_{-\infty}^{\infty} f^{(n)}(t) e^{izt}\, dt .$$

The previous estimate implies

$$\begin{aligned} |\tilde{f}(x+iy)| &\le \frac{CB^n m_n}{|z|^n} \exp\left[\sup_{\tau \ge 0} (\tau|y| - \log L_1(\tau))\right] \int_{-\infty}^{\infty} \frac{dt}{1+t^2} \\ &\le C \exp[l_1^*(|y|) - m(|z|/B)] , \end{aligned}$$

where l_1^* denotes the Legendre transform of $\log L_1(t) = l(t/A) - \log(1+t^2)$. Then, for sufficiently large s, we have

$$\begin{aligned} l_1^*(s) &= \sup_{t\ge 0}[st - l(t/A) + \log(1+t^2)] \\ &= \sup_{t \ge 0}[2Ast - l(t) + (\log(1+t^2) - Ast)] \\ &\le l^*(2As) + O(1) , \qquad s \to \infty . \end{aligned}$$

Hence, $\tilde{f} \in S_{l^*}^m$.

Let us prove the inclusion $S_{l^*}^m \subset \tilde{C}(l_k, m_{n+2})$. We shall show that if $\tilde{f}$ is a nontrivial function from $S_{l^*}^m$, then $f \in C(l_k, m_{n+2})$.

Let us set

$$f(t) = \int_{-\infty}^{\infty} \tilde{f}(z) e^{izt}\, dz .$$

According to the definition of the class $S_{l^*}^m$, we may integrate along any straight line $\{z : \operatorname{Im} z = y\}$ instead of the real axis. After such a substitution we obtain

$$t^k f^{(n)}(t) = t^k i^n \int_{-\infty}^{\infty} \tilde{f}(z) e^{ixt - yt} z^n\, dx , \qquad z = x + iy .$$

If y has the same sign as t, it follows that

$$|t^k f^{(n)}(t)| = C|t|^k \int_{-\infty}^{\infty} |z|^n \exp[l^*(A|y|) - m(B|z|) - yt]\, dx . \tag{37}$$

Since $\{l_k\}$ and $\{m_n\}$ are logarithmically convex sequences, we have

$$|z|^{n+2}e^{-m(B|z|)} = \exp[(n+2)\log|z| - m(B|z|)]$$
$$\leq \sup_{t>0} \exp\left[(n+2)\log\frac{t}{B} - m(t)\right] = \frac{m_{n+2}}{B^{n+2}}\,,$$

for sufficiently large $|z|$ and $|y|$, and (37) implies

$$|t^k f^{(n)}(t)| \leq C_1|t|^k \frac{m_{n+2}}{B^{n+2}} \exp\{l^*(A|y|) - yt\}\,.$$

Minimizing the right-hand side with respect to $|y|$, we obtain

$$(38)\qquad |t^k f^{(n)}(t)| \leq C_1|t|^k \frac{m_{n+2}}{B^{n+2}} \exp\Big\{-l\Big(\frac{|t|}{A}\Big)\Big\}.$$

But

$$|t|^k \exp\Big\{-l\Big(\frac{|t|}{A}\Big)\Big\} = \frac{|t|^k}{L(|t|/A)} \leq |A|^k \sup_{\tau\geq 0}\frac{\tau^k}{L(\tau)} = |A|^k l_k\,.$$

Substituting this bound in (38), we find $f \in C(l_k, m_{n+2})$ proving the Theorem.

REMARK. For Shilov's classes $l_k = k^{\alpha k}$, $m_n = n^{\beta n}$, it is readily seen that

$$\tilde{C}(l_k, m_n) = C(m_n, l_k).$$

Gelfand and Shilov [**39**, Chapter IV, §6], Babenko [**8**], and Mandelbrojt [**93**] proved this relation under various restrictions imposed on the sequences $\{l_k\}$ and $\{m_n\}$.

Combining Theorems 3, 4 and 5 we obtain the main theorem of the present section.

THEOREM 6 (Babenko, Dzhrbashyan). *Assume that $\{l_k\}$ and $\{m_n\}$ are logarithmically convex sequences, $L(x)$ and $M(x)$ are their Ostrowski functions, $m(x) = \log M(x)$, and $l^*(x)$ is the Legendre transform of $\log L(x)$. Let, for some k, condition* (16) *be fulfilled.*

Then the class $C(l_k, m_n)$ is nontrivial if and only if, for some $\theta > 0$, condition (17) *is fulfilled with l replaced by l^*.*

PROOF. If (17) fails, then, according to Theorem 3, the class $S^m_{l^*}$ is trivial, and Theorem 5 implies the same for $C(l_k, m_n)$.

Let (17) be fulfilled. Since

$$xm'(x) = \frac{d\log M(x)}{d\log x} = \frac{d}{dt}\log M(e^t)\,, \qquad t = \log x\,,$$

and the function

$$\log M(e^t) = \sup_{n\geq 0}\log\frac{e^{tn}}{m_n} = \sup_{n\geq 0}\{tn - \log m_n\}$$

is convex, $xm'(x)$ is an increasing function. If $xm'(x) \leq C < \infty$, then $M(x) \leq x^C$ and $x^n m_n^{-1} \leq x^C$ for any $n \geq 0$, which is a contradiction if $n > C$. Hence,

$$\lim_{x\to\infty} xm'(x) = +\infty\,.$$

In the same way, $l^*(x)/x$ is monotonic, nonbounded, and hence tends to infinity. We conclude that the conditions of Theorem 4 are fulfilled for the functions $l^*(x)$

and $p(x) = m(x) - 2\log x$. Hence, the class $S_{l^*}^{p}$ is nontrivial. Since the function $p(x)$ corresponds to the sequence m_{n-2}, the class $C(l_k, m_n)$ is also nontrivial, which proves the theorem.

REMARK. The necessity of (17) for the class $C(l_k, m_n)$ to be nontrivial was proved above (as in the paper of Dzhrbashyan) only if the sequences $\{l_k\}$ and $\{m_n\}$ satisfy (35).

LECTURE 26

The Matsaev Theorem on the Growth of Entire Functions Admitting a Lower Bound

In this lecture we will prove[26]

THEOREM 1 (Matsaev). *If an entire function $f(z)$ has the lower bound*

$$(1)\qquad |f(z)| \geq \exp\Big\{ -Cr^{\rho}\frac{1}{|\sin\varphi|^{k}}\Big\}, \qquad z = re^{i\varphi}, \quad \rho > 1, \quad k > 0,$$

then the function $f(z)$ is of order ρ and finite type.

The proof of this theorem consists of two parts, each of which is of independent interest. In the first part the upper bound

$$(2)\qquad |f(z)| \leq C\exp\Big\{Cr^{\rho}\frac{1}{|\sin\varphi|^{k+1}}\Big\},$$

is derived from the lower bound (1) using Carleman's and R. Nevanlinna's formulas. In the second part the assertion of Matsaev's theorem is derived from this upper bound.

In what follows we shall denote by C various values which may depend on ρ or k, but not on z or $f(z)$.

26.1. A lower bound for harmonic functions of order greater than one in the upper half-plane

First, we prove

THEOREM 2. *Let $u(z)$ be a harmonic function in the upper half-plane with continuous boundary values on the real axis. Suppose that*

$$(3)\qquad u(z) \leq Kr^{\rho}, \qquad z \in \mathbb{C}_+, \quad r = |z| \geq 1, \quad \rho > 1,$$

and

$$(4)\qquad |u(z)| \leq K, \qquad |z| \leq 1, \qquad \operatorname{Im} z \geq 0.$$

Then

$$(5)\qquad u(re^{i\varphi}) \geq -CK\frac{1+r^{\rho}}{\sin\varphi}, \qquad re^{i\varphi} \in \mathbb{C}_+,$$

where C does not depend on K, r, φ and the function $u(z)$.

[26] Following [**96, 97**].

PROOF. To begin with, we apply Carleman's formula (6) from Section 24.2 to the harmonic function $u(z)$:

$$\frac{1}{\pi R}\int_0^\pi u^-(Re^{i\theta})\sin\theta\,d\theta + \frac{1}{2\pi}\int_1^R\left(\frac{1}{t^2}-\frac{1}{R^2}\right)[u^-(t)+u^-(-t)]\,dt$$

$$(6)\qquad = \frac{1}{\pi R}\int_0^\pi u^+(Re^{i\theta})\sin\theta\,d\theta$$

$$+\frac{1}{2\pi}\int_1^R\left(\frac{1}{t^2}-\frac{1}{R^2}\right)[u^+(t)+u^+(-t)]\,dt + A_u(R)\,, \qquad R\ge 2\,.$$

Here and further on, $u^- = (-u)^+$, i.e., $u = u^+ - u^-$, and the remainder term is

$$A_u(R) = \frac{1}{\pi}\int_0^\pi u(e^{i\varphi})\sin\varphi\,d\varphi + \frac{1}{2\pi}\int_0^1\left(1-\frac{1}{R^2}\right)[u(t)+u(-t)]\,dt.$$

Using estimate (4), we obtain

$$(7)\qquad |A_u(R)| \le 2K\,.$$

Other terms on the right-hand side of (6) can be estimated using (3):

$$\frac{1}{\pi R}\int_0^\pi u^+(Re^{i\theta})\sin\theta\,d\theta \le KR^{\rho-1}\,,$$

$$\frac{1}{2\pi}\int_1^R\left(\frac{1}{t^2}-\frac{1}{R^2}\right)[u^+(t)+u^+(-t)]\,dt \le \frac{2K}{2\pi}\int_1^R\left(\frac{1}{t^2}-\frac{1}{R^2}\right)t^\rho\,dt$$

$$\le \frac{K}{\pi(\rho-1)}R^{\rho-1}\,.$$

Thus, for $R\ge 2$ we obtain from (6)

$$(8)\qquad \frac{1}{\pi R}\int_0^\pi u^-(Re^{i\theta})\sin\theta\,d\theta \le CKR^{\rho-1}\,,$$

and

$$(9)\qquad \begin{aligned}\frac{1}{2\pi}\int_1^R \frac{u^-(t)+u^-(-t)}{t^2}\,dt &\le \frac{4}{3}\frac{1}{2\pi}\int_1^R\left(\frac{1}{t^2}-\frac{1}{(2R)^2}\right)[u^-(t)+u^-(-t)]\,dt\\ &\le \frac{4}{3}\frac{1}{2\pi}\int_1^{2R}\left(\frac{1}{t^2}-\frac{1}{(2R)^2}\right)[u^-(t)+u^-(-t)]\,dt \le CKR^{\rho-1}\,.\end{aligned}$$

R. Nevanlinna's representation for the half-disk (formula (14), Section 24.3), and estimates (8) and (9) allow us to find a lower bound for the function $u(z)$. We have

$$\begin{aligned} -u(z) &= \frac{R^2-|z|^2}{2\pi}\int_0^{\pi}\Big(\frac{1}{|Re^{i\theta}-z|^2}-\frac{1}{|Re^{i\theta}-\overline{z}|^2}\Big)\big(-u(Re^{i\theta})\big)\,d\theta \\ &\quad+\frac{\operatorname{Im} z}{\pi}\int_{-R}^{R}\Big(\frac{1}{|t-z|^2}-\frac{R^2}{|R^2-tz|^2}\Big)\big(-u(t)\big)\,dt\,. \end{aligned} \tag{10}$$

Both integral kernels are positive, since they are derivatives of the Green function with respect to the inner normal. This permits us to replace $-u$ by u^- in the integrals and simultaneously to transform identity (10) into the inequality. Now we shall estimate the kernels which appear in the integrals. Let $r=|z|\geq 2$, $R=2r$. Then, using the inequalities

$$|t-z|^2\geq\begin{cases} t^2\sin^2\varphi\,, & |t|\geq 1,\\ \dfrac{r^2}{4}\,, & |t|<1\,,\end{cases}$$

we obtain the estimate of the kernel in the second integral:

$$\frac{\operatorname{Im} z}{\pi}\Big(\frac{1}{|t-z|^2}-\frac{R^2}{|R^2-tz|^2}\Big)\leq\begin{cases}\dfrac{1}{\pi}\dfrac{r}{t^2\sin\varphi}\,, & |t|\geq 1\,,\\ \dfrac{4\sin\varphi}{\pi r}\,, & |t|<1\,.\end{cases} \tag{11}$$

For the first integral kernel we have

$$\begin{aligned} &\frac{R^2-r^2}{2\pi}\Big(\frac{1}{R^2-2Rr\cos(\theta-\varphi)+r^2}-\frac{1}{R^2-2Rr\cos(\theta+\varphi)+r^2}\Big)\\ &\qquad=\frac{1}{2\pi}\frac{4Rr(R^2-r^2)\sin\theta\sin\varphi}{(R^2-2Rr\cos(\theta-\varphi)+r^2)(R^2-2Rr\cos(\theta+\varphi)+r^2)}\\ &\qquad\leq\frac{4Rr(R+r)\sin\theta}{2\pi(R-r)^3}\leq\frac{12\sin\theta}{\pi}\,. \end{aligned} \tag{12}$$

Substituting inequalities (11) and (12) into (10), we obtain

$$\begin{aligned} -u(z)&\leq\frac{12}{\pi}\int_0^{\pi}u^-(2re^{i\theta})\sin\theta\,d\theta\\ &\quad+\frac{r}{\pi\sin\varphi}\int_1^{2r}\frac{u^-(t)+u^-(-t)}{t^2}\,dt+\frac{4\sin\varphi}{\pi r}\int_{-1}^{1}[u^-(t)+u^-(-t)]\,dt\,. \end{aligned}$$

To estimate the first and the second terms in the right-hand side we use inequalities (8) and (9). The third integral is bounded uniformly with respect to z, by (4). Finally, for $|z|\geq 2$,

$$-u(z)\leq CK\frac{r^{\rho}}{\sin\varphi}\,,$$

and the theorem is proved.

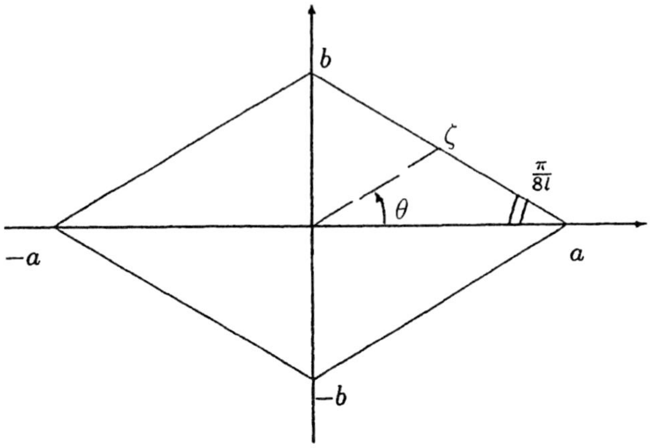

FIGURE 7

REMARK. For $\rho = 1$ the assertion of Theorem 2 is false as is shown by the example of the function

$$u(z) = -\operatorname{Im}[(z + i \log(z + i))], \qquad z \in \mathbb{C}_+ .$$

Further versions and refinements of Theorem 2 may be found in the monograph Nikol'skii [**103,** Chapter I], and in the papers Krasichkov-Ternovskii [**74**], and Matsaev and Mogulskii [**95**]. These results have applications in the study of the division problem for analytic functions.

26.2. Refinement of the upper bound

In this section we shall prove the theorem which plays the basic role in proving Matsaev's theorem.

THEOREM 3. *Let $u(z)$ be a subharmonic function in the complex plane which satisfies the estimate*

$$u(z) \le C \frac{1 + r^\rho}{|\sin \varphi|^l}, \quad z = re^{i\varphi},\ \rho > 1,\ l > 0. \tag{13}$$

Then $u(z)$ is of order ρ and finite type.

PROOF. It suffices to prove that $u(z)$ is of finite order. If that is proved, then, applying the Phragmén-Lindelöf theorem within angles of sufficiently small opening containing the positive and negative rays, we obtain the assertion of the theorem.

Without loss of generality we assume that $l > 1/4$. Then $\pi/8l < \pi/2$. We consider the rhombus L with the vertices at the points

$$\pm a = \pm 2 \csc \frac{\pi}{8l}, \qquad \pm b = \pm 2i \sec \frac{\pi}{8l},$$

and denote by Γ its boundary, see Figure 7.

The function

$$h(\zeta) = \operatorname{Re} \frac{1}{(a^2 - \zeta^2)^{2l}}$$

is harmonic within L and has the lower bound

(14) $$h(\zeta) \geq \frac{C}{|\sin\theta|^{2l}}\,, \qquad \zeta = te^{i\theta}\,,$$

on Γ. Let us prove (14), for example, for $\operatorname{Re}\zeta \geq 0$. We have

$$h(\zeta) = \operatorname{Re}\frac{1}{(a+\zeta)^{2l}(a-\zeta)^{2l}} = \frac{\cos 2l(\psi_1+\psi_2)}{|a-\zeta|^{2l}|a+\zeta|^{2l}}\,,$$

where $\psi_1 = \arg(a+\zeta)$, $\psi_2 = \arg(a-\zeta)$. Inspection of Figure 7 shows that $\psi_1+\psi_2 \in [-\pi/4l, \pi/4l]$, and the estimate (14) is valid. But L contains the unit disk, and we can apply the maximum principle to the function $u(r\zeta) - h(\zeta)$. This function is bounded above inside L and

$$u(z) \leq \max_{|\zeta|=1} h(\zeta) + \max_{\zeta\in\Gamma}[u(r\zeta) - h(\zeta)] \leq C + \max_{0\leq\theta\leq 2\pi}\left(\frac{Cr^\rho}{|\sin\theta|^l} - \frac{C}{|\sin\theta|^{2l}}\right).$$

Further, the maximum of the right-hand side is equal to $C(1+r^{2\rho})$, and hence

$$u(z) \leq C(1+r^{2\rho})\,,$$

proving the theorem.

In this theorem we encounter the situation in which some upper bound of an analytic function yields more precise upper bound. It looks as one of the first theorems of this kind was obtained in the paper [**22**] by Carleman. It states that *if an entire function $f(z)$ satisfies the estimate*

$$|f(re^{i\varphi})| \leq H(\varphi)$$

in the whole complex plane, with $H(\varphi) \geq e$ being an arbitrary function such that

$$\int_0^{2\pi} \log\log H(\varphi)\, d\varphi < \infty\,,$$

then $f(z)$ is bounded, and hence constant.

Other results of this kind may be found in many sources; among them we mention here the monograph Levinson [**84**], and Beurling's lectures published in his Collected Works, Beurling [**13,** vol. I]. See also Domar [**31**] and Rashkovskii [**113, 114**].

26.3. Proof of Matsaev's theorem

Let $f(z)$ be a function satisfying the assumptions of Theorem 1. Evidently, $f(z)$ has only real zeros and $f(0) \neq 0$. Without loss of generality we suppose that $f(z) \neq 0$ for $|z| \leq 1$, and hence

(15) $$|\log|f(z)|| \leq C\,, \qquad |z| \leq 1\,.$$

We choose γ and ψ such that $1 < \gamma < \min(2,\rho)$, $0 < \psi < \pi\left(1 - \frac{1}{\gamma}\right)$, and consider the function

(16) $$u_{\gamma,\psi}(\zeta) = \log|f(\zeta^{1/\gamma}e^{i\psi/2})|.$$

This is a harmonic function in the closed upper half-plane satisfying the estimate

$$u_{\gamma,\psi}(\zeta) \geq -C\tau^{\rho/\gamma} \csc^k \left(\frac{\psi}{2} + \frac{\theta}{\gamma}\right) \geq -C\tau^{\rho/\gamma} \csc^k \left(\frac{\psi}{2}\right), \qquad \zeta = \tau e^{i\theta}. \tag{17}$$

By Theorem 2 applied to $-u_{\gamma,\psi}(\zeta)$ the estimates (15) and (17) yield the upper bound

$$u_{\gamma,\psi}(\zeta) \leq C\tau^{\rho/\gamma} \csc^k \left(\frac{\psi}{2}\right) \csc\theta. \tag{18}$$

In order to estimate the function $f(z)$, we note that $0 < \gamma\psi/2 < \pi$, and set $\zeta = r^\gamma e^{i\gamma\psi/2}$ in (17). By (16) we obtain the estimate

$$\begin{aligned} \log|f(re^{i\psi})| &= u_{\gamma,\psi}(r^\gamma e^{i\gamma\psi/2}) \\ &\leq Cr^\rho\Big(\csc\frac{\psi}{2}\Big)^k \csc\left(\frac{\gamma\psi}{2}\right) \leq Cr^\rho\Big(\csc\frac{\psi}{2}\Big)^{k+1}, \quad r \geq 2, \end{aligned} \tag{19}$$

valid within the angle $\{z = re^{i\psi} : 0 < \psi < \pi(1-1/\gamma)\}$. In the same way the estimate is proved for $0 > \psi > -\pi(1-1/\gamma)$. If we change $f(z)$ to $f(-z)$, we find that (19) holds within the angle $\{z = re^{i\psi} : |\pi - \psi| < \pi(1-1/\gamma)\}$. It remains to obtain the estimate of $f(z)$ within the angle $\{z = re^{i\psi} : \pi(1-1/\gamma) < \psi < \pi/\gamma\}$. For this purpose we set $\zeta = r^\gamma e^{i\gamma(\varphi-\psi/2)}$ in (17) and obtain

$$\log|f(re^{i\varphi})| = u_{\gamma,\psi}(r^\gamma e^{i\gamma(\varphi-\psi/2)}) \leq C(1+r^\rho).$$

The same estimate is evidently valid for $-\pi/\gamma < \varphi < -\pi(1-1/\gamma)$. Combining all these estimates, we obtain (1) with $l = k+1$. Thus, the assertion of Theorem 1 follows immediately from Theorem 3.

26.4. Entire functions admitting a lower bound for $\rho \leq 1$

In Theorem 1 we assumed that $\rho > 1$. Evidently, for $\rho \leq 1$ this theorem implies that an entire function $f(z)$ admitting the lower bound is of order at most one. A more precise estimate of the growth of $f(z)$ is given by the following proposition.

THEOREM 4 (Matsaev). *If an entire function $f(z)$ has a lower bound*

$$|f(z)| \geq \exp\Big\{-M\frac{1+|z|^\rho}{|\operatorname{Im} z|^k}\Big\}, \qquad 0 \leq \rho < 1, \quad k > 0, \tag{20}$$

then it belongs to Cartwright's class C.

PROOF. We shall denote by M different positive constants depending on ρ and k only.

To begin with, we consider the harmonic function $\log\dfrac{1}{|f(z)|}$ in the half-plane $\{z : \operatorname{Im} z > 1\}$. By (20)

$$\log\frac{1}{|f(z)|} \leq M(1+|z|^\rho), \qquad \rho < 1, \quad \operatorname{Im} z \geq 1,$$

and hence this function has a positive harmonic majorant in the upper half-plane. According to the Remark to Theorem 3, Section 14.2, the function $\log|f(z)|$ can be represented in the form

$$\log|f(z)| = \frac{\operatorname{Im} z - 1}{\pi}\int_{-\infty}^{\infty} \frac{\log|f(t+i)|}{|t-z+i|^2}\,dt + M(\operatorname{Im} z - 1), \quad \operatorname{Im} z > 1, \tag{21}$$

where

$$\int_{-\infty}^{\infty} \frac{\log^+ |f(t+i)|}{1+t^2}\, dt < \infty \,. \tag{21}$$

By Hayman's theorem, Lecture 15, we conclude that $f(z)$ is of finite exponential type in the half-planes $\{z : \operatorname{Im} z \ge 1\}$ and $\{z : \operatorname{Im} z \le -1\}$.

By Theorem 1 the entire function $f(z)$ has finite order. To conclude that $f(z)$ is an entire function of exponential type we apply the Phragmén-Lindelöf principle in the strip $\{|\operatorname{Im} z| \le 1\}$.

By Krein's theorem on functions of Cartwright's class (Section 16.1) the condition (21) implies that $\log|f(z)|$ has a positive harmonic majorant in the half-plane $\{z : \operatorname{Im} z < 1\}$. Similarly, $\log|f(z)|$ has a positive harmonic majorant in the half-plane $\{z : \operatorname{Im} z > -1\}$. Therefore, this function has positive harmonic majorant in both half-planes $\mathbb{C}_-$ and $\mathbb{C}_+$. Applying M. G. Krein's theorem once more, we conclude that $f \in C$. The theorem is proved.

REMARK. Theorem 4 immediately implies M. G. Krein's theorem on entire functions $f(z)$ such that $1/f(z)$ is represented by an absolutely convergent series of simple fractions with real poles (Theorem 3, Section 16.1). Indeed, if

$$\frac{1}{f(z)} = \sum_{n=-\infty}^{\infty} \frac{c_n}{z-\lambda_n}$$

where $\{\lambda_n\}$ are real and if

$$\sum_{n=-\infty}^{\infty} \frac{|c_n|}{1+|\lambda_n|} < \infty \,,$$

then

$$\frac{1}{|f(x+iy)|} \le \frac{M}{|y|} \,.$$

Now Theorem 4 may be applied.

Both Matsaev's theorems have important applications to the spectral theory of operators; see Gohberg and Krein [**40**].

LECTURE 27

Entire Functions of Class P

DEFINITION. An entire function $\omega(z)$ of exponential type belongs to the class P if

(a) $\omega(z)$ has no zeros in the lower half-plane $\mathbb{C}_-$;
(b) the number $h_\omega(-\pi/2) - h_\omega(\pi/2)$ is nonnegative.

It is evident that every polynomial without zeros in $\mathbb{C}_-$ belongs to the class P.

In this and the following lectures we will discuss some properties of functions of class P as well as some of their applications.

27.1. Properties of functions of class P

We start with a lemma which may be regarded as a version of the Phragmén-Lindelöf principle.

LEMMA 1. *Let $f(z)$ and $\omega(z)$ be EFET, let $\omega(z)$ have no zeros in the upper half-plane and also*

$$|f(x)| \le |\omega(x)| , \qquad -\infty < x < \infty .$$

Then $\psi(z) = \dfrac{f(z)}{\omega(z)}$ is an analytic function of exponential type in the upper half-plane. Further,

$$h_f(\theta) = h_\omega(\theta) + h_\psi(\theta) = h_\omega(\theta) + k\sin\theta , \qquad 0 \le \theta \le \pi , \tag{1}$$

and

$$|\psi(z)| \le e^{ky} , \qquad y = \operatorname{Im} z \ge 0 , \tag{2}$$

where $k = h_\psi(\pi/2)$.

PROOF. According to the theorem on division of analytic functions (Theorem 5, Lecture 11), the function $\psi(z)$ is of exponential type in $\mathbb{C}_+$. Since the function $\psi(z)$ is bounded on the real axis, $h_\psi(0) \le 0$ and $h_\psi(\pi) \le 0$, and by properties 4 and 5 of the indicator function (see Section 8.2)

$$h_\psi(\theta) = k\sin\theta , \qquad k = h_\psi(\pi/2) , \qquad 0 \le \theta \le \pi .$$

By the theorem on addition of indicator functions, Section 16.1, we obtain inequality (1). It is evident that the type of the function $\psi(z)$ in $\mathbb{C}_+$ equals k, and estimate (2) follows from the Phragmén-Lindelöf theorem, proving the lemma.

COROLLARY 1 (M. G. Krein). *The indicator diagram of EFET $\omega(z)$ without zeros in the lower half-plane is symmetric with respect to a line which is parallel to the real axis.*

To prove the corollary, it suffices to set $f(z) = \omega(z)$ in Lemma 1 and to replace $\omega(z)$ by $\overline{\omega}(z)$.[27] Since $h_{\overline{\omega}}(\theta) = h_\omega(-\theta)$, (1) has the form

$$h_\omega(\theta) = h_\omega(-\theta) + k \sin\theta\,, \qquad 0 < \theta < \pi\,.$$

Thus, the upper part of the indicator diagram can be obtained from the lower part by the reflection in the real axis followed by the translation along the imaginary axis by k. It is evident that the line $\{z = x + iy : y = k/2\}$ is the axis of symmetry for the indicator diagram. The corollary is proved.

DEFINITION. The number

$$d_\omega = \frac{1}{2}[h_\omega(-\pi/2) - h_\omega(\pi/2)] = -\frac{k}{2}$$

is called the defect of the EFET $\omega(z)$ without zeros in $\mathbb{C}_-$.

PROBLEM 1. Prove that

$$d_\omega = \lim_{r\to\infty} \frac{1}{\pi r}\int_0^\pi \log\left|\frac{\omega(re^{i\varphi})}{\overline{\omega}(re^{i\varphi})}\right| \sin\varphi\, d\varphi\,, \tag{3}$$

where the function under the limit sign does not decrease.

HINT. Use the Carleman formula.

The theorem on addition of indicator functions in the upper half-plane combined with the definition of the defect immediately yields:

COROLLARY 2. *The defect of the product of EFET without zeros in the lower half-plane equals the sum of defects of the factors.*

COROLLARY 3. *In order that an EFET $\omega(z)$ without zeros in the lower half-plane be a function of class P it is necessary and sufficient that*

$$\left|\frac{\omega(z)}{\overline{\omega}(z)}\right| \le 1\,, \qquad \operatorname{Im} z > 0\,. \tag{4}$$

The sufficiency of condition (4) is evident, and the necessity follows from inequality (2) in the statement of Lemma 1.

REMARK. Let $\omega(z)$ be an EFET. Represent it in the form

$$\omega(z) = P(z) + iQ(z)\,,$$

where $P(z)$ and $Q(z)$ are real entire functions.[28] Condition (4) is equivalent to the fact that the function

$$\theta(z) = \frac{Q(z)}{P(z)}$$

---

[27] $\overline{\omega}(z)$ denotes the entire function obtained from $\omega(z)$ by conjugating all Taylor coefficients.

[28] An entire function is called *real* if it takes real values on the real axis.

maps the upper half-plane into itself. Indeed, if $\psi(z) = \dfrac{\omega(z)}{\overline{\omega}(z)}$, then

$$\psi(z) = \frac{1 + i\theta(z)}{1 - i\theta(z)},$$

and the unit disk in the ψ-plane corresponds to the upper half-plane of the θ-plane. Note that the function $\theta(z)$ is constant if and only if $\omega(z)$ is real on the real axis up to a constant factor.

COROLLARY 4. *If a sequence $\omega_n(z)$ of functions of class P converges uniformly on each compact set to an EFET $\omega(z)$, then $\omega(z)$ is a function of class P.*

Indeed, $\omega(z)$ has no zeros in $\mathbb{C}_-$, and the inequalities $\left|\dfrac{\omega_n(z)}{\overline{\omega}_n(z)}\right| \le 1$ for $\operatorname{Im} z \ge 0$ yield $\left|\dfrac{\omega(z)}{\overline{\omega}(z)}\right| \le 1$ for $\operatorname{Im} z \ge 0$.

COROLLARY 5 *Each function of class P can be represented in the form*

(5) $$\omega(z) = cz^m e^{az} \prod_k \left(1 - \frac{z}{\alpha_k}\right) e^{z \operatorname{Re} \frac{1}{\alpha_k}},$$

where $\operatorname{Im} a = d_\omega \ge 0$, $\operatorname{Im} \alpha_k \ge 0$ *and*

(6) $$\sum_k \left| \operatorname{Im} \frac{1}{\alpha_k} \right| < \infty .$$

PROOF. The function $\psi(z) = \dfrac{\omega(z)}{\overline{\omega}(z)}$, $z \in \mathbb{C}_+$, has the same zeros α_k as the function $\omega(z)$. Since the function $\psi(z)$ is bounded on the real axis, its zeros satisfy the Blaschke condition (6), as proved in Theorem 2, Section 14.2. Representation (5) follows from the Hadamard theorem and (6). Further,

$$\psi(z) = \frac{\omega(z)}{\overline{\omega}(z)} = \frac{c}{\overline{c}}\, e^{(a-\overline{a})z} \prod_k \frac{1 - z/\alpha_k}{1 - z/\overline{\alpha}_k},$$

where the product converges by virtue of (6). According to the Hayman theorem (Lecture 15), we have

$$h_\psi(\theta) = -2(\operatorname{Im} a)\sin\theta .$$

Thus, $\operatorname{Im} a = d_\omega \ge 0$.

COROLLARY 6. *Each function $\omega(z)$ of class P can be approximated uniformly on compact sets by polynomials $\omega_n(z)$ without zeros in the lower half-plane.*

PROOF. We will use representation (5). The sequence of functions

$$cz^m e^{\kappa_n z} \prod_{k=1}^{n} \left(1 - \frac{z}{\alpha_k}\right)$$

with $\kappa_n = a + \sum_{k=1}^n \operatorname{Re} \dfrac{1}{\alpha_k}$ approximates the function $\omega(z)$ uniformly on every compact set. Choosing positive integers p_n, large enough, we obtain the sequence of approximating polynomials

$$\omega_n(z) = cz^m\left(1+\frac{\kappa_n z}{p_n}\right)^{p_n}\prod_{k=1}^{n}\left(1-\frac{z}{\alpha_k}\right)$$

without zeros in $\mathbb{C}_-$. The corollary is proved.

The well-known Laguerre-Pólya theorem describes the general form of entire functions which can be approximated on every compact set by polynomials without zeros in the lower half-plane. Such functions have the form $f(z) = e^{-\gamma z^2}\Phi(z)$ where $\gamma \geq 0$, and $\Phi(z)$ is an entire function of genus one satisfying condition (6). For more details related to this subject we refer the reader to Chapter 8 of the book [**82**].

Entire functions of class P have important applications in the framework of the theory of Hilbert spaces of entire functions developed by de Branges. This theory is exposed in the monograph de Branges [**20**].

27.2. Meromorphic functions with interlacing zeros and poles

In this section we prove a theorem which will be used in the sequel.

We say that two sequences of real numbers $\{a_k\}$ and $\{b_k\}$ *are interlacing* if exactly one member of one of them lies between each pair of neighboring terms of the other.

THEOREM 1 (M. G. Krein). *In order that a real meromorphic function $\theta(z)$ in $\mathbb{C}$ map the upper half-plane onto itself, it is necessary and sufficient that $\theta(z)$ be represented in the form*

$$\theta(z) = c\frac{z-a_0}{z-b_0}\prod_k{}'\left(1-\frac{z}{a_k}\right)\left(1-\frac{z}{b_k}\right)^{-1}, \tag{7}$$

where $b_k < a_k < b_{k+1}$, $k \in \mathbb{Z}$, $a_{-1} < 0 < b_1$, $c > 0$.

The prime here means that the index k takes all integer values except zero.

SUFFICIENCY. Since a_k and b_k are interlacing, the series

$$\sum_k{}'\left(\frac{1}{b_k}-\frac{1}{a_k}\right),$$

and hence the series

$$\sum_k{}'\left[\left(1-\frac{z}{a_k}\right)\left(1-\frac{z}{b_k}\right)^{-1}-1\right] = z\sum_k{}'\left(\frac{1}{b_k}-\frac{1}{a_k}\right)\left(1-\frac{z}{b_k}\right)^{-1}$$

are convergent. Thus the infinite product (7) converges uniformly on every compact set not containing points b_k.

For all values of the index k we have

$$\arg\frac{1-\dfrac{z}{a_k}}{1-\dfrac{z}{b_k}} = \arg(z-a_k)-\arg(z-b_k)\,.$$

For $z \in \mathbb{C}_+$, the left-hand side is the angle at which the segment $[b_k, a_k]$ is viewed from the point z. The equation

$$\arg\theta(z) = \sum_k [\arg(z - a_k) - \arg(z - b_k)]$$

implies that $0 < \arg\theta(z) < \pi$, $z \in \mathbb{C}_+$.

NECESSITY. Suppose that the real meromorphic function $\theta(z)$ maps the upper half-plane into itself, i.e.,

$$0 \le \arg\theta(z) \le \pi\,, \qquad z \in \mathbb{C}_+\,.$$

If $\theta(z)$ has either zeros or poles in the upper half-plane, then, going around a contour lying in the upper half-plane and enclosing one pole (or zero) of $\theta(z)$, we find that the absolute value of the increment of $\arg\theta(z)$ is not less than 2π. Since this is impossible, all zeros and poles of the function $\theta(z)$ lie on the real axis.

When $\operatorname{Im} z < 0$, we have $-\pi < \arg\theta(z) < 0$, and when the point z runs over any circumference, the absolute value of the increment of $\arg\theta(z)$ does not exceed 2π. Therefore, all zeros and poles of the function $\theta(z)$ are simple, and the number of zeros on any interval of the real axis differs from the number of its poles by at most one. This means that the zeros and poles of $\theta(z)$ interlace. The infinite product

$$\Phi(z) = \frac{z - a_0}{z - b_0} \prod_k{}' \left(1 - \frac{z}{a_k}\right)\left(1 - \frac{z}{b_k}\right)^{-1}$$

corresponding to the zeros $\{a_k\}$ and poles $\{b_k\}$ of the function $\theta(z)$ converges uniformly on every compact set not containing points of $\{b_k\}$ and maps the upper half-plane into itself.

The function

$$\chi(z) = \frac{\theta(z)}{\Phi(z)}$$

is entire, has no zeros in the complex plane, and $|\arg\chi(z)| \le 2\pi$ for all $z \in \mathbb{C}$. Thus the entire function $u = \log\chi(z)$ maps the whole complex plane into the strip $|\operatorname{Im} u| \le 2\pi$, and therefore $u \equiv \text{const}$. The theorem is proved.

This theorem immediately yields the following statement.

REMARK 1 (N. G. Chebotarev, P. Montel). In order that a real meromorphic function $\theta(z)$ can be uniformly approximated on every compact set in the complex plane by real rational functions with real interlacing zeros and poles, it is necessary and sufficient that $\theta(z)$ maps the upper half-plane into itself.

PROBLEM 1 (N. G. Chebotarev, P. Montel). Prove that a real meromorphic function $\theta(z)$ which maps the upper half-plane into itself can be represented in the form

$$\theta(z) = az + b + \frac{c_0}{z} + \sum_{n=\omega_1}^{\omega_2} c_n\left\{\frac{1}{a_n - z} - \frac{1}{a_n}\right\}, \qquad -\infty \le \omega_1 \le \omega_2 \le \infty\,,$$

where $a \ge 0$, b is real, $c_n \ge 0$, $\dots a_{n-1} < a_n < a_{n+1} \dots$, and the series

$$\sum_{n=\omega_1}^{\omega_2} \frac{c_n}{a_n^2}$$

converges.

In fact, this representation is a particular case of the representation of a holomorphic function mapping the upper half-plane into itself, due to R. Nevanlinna, given by relation (2), Section 14.1.

REMARK 2 (N. G. Chebotarev). Two real entire functions $P(z)$ and $Q(z)$ are said to be a real pair if they have no common zeros, and if any linear combination $\mu P(z) + \nu Q(z)$ with real coefficients has no complex zeros. If entire functions $P(z)$ and $Q(z)$ form a real pair, then the meromorphic function

$$\theta(z) = \frac{Q(z)}{P(z)} \tag{8}$$

does not take on real values for $\operatorname{Im} z \neq 0$, so that it maps $\mathbb{C}_+$ either onto itself or onto $\mathbb{C}_-$. If, in addition,

$$Q'(x_0)P(x_0) - Q(x_0)P'(x_0) > 0 \tag{9}$$

at some real point x_0, then $\theta'(x_0) > 0$, and $\theta(z)$ maps the upper half-plane onto itself.

Conversely, if a function $\theta(z)$ of the form (8) maps the upper half-plane onto itself, then it takes on real values on the real axis only, and hence any real combination $\mu P(z) + \nu Q(z)$ with real coefficients has only real zeros. The condition (9), is evidently fulfilled.

27.3. Theorem of Hermite and Biehler for entire functions of exponential type

In many applications it is necessary to know whether a certain half-plane does not contain zeros of a given entire function. For polynomials the criterion is given by the well-known theorem due to Hermite and Biehler:

In order that the polynomials

$$\omega(z) = P(z) + iQ(z)$$

where $P(z)$ and $Q(z)$ are real polynomials, have no zeros in the closed lower half-plane $\{z : \operatorname{Im} z \leq 0\}$, it is necessary and sufficient that the following conditions be satisfied:

1) *the polynomials $P(z)$ and $Q(z)$ have simple interlacing zeros;*
2) *at some real point x_0*

$$Q'(x_0)P(x_0) - Q(x_0)P'(x_0) > 0\,.$$

In this section we will prove the analog of the Hermite-Biehler theorem for EFET. We call an EFET $\omega(z)$ of class P *trivial* if it is real up to a constant factor.

THEOREM 2 (Generalized Hermite-Biehler Theorem)[29]. *In order that an EFET $\omega(z) = P(z) + iQ(z)$ be a nontrivial function of class P, it is necessary and sufficient that real entire functions $P(z)$ and $Q(z)$ satisfy the following conditions:*

[29]This theorem was proved independently by M. G. Krein and the author, but the work of M. G. Krein was not published. Similar results were proved by N. N. Meiman.

(A) *the functions* $P(z)$ *and* $Q(z)$ *can be represented in the form*

$$P(z) = R(z)P_1(z)\ , \qquad Q(z) = R(z)Q_1(z)\ ,$$

where the zeros of R *are real, and the zeros of* P_1 *and* Q_1 *are real, simple and interlacing;*

(B) *EFET* P *and* Q *have coinciding indicator diagrams;*

(C) *at some real point* x_0,

$$Q'(x_0)P(x_0) - Q(x_0)P'(x_0) > 0\ .$$

To prove this theorem we need the following result.

LEMMA 2. *Let* $P(z)$ *and* $Q(z)$ *be EFET with real, simple and interlacing zeros, and let*

$$(10)\qquad \begin{aligned} P(z) =& z^m e^{az+c}\prod_k\Big(1-\frac{z}{a_k}\Big)e^{z/a_k}\ , \\ Q(z) =& z^n e^{bz+d}\prod_k\Big(1-\frac{z}{b_k}\Big)e^{z/b_k} \\ &(a_{-1}<0<b_1) \end{aligned}$$

be their canonical representations. Then

$$(11)\qquad h_Q(\theta) = h_P(\theta) + \delta\cos\theta\ , \qquad 0\le\theta\le 2\pi\ ,$$

where

$$(12)\qquad \delta = b - a + \sum_k\Big(\frac{1}{b_k}-\frac{1}{a_k}\Big)\ .$$

In other words, the indicator diagram of the function $Q(z)$ *can be obtained from the indicator diagram of the function* $P(z)$ *upon translation of the latter by* δ *along the real axis. If* $P(z)$ *and* $Q(z)$ *are real entire functions, then the vector* δ *is real as well.*

PROOF. It follows from representations (10) that

$$(13)\qquad \frac{Q(z)}{P(z)} = ce^{\delta z}\frac{z-b_0}{z-a_0}\prod{}'\Big(1-\frac{z}{b_k}\Big)\Big(1-\frac{z}{a_k}\Big)^{-1}\ .$$

Denote by $\Phi(z)$ the infinite product on the right-hand side of (13). The convergence of this product was proved while proving Theorem 1. By Theorem 1 the real function $\Phi(z)$ maps the upper half-plane either onto itself or onto the lower half-plane. Using its representation as described in Problem 1 above we obtain the estimate

$$|\Phi(z)| \le a|z| + b + \frac{c_0}{|z|} + \frac{|z|^2}{|y|}\sum_{n\ne 0}\frac{c_n}{a_n^2} \le C\frac{|z|^2}{|y|}\ , \qquad |z|\ge 1\ .$$

Since the function $-1/\Phi(z)$ has positive imaginary part in the upper half-plane, the similar lower estimate is valid for $\Phi(z)$ and we arrive at the estimates[30]

$$C_1\frac{|\sin\theta|}{r} \le |\Phi(re^{i\theta})| \le C_2\frac{r}{|\sin\theta|}\ , \qquad r\ge 1, \qquad \theta\ne 0,\pi$$

[30]Compare with Carathéodory's inequality (Section 11.1, Problem 2).

with some positive constants C_1 and C_2. It follows from these inequalities that, for $\theta \neq 0, \pi$,

$$\lim_{r\to\infty} \frac{\log|\Phi(re^{i\theta})|}{r} = 0 . \tag{14}$$

Relation (11) follows from (13) and (14), and the lemma is proved.

COROLLARY. *The relation*

$$b - a + \sum_k \Big(\frac{1}{b_k} - \frac{1}{a_k}\Big) = 0 \tag{15}$$

is necessary and sufficient in order that the entire functions $P(z)$ and $Q(z)$ have coinciding indicator diagrams.

PROOF OF THEOREM 2. SUFFICIENCY. Let us consider the function

$$\theta(z) = \frac{Q(z)}{P(z)} = \frac{Q_1(z)}{P_1(z)} .$$

The previous corollary implies that this function is represented in the form (7) with a real constant c. Then by Theorem 1 and by Remark 2 to this theorem either $\omega(z) \in P$ $(c > 0)$, or $\overline{\omega}(z) \in P$ $(c < 0)$. Using condition (C) we conclude that $\omega(z) \in P$.

NECESSITY. If $\omega(z) \in P$, then the function $\theta(z)$ maps the upper half-plane onto itself. Evidently, condition (C) is fulfilled. Condition (B) follows from Theorem 1 and the Corollary from Lemma 2. The theorem is proved.

REMARK 1. Suppose that assumption (A) of Theorem 2 is fulfilled but assumption (B) is violated. Then by Lemma 2 the indicator diagram I_P of $P(z)$ can be obtained from the indicator diagram I_Q of $Q(z)$ translating the latter by a real number $\delta \neq 0$. According to the property of the indicator diagrams (Section 9.1, Problem 1), the diagram I_ω of the function $\omega = P + iQ$ coincides with the smallest convex compact set which contains I_P and I_Q. The functions $P(z)$ and $Q(z)$ are real; hence their indicator diagrams are symmetric with respect to the real axis. Therefore, I_ω is also symmetric with respect to the real axis, and $d_\omega = 0$. In addition, in this case the function $\omega(z)$ has zeros in both half-planes $\mathbb{C}_+$ and $\mathbb{C}_-$, since otherwise condition (B) would be satisfied.

Thus, if the function $\omega(z)$ satisfies condition (A) and also at least one of the following two conditions:

(a) $\omega(z) \neq 0 , \qquad z \in \mathbb{C}_- ;$

(b) $d_\omega > 0,$

then $\omega(z) \in P$.

REMARK 2. If the function $\omega(z)$ is of zero exponential type (in particular, if $\omega(z)$ is a polynomial), then the indicator diagrams of the functions $\omega(z)$, $P(z)$ and $Q(z)$ coincide with the origin. In this case, if the function $\omega(z)$ has no zeros in the lower half-plane, then $\omega(z) \in P$ and condition (B) holds automatically.

Thus, if $P(z)$ and $Q(z)$ are entire functions of zero exponential type, the function $\omega(z) = P(z) + iQ(z)$ has no zeros in the closed lower half-plane if and only if the zeros of $P(z)$ and $Q(z)$ are real and interlacing, and condition (C) is satisfied.

Now we present an example of application of Theorem 2. Let us consider an exponential polynomial

$$\omega(z) = \sum_{k=1}^{n} e^{i\lambda_k z} P_k(z) ,$$

where $\lambda_1 < \lambda_2 < \cdots < \lambda_n$, $-\lambda_1 < \lambda_n$, and $P_k(z)$ are polynomials. The indicator diagram of $\omega(z)$ coincides with the segment $[-i\lambda_n, -i\lambda_1]$ of the imaginary axis. Hence

$$d_\omega = \frac{1}{2}(\lambda_n - \lambda_1) > 0 .$$

Let $\omega(z) = P(z) + iQ(z)$, where $P(z)$ and $Q(z)$ are exponential polynomials. By Remark 2 to Theorem 1 the interlacing property of the zeros of $P(z)$ and $Q(z)$ is equivalent to the condition that all zeros of $\omega(z)$ lie in the open upper half-plane.

More detailed account of this circle of problems can be found in the monographs Chebotarev and Meiman [**25**], Levin [**82**] and in the paper Levin [**80**].

LECTURE 28

S. N. Bernstein's Inequality for Entire Functions of Exponential Type and its Generalizations

One of the most significant properties of entire functions of exponential type was discovered by S. N. Bernstein:

If $f(z)$ is an entire function of exponential type $\sigma > 0$ and

$$|f(x)| \leq M \,, \qquad -\infty < x < \infty \,,$$

then

$$|f'(x)| \leq M\sigma \,, \qquad -\infty < x < \infty \,.$$

The equality sign is attained here for some x if and only if

$$f(z) = c_1 \cos \sigma z + c_2 \sin \sigma z \,.$$

This theorem was generalized by many authors in various directions.

The inequalities in the statement of the Bernstein theorem may be written in the form

$$|f(x)| \leq |Me^{i\sigma x}| \quad \text{and} \quad |f'(x)| \leq |(Me^{i\sigma x})'| \,.$$

In other words, the Bernstein theorem states that under some conditions the inequality between the functions yields the same inequality between their derivatives.

Taking the properties of functions of the class P as our starting point, we will prove that if $f(z)$ is an EFET, $\omega(z) \in P$, and if

$$\sigma_f \leq \sigma_\omega \,, \tag{1}$$

then the inequality between the functions

$$|f(x)| \leq |\omega(x)| \,, \qquad -\infty < x < \infty \,, \tag{2}$$

yields the same inequality between their derivatives:

$$|f'(x)| \leq |\omega'(x)| \,, \qquad -\infty < x < \infty \,.$$

S. N. Bernstein's inequality has many applications, mainly in approximation theory (see Akhiezer [**4**], Timan [**122**]). In this lecture we shall discuss only one surprising application to the theory of Banach algebras.

28.1. P-majorants

DEFINITION. A function $\omega(z) \in P$ is a P-majorant of an EFET $f(z)$ if inequalities (1) and (2) are fulfilled.

We shall denote this relation by the symbol $f \prec \omega$.

In this section we shall prove several important properties of P-majorants.

LEMMA 1. *If $f \prec \omega$, then*

$$|f(z)| \le |\omega(z)|\,, \qquad z \in \mathbb{C}_-\,, \tag{3}$$

$$|f(z)| \le |\overline{\omega}(z)|\,, \qquad z \in \mathbb{C}_+\,. \tag{4}$$

PROOF. By Lemma 1 from the previous lecture we have

$$h_f(\theta) = h_\omega(-\theta) + k|\sin\theta|\,, \qquad -\pi \le \theta \le 0\,. \tag{5}$$

To prove the lemma we shall show that $k \le 0$.

Since $\omega(z) \in P$, then

$$h_\omega(\theta) \ge h_\omega(-\theta)\,, \qquad -\pi \le \theta \le 0\,,$$

and hence $h_\omega(\theta)$ attains its maximum σ_ω at some point θ_0, $-\pi \le \theta_0 \le 0$. If $\theta_0 \ne 0, \pi$, then, setting $\theta = \theta_0$ in (5), we obtain

$$k|\sin\theta_0| = h_f(\theta_0) - \sigma_\omega \le \sigma_f - \sigma_\omega \le 0\,.$$

Thus, in this case, $k \le 0$.

Now let $\theta_0 = 0$ (or $\theta_0 = \pi$). It follows from (5), that the part of the indicator diagram I_f corresponding to $\theta \in [-\pi, 0]$ can be obtained from the respective part of the indicator diagram I_ω translating the latter by $-ki$. In this case, the point $\sigma_\omega - ki$ belongs to the indicator diagram I_f, and

$$\sigma_f \ge \sqrt{\sigma_\omega^2 + k^2}\,,$$

i.e., $k = 0$. The lemma is proved.

REMARK 1. If $\omega(z)$ is an EFET such that for each EFET $f(z)$ inequalities (1) and (2) yield (3) and (4), then $\omega(z) \in P$.

REMARK 2. If an EFET $f(z)$ possesses a trivial (i.e., real up to a constant factor) majorant $\omega(z)$, then

$$f(z) = c\omega(z), \qquad |c| \le 1\,. \tag{6}$$

Indeed, in such a case $\overline{\omega}(z) = c\omega(z)$, and by Lemma 1 the inequality

$$|f(z)| \le |\omega(z)|$$

holds in the whole complex plane. By the Liouville theorem we obtain (6).

PROBLEM 1. Let $f(z)$ be an EFET. Prove that the following conditions are equivalent:

(a) $f(z)$ has a P-majorant.

(b) There is a function $\omega(z) \in P$ such that $f \prec \omega$,

$$|f(x)| = |\omega(x)|\,, \qquad -\infty < x < \infty\,,$$

and either the indicator diagrams of $f(z)$ and $\omega(z)$ coincide or one can be obtained from the other by reflection in the real axis.[31]

[31] Such a majorant $\omega(z)$ is the best possible.

(c) The zeros $\{\alpha_k\}$ of the function $f(z)$ satisfy the condition

$$\sum_k \left| \operatorname{Im} \frac{1}{\alpha_k} \right| < \infty .$$

(d) $\displaystyle \sup_r \left| \int_{-r}^{r} \frac{\log|f(x)|}{1+x^2}\, dx \right| < \infty.$

HINT. The equivalence of conditions (c) and (d) follows from Carleman's formula (see Problem 1, Section 24.1). If the zeros of an EFET

$$f(z) = ce^{(a+ib)z} \prod_k \left(1 - \frac{z}{\alpha_k}\right) \exp \frac{z}{\alpha_k}$$

satisfy condition (c), then a majorant $\omega(z)$ satisfying condition (b) can be found in the form

$$\omega(z) = ce^{(a+ip)z} \prod_k \left(1 - \frac{z}{\beta_k}\right) \exp z \operatorname{Re} \frac{1}{\beta_k} ,$$

where $\beta_k = \alpha_k$ for $\operatorname{Im} \alpha_k \geq 0$, $\beta_k = \alpha_k$ for $\operatorname{Im} \alpha_k < 0$, and $p \geq 0$ is appropriately chosen.

LEMMA 2. *Let $\omega(z)$ and $f(z)$ be entire functions. In order that the function*

$$\varphi_u(z) = f(z) - u\omega(z)$$

belong to the class P for an arbitrary complex value u satisfying $|u| \geq 1$, it is necessary and sufficient that $\omega \in P$ and $f \prec \omega$.

NECESSITY. Let $\varphi_u \in P$ for $|u| \geq 1$. The equations

$$\omega(z) = \frac{\varphi_u(z) - \varphi_v(z)}{v - u} ,$$
$$f(z) = \frac{v\varphi_u(z) - u\varphi_v(z)}{v - u}$$

imply that the functions $\omega(z)$ and $f(z)$ are of finite exponential type. As $u \to \infty$, the function $\frac{1}{u}\varphi_u(z)$ approaches $-\omega(z)$ uniformly on every compact set, and by Corollary 4 to Lemma 1 from the previous Lecture we have $\omega(z) \in P$. Further, if $|u| \geq 1$, then the holomorphic function

$$\frac{\varphi_u(z)}{\omega(z)} = \frac{f(z)}{\omega(z)} - u$$

does not vanish in the half-plane $\mathbb{C}_-$. In other words,

$$\left| \frac{f(z)}{\omega(z)} \right| \leq 1 , \qquad \operatorname{Im} z \leq 0 . \tag{7}$$

To estimate $f(z)$ in the upper half-plane, we note that, according to Corollary 3, Section 27.1, the condition $\varphi_u \in P$ and estimate (7) imply the inequality

$$|f(\overline{z}) - u\omega(\overline{z})| \leq |f(z) - u\omega(z)| \leq (1 + |u|)|\omega(z)| , \qquad z \in \mathbb{C}_- ,$$

whence

$$|f(\overline{z})| \leq (1 + 2|u|)|\omega(z)| . \tag{8}$$

Evidently, estimates (7) and (8) yield (1) and (2); i.e., $f \prec \omega$.

SUFFICIENCY. If $f \prec \omega$, then by virtue of Lemma 1 the estimate (3) holds, the function $\varphi_u(z)$ has no zeros in $\mathbb{C}_-$ for $|u| > 1$, and hence for $|u| \geq 1$.

Further, for $z \in \mathbb{C}_-$ and $|u| > 1$, we have

$$\left|\frac{\varphi_u(\overline{z})}{\varphi_u(z)}\right| = \left|\frac{f(\overline{z}) - u\omega(\overline{z})}{f(z) - u\omega(z)}\right| = \left|\frac{f(\overline{z})}{\omega(z)} - u\frac{\omega(\overline{z})}{\omega(z)}\right| : \left|\frac{f(z)}{\omega(z)} - u\right| \leq \frac{1+|u|}{|u|-1} .$$

Also, $\left|\dfrac{\varphi_u(\overline{z})}{\varphi_u(z)}\right| = 1$ if z is real.

Applying the Phragmén-Lindelöf principle to the function $\left|\dfrac{\overline{\varphi}_u(z)}{\varphi_u(z)}\right|$ in the half-plane $\mathbb{C}_-$, we obtain

$$|\varphi_u(\overline{z})| = |\overline{\varphi}_u(z)| \leq |\varphi_u(z)| , \qquad \operatorname{Im} z \leq 0 .$$

It is clear that the last inequality is valid for $|u| = 1$ as well. Thus $\varphi_u \in P$ for each u satisfying $|u| \geq 1$. The lemma is proved.

For an arbitrary function $\omega \in P$ and an arbitrary value $\alpha \in [0, 2\pi]$ we set

$$\Omega(z;\alpha) = \frac{1}{2}\left(\omega(z)e^{i\alpha} + \overline{\omega}(z)e^{-i\alpha}\right) .$$

If $\omega(z) = P(z) + iQ(z)$, then $\Omega(z;\alpha) = P(z)\cos\alpha - Q(z)\sin\alpha$, and according to part (A) of the generalized Hermite-Biehler theorem (Theorem 2, Section 27.3) and Remark 2, Section 27.2, $\Omega(z;\alpha)$ has real zeros only. Multiple zeros of this function are simultaneously zeros of the function $\omega(z)$.

PROBLEM 2 (N. I. Akhiezer). Let $f(z)$ be a real EFET and let $f \prec \omega$. Then the difference $\Omega(z;\alpha) - f(z)$ either vanishes identically or has only real zeros, and these zeros are simple except when $f(x) = \pm|\omega(x)|$.

HINT. By virtue of Lemma 2, the function $f(z) - e^{i\alpha}\omega(z)$ belongs to the class P. It suffices to apply Theorem 2, Section 27.3 to this function.

PROBLEM 3. [32] Let $f(z)$ be a real EFET and let $f \prec \omega$. Then for every complex z

$$|f(z)| \leq \frac{1}{2}(|\omega(z)| + |\overline{\omega}(z)|) ,$$

where the equality holds for at least one nonreal point if and only if $f(z) \equiv \Omega(z;\alpha)$ for some $\alpha \in [0, 2\pi]$.

HINT. Use the previous problem.

28.2. Operators preserving inequalities

DEFINITION. A linear operator K defined on the set of all entire functions of exponential type and mapping the class P into itself is said to be a $\mathfrak{B}$-operator.

[32] In the case $\omega(z) = e^{i\sigma z}$ this theorem is due to R. J. Duffin and A. C. Schaeffer. The general statement was obtained by N. I. Akhiezer.

THEOREM 1. *$\mathfrak{B}$-operators preserve the relation of majorization.*

PROOF. Let $f \prec \omega$ and let K be an arbitrary $\mathfrak{B}$-operator. By Lemma 2 we have

$$f(z) - u\omega(z) \in P\,, \qquad |u| \geq 1\,.$$

Then

$$K[f(z) - u\omega(z)] = K[f] - uK[\omega] \in P\,, \quad |u| \geq 1\,.$$

Applying Lemma 2 once more, we obtain $K[f] \prec K[\omega]$, proving the theorem.

THEOREM 2. *The operator of differentiation is a $\mathfrak{B}$-operator.*

PROOF. Let $\omega(z)$ be an entire fuction of exponential type $\sigma \geq 0$. Its derivative $\omega'(z)$ has the same type, which follows, for example, from the formula expressing the type of an entire function via its Taylor coefficients (see Section 1.3).

Let us consider the imaginary part of the logarithmic derivative of a function $\omega(z) \in P$. Representation (5) from the previous lecture yields

$$\frac{\omega'(z)}{\omega(z)} = \frac{m}{z} + a + \sum_k{}' \left[\frac{1}{z - \alpha_k} - \operatorname{Re}\frac{1}{\alpha_k}\right].$$

Since $\operatorname{Im} a = d_\omega \geq 0$, we obtain

$$\operatorname{Im}\frac{\omega'(z)}{\omega(z)} = d_\omega + \sum_k \frac{-(\operatorname{Im} z - \operatorname{Im}\alpha_k)}{|z - \alpha_k|^2} > 0\,, \qquad \operatorname{Im} z < 0\,. \tag{9}$$

Therefore, $\omega'(z) \neq 0$ for $\operatorname{Im} z < 0$.

By Corollary 6 of Lemma 2 of the previous lecture there is a sequence of polynomials $\omega_n(z) \in P$ without zeros in the lower half-plane converging to $\omega(z)$ uniformly on every compact set. By (9) the polynomials $\omega_n'(z)$ have no zeros in the lower half-plane[33] and hence belong to the class P. Applying Corollary 4 of the same Lemma 1, we conclude that $\omega'(z) \in P$. The theorem is proved.

REMARK. If the function $\omega(z) \in P$ has at least one nonreal zero, then inequality (9) holds on the real axis.

Indeed, if at some real point x

$$\operatorname{Im}\frac{\omega'(x)}{\omega(x)} = d_\omega + \sum_k \frac{\operatorname{Im}\alpha_k}{|x - \alpha_k|^2} = 0\,,$$

then $d_\omega = 0$, and all zeros α_k are real.

PROBLEM 4. Prove that $d_{\omega'} \geq d_\omega$ for $\omega(z) \in P$.

LEMMA 3. *If $f \prec \omega$ and for some real α*

$$\varphi(z) = f(z) - e^{i\alpha}\omega(z)$$

is a trivial function of class P, i.e., it is real up to a constant factor, then

$$f(z) = c_1\omega(z) + c_2\overline{\omega}(z) \tag{10}$$

with $|c_1| + |c_2| = 1$.

[33]These arguments prove the Gauss theorem: zeros of the derivative of a polynomial are located in the closed convex hull of zeros of the polynomial itself.

PROOF. Without loss of generality we assume that the trivial function $\varphi(z)$ of class P is real and fix real values γ_1 and γ_2 such that

$$e^{i\gamma_1} f(z) + e^{i\gamma_2}\omega(z) = i\varphi(z) .$$

Set

$$e^{i\gamma_1} f(z) = P(z) + iQ(z) ,$$
$$e^{i\gamma_2}\omega(z) = -P(z) + iS(z) ,$$

where $P(z)$, $Q(z)$ and $S(z)$ are real EFET. The inequality

$$|P(x) + iQ(x)| \le |P(x) - iS(x)|$$

implies that

$$|Q(x)| \le |S(x)| , \quad x \in \mathbb{R} . \tag{11}$$

Now we shall show that

$$Q(z) = cS(z) , \qquad |c| \le 1 .$$

By the generalized Hermite-Biehler theorem (Theorem 2, Section 27.3), the function $S(z)$ has real zeros only and, according to inequality (11), they are zeros of $Q(z)$. Hence

$$\theta(z) = \frac{Q(z)}{S(z)}$$

is an EFET bounded by 1 on the real axis. Let us consider the entire function

$$te^{i\gamma_1} f(z) + e^{i\gamma_2}\omega(z) , \qquad 0 < t < 1 .$$

By virtue of Lemma 2 this is a function of class P, and, once more, by the generalized Hermite-Biehler Theorem, the function $tQ + S$ has real zeros only. Hence the function

$$t\theta(z) + 1 = \frac{tQ(z) + S(z)}{S(z)}$$

also has real zeros only. Since $|\theta(x)| \le 1$ on the real axis, then, for $0 < t < 1$, the entire function $t\theta(z) + 1$ has no real zeros. If follows that the entire function $\theta(z)$ omits values $\tau < -1$, and therefore $\theta(z) \equiv c$ with real c, $|c| \le 1$.

Thus,

$$\begin{aligned} e^{i\gamma_1} f(z) &= P(z) + icS(z) , \\ e^{i\gamma_2}\omega(z) &= -P(z) + iS(z) , \end{aligned} \tag{12}$$

which is equivalent to (10). The lemma is proved.

REMARK. If an EFET $f(z)$ is represented in the form (10) with $\omega \in P$, then, as follows from relations (12), $|f(x_0)| = |\omega(x_0)|$ at each zero x_0 of the function $S(z)$. We note that if the function $S(z)$ has no zeros, then again by the generalized Hermite-Biehler theorem, $\omega(z)$ is a polynomial of the first degree.

THEOREM 3. *Let $f \prec \omega$. Then for all positive integers k*

$$|f^{(k)}(x)| \le |\omega^{(k)}(x)|\,, \qquad -\infty < x < \infty\,. \tag{13}$$

If $\omega(z)$ is not a polynomial, then the equality sign is attained in (13) at least at some real point x_0 and for some k, if and only if the function $f(z)$ has the form (10).

PROOF. Inequalities (13) follow directly from Theorems 1 and 2. Note that by virtue of Theorem 2 we have $\omega^{(k)} \in P$, and if $f(z)$ has the form (10), then

$$f^{(k)}(z) = c_1\omega^{(k)}(z) + c_2\overline{\omega}^{(k)}(z)$$

is a function of the same form. In this case, according to the remark to the previous lemma, for each k there is x_k such that

$$|f^{(k)}(x_k)| = |\omega^{(k)}(x_k)|\,.$$

Now, let us prove the converse statement. Assume that for some positive integer k there is a real x_0 such that

$$|f^{(k)}(x_0)| = |\omega^{(k)}(x_0)|\,.$$

Then for some real α the derivative of the function

$$\varphi(z) = f^{(k-1)}(z) - e^{i\alpha}\omega^{(k-1)}(z)$$

vanishes at x_0. The function $\varphi(z)$ itself cannot vanish at x_0, since otherwise

$$\frac{f^{(k-1)}(z)}{\omega^{k-1}(z)} = e^{i\alpha} + (z-x_0)^p\theta(z), \qquad \theta(x_0) \ne 0,\ p \ge 2\,,$$

which gives an immediate contradiction, since $f^{(k-1)}$ is majorized by $\omega^{(k-1)}$. Thus

$$\operatorname{Im}\frac{\varphi'(x_0)}{\varphi(x_0)} = 0.$$

By the Remark to Theorem 2 we conclude that all zeros of $\varphi(z)$ are real, and hence it is a trivial function of class P. Hence, Lemma 3 yields

$$f^{(k-1)}(z) = c_1\omega^{(k-1)}(z) + c_2\overline{\omega}^{(k-1)}(z)\,, \qquad |c_1| + |c_2| = 1\,.$$

By virtue of the Remark to Lemma 3 it follows that, for some real point x_1,

$$|f^{(k-1)}(x_1)| = |\omega^{(k-1)}(x_1)|\,.$$

Repeating this argument several times, we obtain that the function $f(z)$ has the form (10). The theorem is proved.

Now we will show that the class P is the largest class of EFET for which Theorem 3 holds. More precisely, we will prove

THEOREM 4. *Let $\omega(z)$ be an EFET such that the condition*

$$|f(x)| \le |\omega(x)|\,, \qquad -\infty < x < \infty\,,$$

where f is any EFET with $\sigma_f \le \sigma_\omega$, implies that

$$|f'(x)| \le |\omega'(x)|\,, \qquad -\infty < x < \infty\,.$$

Then either $\omega(z) \in P$ or $\overline{\omega}(z) \in P$.

PROOF. To begin with, we assume that the function $\omega(z)$ has zeros in both half-planes $\mathbb{C}_+$ and $\mathbb{C}_-$. Choose a real point x_0 such that $\omega(x_0) \ne 0$ and $\omega'(x_0) \ne 0$. For the sake of definiteness we assume that

$$\operatorname{Im}\frac{\omega(x_0)}{\omega'(x_0)} \le 0\,.$$

Set

$$f(z) = \frac{z-\overline{\alpha}}{z-\alpha}\omega(z)\,,$$

where α is a zero of $\omega(z)$ lying in $\mathbb{C}_+$ (if

$$\operatorname{Im}\frac{\omega(x_0)}{\omega'(x_0)} > 0\,,$$

then we choose $\alpha \in \mathbb{C}_-$). Evidently, $\sigma_f = \sigma_\omega$ and

$$|f(x)| = |\omega(x)|\,, \qquad -\infty < x < \infty\,.$$

On the other hand,

$$|f'(x_0)| = \left|\omega'(x_0)\frac{x_0-\overline{\alpha}}{x_0-\alpha} - \omega(x_0)\frac{\alpha-\overline{\alpha}}{(x_0-\alpha)^2}\right| = |\omega'(x_0)|\left|1 - 2i\frac{\omega(x_0)}{\omega'(x_0)}\frac{\operatorname{Im}\alpha}{|x_0-\alpha|^2}\right|\,.$$

According to our choice of α, we have

$$\operatorname{Re}\left[-2i\frac{\omega(x_0)}{\omega'(x_0)}\frac{\operatorname{Im}\alpha}{|x_0-\alpha|^2}\right] \ge 0\,.$$

If the left-hand side vanishes, then the imaginary part of the expression in the brackets does not vanish. In either case

$$\left|-2i\frac{\omega(x_0)}{\omega'(x_0)}\frac{\operatorname{Im}\alpha}{|x_0-\alpha|^2}\right| > 1\,,$$

and hence

$$|f'(x_0)| > |\omega(x_0)|\,.$$

Thus we arrive at a contradiction.

Let us assume now that all zeros of the function $\omega(z)$ lie in a closed half-plane which, for the sake of definiteness, is $\overline{\mathbb{C}}_+$. If all zeros are real, then $\omega(z)$ is a trivial function of class P.

Assume that some zeros of $\omega(z)$ are located in $\mathbb{C}_+$. If the defect of ω is non-negative ($d_\omega \ge 0$), then $\omega \in P$. Suppose that $d_\omega < 0$ and consider the function

$$f(z) = e^{2i\delta z}\omega(z)$$

where $\delta = -d_\omega > 0$. Then $f \in P$. It is evident that with such a choice of δ we have $\sigma_f = \sigma_\omega$ and

$$|f(x)| = |\omega(x)|\,. \tag{14}$$

Further,

$$\operatorname{Re}\frac{f'(x)}{f(x)}=\operatorname{Re}\frac{\omega'(x)}{\omega(x)}\,. \tag{15}$$

Corollary 5, Section 27.1 yields

$$\omega(z)=cz^me^{-i\delta z}\prod_k\left(1-\frac{z}{\alpha_k}\right)e^{z\operatorname{Re}1/\alpha_k}.$$

Therefore

$$\operatorname{Im}\frac{\omega'(x)}{\omega(x)}=-\delta+\sum_{\alpha_k\in\mathbb{C}_+}\frac{\operatorname{Im}\alpha_k}{|x-\alpha_k|^2}\,,$$

and

$$\operatorname{Im}\frac{f'(x)}{f(x)}=\delta+\sum_{\alpha_k\in\mathbb{C}_+}\frac{\operatorname{Im}\alpha_k}{|x-\alpha_k|^2}\,.$$

Since there is at least one $\alpha_k\in\mathbb{C}_+$, we obtain

$$\left|\operatorname{Im}\frac{\omega'(x)}{\omega(x)}\right|<\left|\operatorname{Im}\frac{f'(x)}{f(x)}\right|\,. \tag{16}$$

Examining (14), (15) and (16), we conclude that

$$|f'(x)|>|\omega'(x)|\,.$$

Hence the assumption $d_\omega<0$ leads us to a contradiction, which proves that $\omega(z)\in P$.

PROBLEM 5. Let $\omega(z)$ be a nontrivial function of class P, and let

$$\Omega(z;\alpha)=\frac{1}{2}\{\omega(z)e^{i\alpha}+\overline{\omega}(z)e^{-i\alpha}\}\,,\qquad \Phi(x)=\frac{1}{2}\arg\frac{\omega(x)}{\overline{\omega}(x)}\,.$$

Then the function $y=y(x)=\dfrac{\Omega(x;\alpha)}{|\omega(x)|}$ satisfies the identity

$$(y')^2+\{\Phi'(x)\}^2y^2=\{\Phi'(x)\}^2\,.$$

PROBLEM 6 (N. I. Akhiezer). Let f be a real EFET, and let $f\prec\omega$ where ω is a nontrivial function of class P. Then

$$\left|\frac{d}{dx}\frac{f(x)}{|\omega(x)|}\right|\le\Phi'(x)\sqrt{1-\left|\frac{f(x)}{\omega(x)}\right|^2}\,, \tag{17}$$

where $\Phi(x)=\dfrac{1}{2}\arg\dfrac{\omega(x)}{\overline{\omega}(x)}$. If there is at least one real point x_0 at which $|f(x_0)|<|\omega(x_0)|$ and the equality holds in (17), then $f(z)\equiv\Omega(z;\alpha)$ for some real α.

HINT. For each real x one can find values L, $0 \le L \le 1$, and α, $0 \le \alpha \le 2\pi$, such that

$$f(x) = L\Omega(x;\alpha)\,, \qquad f'(x) = L\Omega'(x;\alpha)\,,$$

and then use the previous problem. If the equality holds in (17), then it is necessary to use Problem 2.

For $\omega(z) = e^{i\sigma z}$ the assertion of the previous problem gives a useful inequality

$$|f'(x)| \le \sigma_f \sqrt{1 - |f(x)|^2}\,,$$

which is valid for any arbitrary EFET $f(z)$ such that $|f(x)| \le 1$.

Let E be a Banach space of functions defined on the real axis whose norm is invariant with respect to the real shift. This means that, for every real t,

$$\|g(x+t)\| = \|g(x)\|\,.$$

Examples of such spaces are provided by the space $L^p(-\infty,\infty)$, $p \ge 1$, and by the space of locally integrable functions with the norm

$$\sup_{-\infty < s < \infty} \left(\int_s^{2\pi + s} |g(t)|^p \, dt \right)^{1/p}.$$

PROBLEM 7. Let $f(z)$ be an EFET σ belonging to the space E. Then $\|f'\| \le \sigma \|f\|$.

HINT. Apply S. N. Bernstein's inequality to the function $\chi(t) = \varphi[f(z+t)]$, where $\varphi \in E^*$, $\|\varphi\| = 1$.

A description of $\mathfrak{B}$-operators and other results connected with Theorems 1–4 can be found in Levin [**80**] as well as in Chapter IX of Levin [**82**]. A somewhat different approach, based on interpolation formulas, to this circle of problems can be found in Akhiezer [**4,** pages 336–347].

28.3. S. N. Bernstein's inequality and Banach algebras

V. E. Katsnelson discovered a direct relation between S. N. Bernstein's inequality and the theory of Banach algebras.

DEFINITION. An element a of a Banach algebra A is called Hermitian if $\|e^{ita}\| = 1$ for all real t.

For example, Hermitian elements of the algebra of all bounded operators of a Hilbert space are selfadjoint operators. It is well known that the operator norm of a selfadjoint operator in a Hilbert space coincides with its spectral radius. Using S. N. Bernstein's inequality, V. E. Katsnelson showed[34] that *for every Hermitian element in an arbitrary Banach algebra the norm coincides with the spectral radius* $\rho(a)$. Moreover, this statement appeared to be equivalent to S. N. Bernstein's inequality.

First, using this inequality, we shall prove that

$$\|a\| = \rho(a) \tag{18}$$

for the Hermitian element $a \in A$.

[34] Independently and almost simultaneously this fact was proved by a number of other authors (F. F. Bonsall, M. T. Grabb, A. M. Sinclair, A. Browder).

Let $\psi \in A^*$ be an arbitrary linear functional whose norm equals 1. Consider the entire function $f(z) = \psi(e^{iaz})$. By Gelfand's formula for the spectral radius $\rho(a)$ (see Section 6.3) $f(z)$ is an EFET and

$$\sigma_f = \frac{1}{e} \limsup_{n\to\infty} n\Big[\frac{|\psi(a^n)|}{n!}\Big]^{1/n} \le \rho(a).$$

Since the element a is Hermitian, we have $|f(x)| \le 1$, $x \in \mathbb{R}$. Applying S. N. Bernstein's inequality, we obtain

$$|\psi(a)| = |f'(0)| \le \sigma_f \sup\{|f(x)| : x \in \mathbb{R}\} \le \rho(a) .$$

The Hahn-Banach theorem yields $\|a\| \le \rho(a)$. Since the converse inequality is evident, (18) is proved.

Now, we shall deduce from (15) S. N. Bernstein's inequality. To this end we consider the Banach space B_σ of EFET not exceeding σ and bounded on the real axis and assume that B_σ is endowed with the norm

$$\|f\| = \sup\{|f(x)| : x \in \mathbb{R}\} .$$

Let $\{f_n\} \in B_\sigma$ be a Cauchy sequence. The inequality

$$|f_n(x+iy) - f_m(x+iy)| \le \sup_{x\in\mathbb{R}} |f_n(x) - f_m(x)| e^{\sigma|y|}$$

implies that the limit function belongs to B_σ proving its completeness.

Let $a = \dfrac{1}{i}\dfrac{d}{dz}$ be the operator of differentiation in this space. If $t \in \mathbb{R}$, then

$$(e^{ita})[f](z) = \Big(\sum_{n=0}^{\infty} \frac{i^n t^n a^n}{n!}\Big)[f](z) = \sum_{n=0}^{\infty} \frac{i^n}{n!} t^n \Big(\frac{1}{i}\Big)^n \frac{d}{dz^n} f(z) = f(z+t) .$$

Hence, for $f \in B_\sigma$, we have $\|(e^{ita})[f]\| = \|f\|$. Thus $\|e^{ita}\| = 1$, and a is an Hermitian element in the Banach algebra of all bounded operators in B_σ.

Now we shall evaluate the spectral radius of a. Let $x \in \mathbb{R}$. We have

$$f^{(n)}(x) = \frac{n!}{2\pi i} \int_{|\zeta|=r} \frac{f(x+\zeta)}{\zeta^{n+1}}\, d\zeta .$$

Using the inequality $|f(x+\zeta)| \le e^{\sigma|\zeta|}\|f\|$, we obtain

$$|f^{(n)}(x)| \le \frac{n!}{2\pi}\frac{2\pi r}{r^{n+1}} e^{\sigma r}\|f\| = \frac{n!}{r^n} e^{\sigma r}\|f\| .$$

Minimizing this estimate with respect to r, we conclude that

$$|f^{(n)}(x)| \le \|f\| n! e^{n - n\log n + n\log\sigma} .$$

In other words

$$\|a^n\| \le n! e^{n - n\log n + n\log\sigma} .$$

By Gelfand's formula for the spectral radius and by the Stirling formula the last inequality yields

$$\rho(a) = \lim_{n\to a} \|a^n\|^{1/n} \le \lim_{n\to\infty} \{n! e^{n-n\log n + n\log\sigma}\}^{1/n} = \sigma .$$

Taking into account condition (18), we obtain now for an arbitrary function $f \in B_\sigma$,

$$|f'(x)| = |a[f](x)| \le \|a\|\,\|f\| = \rho(a)\|f\| \le \sigma\|f\| ,$$

proving S. N. Bernstein's inequality.

REMARK. F. F. Bonsall and T. Duncan, somewhat later than V. E. Katsnelson, found the following simple and direct proof of the coincidence of the norm and spectral radius of Hermitian elements in Banach algebras. This proof uses only basic facts of the theory of Banach algebras, and leads one to another proof of the inequality of S. N. Bernstein.

Let us consider the function $f(\zeta) = \arcsin\zeta$ analytic in the disk $|\zeta| < 1$. The Taylor coefficients f_k of this function are nonnegative and their sum equals $f(1) = \pi/2$. Let a be an arbitrary Hermitian element of a Banach algebra such that $\rho(a) < \pi/2$. Then the identity $a = f(\sin a)$ holds, whence

$$\|a\| \leq \sum_{k=1}^{\infty} f_k \|\sin a\|^k .$$

Since the element a is Hermitian, we have $\|\sin a\| \leq 1$, and therefore,

$$\|a\| \leq \sum_{k=1}^{\infty} f_k = \pi/2 .$$

This allows us to conclude that for an arbitrary Hermitian element and an arbitrary $\varepsilon > 0$, $\|a\| \leq (1+\varepsilon)\rho(a)$. Hence $\|a\| \leq \rho(a)$.

Further results connecting inequalities of S. N. Bernstein type (including Theorem 3) with the theory of Banach algebras can be found in Gorin [**44**].

Added in Proof

page 187, lines 1–2 below must be

$$= \frac{2}{R}\int_0^{\pi} \log|f(Re^{i\varphi})|\sin\varphi\, d\varphi + \int_\rho^R \left(\frac{1}{t^2} - \frac{1}{R^2}\right)\log|f(t)f(-t)|\, dt$$
$$+ 2\pi \sum_{a_n\in S} \operatorname{Im}\left(\frac{1}{a_n} + \frac{a_n}{R^2}\right) - 2\pi \sum_{b_n\in S} \operatorname{Im}\left(\frac{1}{b_n} + \frac{b_n}{R^2}\right) - 2\pi A_f(\rho, R) ,$$

page 200, line 3 below must be

$$H(z) = \left(\frac{\sin z}{z}\right)^{2p} h(z)$$

page 201, line 3 above, instead of $\sigma = 2 + 2p$ must be $\sigma = 4p$

page 213, lines 5–6 above, instead $\psi_1 + \psi_2 \in [-\pi/4l, \pi/4l]$ must be $\psi_1 + \psi_2 \in (-\pi/4l, \pi/4l)$

page 214, line 9 above must be

$$\leq Cr^\rho\left(\csc\frac{\psi}{2}\right)^k \csc\left(\frac{\gamma\psi}{2}\right) \leq Cr^\rho|\csc\psi|^{k+1}, \qquad r \geq 2 ,$$

page 215, line 4 below must be

$$\frac{1}{|f(x+iy)|} \leq M\left(1 + \frac{|z|}{|y|}\right) .$$

page 223, Eq.(11) must read

$$h_Q(\theta) = h_P(\theta) + \operatorname{Re}(\delta \exp i\theta) , \qquad 0 \leq \theta \leq 2\pi ,$$

page 223, line 10 below omit the words: along the real axis.

Bibliography

1. S. Agmon, *Functions of exponential type in an angle and singularities of Taylor series*, Trans. Amer. Math. Soc. **70** (1951), 492–508.
2. L. Ahlfors, *On Phragmén-Lindelöf's principle*, Trans. Amer. Math. Soc. **41** (1937), 1–8.
3. ———, *Remarks on Carleman's formula for functions in a half-plane*, SIAM J. Numer. Anal. **3** (1966), 183–187.
4. N. I. Akhiezer, *Lecture on approximation theory*, 2nd rev. ed., "Nauka", Moscow, 1965. (Russian)
5. V. S. Azarin, *Generalization of a theorem of Hayman's on a subharmonic function in an n-dimensional cone*, Mat. Sb. **66** (1965), no. 2, 248–264; English transl. in Amer. Math. Soc. Transl. Ser. 2 **80** (1969).
6. ———, *A characteristic property of functions of completely regular growth in the interior of an angle*, Teor. Funktsiĭ Funktsional. Anal. i Prilozhen. **1966**, no. 2, 55–66. (Russian)
7. ———, *On asymptotic behavior of subharmonic functions of finite order*, Mat. Sb. **108** (1979), no. 2, 147–167; English transl. in Math. USSR-Sb. **36** (1980), 135–154.
8. K. I. Babenko, *On a new problem of quasi-analyticity and on the Fourier transform of entire functions*, Trudy Moskov. Mat. Obshch. **5** (1956), 523–542. (Russian)
9. ———, *On certain classes of spaces of infinitely differentiable functions*, Dokl. Akad. Nauk SSSR **132** (1960), no. 6, 1231–1234; English transl. in Soviet Math. Dokl. **1** (1960).
10. G. Z. Ber, *An interference phenomenon in integral metrics, and approximation by entire functions*, Teor. Funktsiĭ Funktsional. Anal. i Prilozhen. **1980**, no. 34, 11–24. (Russian)
11. C. A. Berenstein and B. A. Taylor, *Mean periodic functions*, Internat. J. Math. Math. Sci. **3** (1980), 199–235.
12. A. Beurling, *Some theorems on boundedness of analytic functions*, Duke Math. J. **16** (1949), 355–359.
13. ———, *Collected works* I, II, Birkhäuser, Basel, 1989.
14. A. Beurling and P. Malliavin, *On Fourier transforms of measures with compact support*, Acta Math. **107** (1962), 291–309.
15. ———, *On the closure of characters and the zeros of entire functions*, Acta Math. **118** (1967), 79–93.
16. L. Bieberbach, *Analytische Fortsetzung*, Springer-Verlag, Berlin, 1955.
17. R. P. Boas, *Interference phenomena for entire functions*, Michigan Math. J. **3** (1956), 123–132.
18. A. A. Borichev, *Generalized Fourier transform, Titchmarsh transform, and almost analytic functions*, Algebra i Analiz **1** (1989), no. 4, 17–53; English transl. in Leningrad Math. J. **1** (1990), no. 4, 825–857.
19. N. Bourbaki, *Thèories spectrales*, Hermann, Paris, 1967.
20. L. de Branges, *Hilbert spaces of entire functions*, Prentice Hall, Englewood Cliffs, NJ, 1968.
21. Yu. A. Brudnyi and E. A. Gorin, *Isoperimetric representations and differential inequalities*, Yaroslavl University, Yaroslavl, 1981. (Russian)
22. T. Carleman, *Extension d'un thèoréme de Liouville*, Acta Math. **48** (1932), 363–366.
23. ———, *Les fonctions quasi-analytiques*, Gauthier-Villars, Paris, 1926.
24. H. Cartan, *Sur les systèmes de fonctions holomorphes à variétés linéaires lacunaires et leur applications*, Ann. Sci. École Norm. Sup. (3) **45** (1928), 255–346.
25. N. G. Chebotarev and N. N. Meiman, *The Routh-Hurwitz problem for polynomials and entire functions*, Trudy Mat. Inst. Steklov. **26** (1949). (Russian)

26. E. Davydova and V. Logvinenko, *On entire functions of exponential type bounded on the real line*, St. Petersburg Math. J. (to appear).
27. M. M. Dzhrbashyan [Djrbashian], *Uniqueness theorems for Fourier transforms and for infinitely differentiable functions*, Izv. Akad. Nauk Armyan. SSR Ser. Mat. **10** (1957), no. 6, 7–24. (Russian)
28. M. M. Dzhrbashyan [Djrbashian] and A. E. Avetisyan, *Integral representation of some classes of functions analytic in an angular region*, Dokl. Akad. Nauk SSSR **120** (1958), 457–460. (Russian)
29. M. M. Dzhrbashyan [Djrbashian] and S. G. Rafaelyan, *Interpolation expansions in classes of entire functions and Riesz' bases generated by them*, Izv. Akad. Nauk Armyan. SSR Ser. Mat. **22** (1987), no. 1, 23–63; English transl. in Soviet J. of Contemporary Analysis **22** (1987).
30. Y. Domar, *Extensions of the Titchmarsh convolution theorem with applications in the theory of invariant subspaces*, Proc. London Math. Soc. (3) **46** (1983), 288–300.
31. ______, *Uniform boundedness in families related to subharmonic functions*, J. London Math. Soc. (2) **38** (1988), 485–491.
32. H. Dym and H. P. McKean, *Gaussian processes, function theory and the inverse spectral problems*, Probab. Math. Statist., vol. 31, Academic Press, New York, 1976.
33. A. Eremenko and M. Sodin, *Parametrization of entire functions of sine type by their critical values*, Adv. Soviet Math., vol. 11, Amer. Math. Soc., Providence, RI, 1992, pp. 237–242.
34. M. Essén, *The* $\cos \pi\lambda$ *theorem*, Lecture Notes in Math., vol. 467, Springer-Verlag, Berlin and New York, 1975.
35. ______, *A superharmonic proof of the M. Riesz conjugate function theorem*, Ark. Mat. **22** (1984), 241–249.
36. M. A. Evgrafov, *Asymptotic estimates and entire functions*, 3rd ed., "Nauka", Moscow, 1979.
37. J. B. Garnett, *Bounded analytic functions*, Pure Appl. Math., vol. 96, Academic Press, New York, 1981.
38. I. M. Gelfand, D. A. Raikov, and G. E. Shilov, *Commutative normed rings*, Chelsea, New York, 1964.
39. I. M. Gelfand and G. E. Shilov, *Generalized functions*. II, *Spaces of fundamental and generalized functions*, Academic Press, New York, 1968.
40. I. Ts. Gohberg and M. G. Krein, *Introduction to the theory of linear nonselfadjoint operators in Hilbert space*, Transl. Math. Monographs, vol. 18, Amer. Math. Soc., Providence, RI, 1969.
41. ______, *Theory and applications of Volterra operators in Hilbert space*, Transl. Math. Monographs, vol. 24, Amer. Math. Soc., Providence, RI, 1970.
42. A. A. Goldberg, B. Ya. Levin, and I. V. Ostrovskiĭ, *Entire and meromorphic functions*, Itogi Nauki i Tekhniki: Sovremennye Problemy Mat.: Fundamental′nye Napravleniya, vol. 81, VINITI, Moscow, 1991, pp. 5–185; English transl. in Complex Analysis 1, Encyclopaedia of Math. Sci., Springer-Verlag, Berlin (to appear).
43. A. A. Goldberg and I. V. Ostrovskiĭ, *Distribution of values of meromorphic functions*, "Nauka", Moscow, 1970. (Russian)
44. E. A. Gorin, *Bernstein inequalities from the operator theory point of view*, Vestnik Kharkov. Univ. Prikl. Mat. Mekh. **45** (1980), 77–105; English transl. in Selecta Math. Soviet. **7** (1988).
45. E. A. Gorin and A. L. Koldobskiĭ, *On potentials of measures with values in a Banach space*, Sibirsk. Mat. Zh. **28** (1987), no. 1, 65–80; English transl. in Siberian Math. J. **28** (1987).
46. N. V. Govorov, *Riemann's boundary problem with infinite index*, Birkhäuser, Basel, 1994.
47. A. F. Grishin, *Regularity of growth of subharmonic functions*. I, II, Teor. Funktsiĭ Funktsional. Anal. i Prilozhen. **1968**, no. 6, 3–29; III, Teor. Funktsiĭ Funktsional. Anal. i Prilozhen. **1968**, no. 7, 59–84. (Russian)
48. ______, *Functions of order one subharmonic in a half-plane, and a Tauberian theorem*, Teor. Funktsiĭ Funktsional. Anal. i Prilozhen. **1990**, no. 53, 87–84; English transl. in J. Soviet Math. **58** (1992).
49. A. F. Grishin and A. M. Russakovskiĭ, *Free interpolation by entire functions*, Teor. Funktsiĭ Funktsional. Anal. i Prilozhen. **1985**, no. 44, 32–42; English transl. in J. Soviet Math. **48** (1990).
50. W. K. Hayman, *Question of regularity connected with the Phragmén-Lindelöf principle*, J. Math. Pures Appl. (9) **35** (1956), 115–126.
51. ______, *Meromorphic functions*, Clarendon Press, Oxford, 1964.
52. ______, *Research problems in function theory*, Athlone Press, London, 1967.

53. ———, *Value distribution and exceptional sets*, Sém. Math. Sup., vol. 79, Presses Univ. Montréal, Montreal, 1982, pp. 79–147.
54. W. K. Hayman and P. B. Kennedy, *Subharmonic functions*. I, Academic Press, New York, 1976.
55. W. K. Hayman and B. Kjellberg, *On the minimum of a subharmonic function on a connected set*, Studies in Pure Mathematics (Paul Erdős, ed.), Birkhäuser, Basel, 1983, pp. 291–322.
56. J. R. Higgins, *Five short stories about the cardinal series*, Bull. Amer. Math. Soc. **12** (1985), 45–89.
57. K. Hoffman, *Banach spaces of analytic functions*, Prentice Hall, Englewood Cliffs, NJ, 1962.
58. L. Hörmander, *A uniqueness theorem of Beurling for Fourier transform pairs*, Ark. Mat. **29** (1991), 237–240.
59. S. V. Hruščev [Khrushchev], N. K. Nikol'skiĭ, and B. S. Pavlov, *Unconditional bases of exponentials and of reproducing kernels*, Complex Analysis and Spectral Theory (Leningrad, 1979/1980), Lecture Notes in Math., vol. 864, Springer-Verlag, Berlin, 1981, pp. 214–336.
60. Ya. I. Khurgin and V. P. Yakovlev, *Compactly supported functions in physics and engineering*, "Nauka", Moscow, 1971. (Russian)
61. V. P. Havin [Khavin] and B. Jöricke, *The uncertainty principle in harmonic analysis*, Springer-Verlag, Berlin, 1994.
62. J.-P. Kahane, *Sur quelques problèmes d'unicité et de prolongement, relatifs aux fonctions approchables par des sommes d'exponentielles*, Ann. Inst. Fourier (Grenoble) **5** (1953–1954), 39–130; *Sur les fonctions moyenne-périodiques bornées*, Ann. Inst. Fourier (Grenoble) **7** (1957), 293–314.
63. ———, *Lectures on mean periodic functions*, Tata Institute of Fundamental Research, Bombay, 1959.
64. Y. Katznelson and S. Mandelbrojt, *Quelques classes de fonctions entières. Le problème de Gelfand et Shilov*, C. R. Acad. Sci. Paris Sér. I Math. **257** (1963), 345–348.
65. V. E. Katsnelson, *Exponential bases in L^2*, Funktsional. Anal. i Prilozhen. **5** (1971), no. 1, 37–47; English transl. in Functional. Anal. Appl. **5** (1971), no. 1, 31–38.
66. B. N. Khabibullin, *Smallness of growth on the imaginary axis of entire functions of exponential type with given zeros*, Mat. Zametki **43** (1988), no. 5, 644–650; English transl. in Math. Notes **43** (1988).
67. ———, *Sets of uniqueness in spaces of entire functions of one variable*, Izv. Akad. Nauk. SSSR **55** (1991), no. 4, 1101–1128; English transl. in Math. USSR-Izv. **39** (1992).
68. ———, *Nonconstructive proofs of the Beurling-Malliavin theorem on the radius of completeness, and nonuniqueness theorems for entire functions*, Izv. Ross. Akad. Nauk Ser. Mat. **58** (1994), 125–146; English transl. in Russ. Acad. Sci. Izv. Math. **45** (1995).
69. A. I. Kheifits, *A characterization of zeros of certain special classes of entire functions of finite degree*, Teor. Funktsiĭ Funktsional. Anal. i Prilozhen. **9** (1969), 3–13. (Russian)
70. B. Kjellberg, *On certain integral and harmonic functions: A study in minimum modulus*, Thesis, Univ. Uppsala, Uppsala, 1948.
71. P. Koosis, *Introduction to H^p spaces*, Cambridge Univ. Press, Cambridge, 1980.
72. ———, *The logarithmic integral*. I, Cambridge Univ. Press, Cambridge, 1988; II, Cambridge Univ. Press, Cambridge, 1992.
73. J. Korevaar, *Zero distribution of entire functions and spanning radius for a set of complex exponentials*, Aspects of Contemporary Complex Analysis (D. A. Brannan and J. G. Clunie, eds.), Academic Press, New York, 1980, pp. 293–312.
74. I. F. Krasichkov-Ternovskiĭ, *An estimate for the subharmonic difference of subharmonic functions*. I, Mat. Sb. **102** (1977), no. 2, 216–247; II, Mat. Sb. **103** (1977), no. 1, 69–111; English transl. in Math. USSR-Sb. **31** (1977), 191–218; **32** (1977), 59–97.
75. ———, *An interpretation of the Beurling-Malliavin theorem on the radius of completeness*, Mat. Sb. **180** (1989), no. 3, 397–423; English transl. in Math. USSR-Sb. **66** (1990), 405–429.
76. M. G. Krein, *Fundamental aspects of the representation theory of Hermitian operators with deficiency indices (m,n)*, Ukrain. Mat. Zh. **1** (1949), no. 2, 3–65; English transl. in Amer. Math. Soc. Transl. Ser. 2 **97** (1971).
77. ———, *On the indefinite case of the Sturm-Liouville boundary problem in the interval $[0,\infty)$*, Izv. Akad. Nauk SSSR **16** (1952), no. 5, 293–324. (Russian)
78. N. S. Landkof, *Foundations of the modern potential theory*, Springer-Verlag, Berlin, 1972.
79. A. F. Leont'ev, *Exponential series*, "Nauka", Moscow, 1976. (Russian)

80. B. Ya. Levin, *On a special class of entire functions and on related extremal properties of entire functions of finite degree*, Izv. Akad. Nauk SSSR **14** (1950), no. 1, 45–84. (Russian)
81. ______, *Generalization of a theorem of Cartwright concerning an entire function of finite degree bounded on a sequence of points*, Izv. Akad. Nauk SSSR **21** (1957), 549–558. (Russian)
82. ______, *Distribution of zeros of entire functions*, Transl. Math. Monographs, vol. 5, Amer. Math. Soc., Providence, RI, 1980.
83. B. Ya. Levin and Din Than Hoa, *Interference operators on entire functions of exponential type*, Funktsional. Anal. i Prilozhen. **3** (1969), no. 1, 48–61; English transl. in Functional Anal. Appl. **3** (1969), 39–50.
84. N. Levinson, *Gap and density theorems*, Amer. Math. Soc. Colloq. Publ., vol. 26, Amer. Math. Soc., New York, 1940.
85. W. A. J. Luxemburg, *On an inequality of Levinson of the Phragmén-Lindelöf type*, Indag. Math. **46** (1984), 421–427.
86. Yu. I. Lyubarskiĭ, *Properties of systems of linear combinations of powers*, Leningrad Math. J. **1** (1990), no. 6, 1297–1370.
87. Yu. I. Lyubich and V. A. Tkachenko, *A new proof of the fundamental theorem on functions periodic in the mean*, Teor. Funktsiĭ Funktsional. Anal. i Prilozhen. **1967**, no. 4, 162–170. (Russian)
88. ______, *The abstract quasianalyticity problem*, Teor. Funktsiĭ Funktsional. Anal. i Prilozhen. **1972**, no. 16, 18–29. (Russian)
89. P. Malliavin, *Sur la croissance radiale d'une fonction méromorphe*, Illinois J. Math. **1** (1957), 259–296.
90. P. Malliavin and L. A. Rubel, *On small entire functions of exponential type with given zeros*, Bull. Soc. Math. France **89** (1961), 175–206.
91. S. Mandelbrojt, *Séries adhérentes, régularisation des suites, applications*, Gauthier-Villars, Paris, 1952.
92. ______, *Sur une problème de Gelfand et Shilov*, Ann. Sci. École Norm. Sup. (3) **77** (1960), no. 2, 145–166.
93. ______, *Theorems of closure and theorems of composition*, Inostr. Liter., Moscow, 1962. (Russian)
94. A. S. Markus, *Introduction to the spectral theory of polynomial operator pencils*, Transl. Math. Monographs, vol. 71, Amer. Math. Soc., Providence, RI, 1988.
95. V. I. Matsaev and E. Z. Mogulskiĭ, *A division theorem for analytic functions with a given majorant, and some of its applications*, Zap. Nauchn. Sem. Leningrad. Otdel. Mat. Inst. Steklov. (LOMI) **56** (1976), 73–89; English transl. in J. Soviet Math. **14** (1980).
96. V. I. Matsaev, *On the growth of entire functions that admit a certain estimate from below*, Dokl. Akad. Nauk SSSR **132** (1960), 283–286; English transl. in Soviet Math. Dokl. **1** (1960).
97. ______, *Volterra operators obtained from selfadjoint operators by perturbation*, Dokl. Akad. Nauk SSSR **139** (1961), 810–813; English transl. in Soviet Math. Dokl. **2** (1961).
98. S. N. Mergelyan, *Uniform approximations to functions of a complex variable*, Uspekhi Mat. Nauk **7** (1952), no. 2, 31–122; English transl. in Amer. Math. Soc. Transl. Ser. 1 **3** (1962).
99. A. M. Minkin, *Reflection of exponents and unconditional bases of exponentials*, St. Petersburg Math. J. **3** (1992), no. 5, 1043–1068.
100. G. W. Morgan, *A note on Fourier transform*, J. London Math. Soc. **9** (1935), 187–192.
101. F. L. Nazarov, *Local estimates for exponential polynomials and their applications to inequalities of uncertainty principle type*, St. Petersburg Math. J. **5** (1994), 663–717.
102. R. Nevanlinna, *Analytic functions*, Springer-Verlag, Berlin, 1970.
103. N. K. Nikol'skiĭ, *Selected problems of the weighted approximation and of spectral analysis*, Trudy Mat. Inst. Steklov. **120** (1974); English transl. in Proc. Steklov Inst. Math. **120** (1976).
104. ______, *Treatise on the shift operator. Spectral function theory*, Springer-Verlag, Berlin, 1986.
105. I. V. Ostrovskiĭ, *Generalization of the Titchmarsh convolution theorem and the complex-valued measures uniquely determined by their restrictions to a half-line*, Lecture Notes in Mathematics, vol. 1155, Springer-Verlag, Berlin, 1985, pp. 256–283.
106. ______, *On a class of entire functions*, Soviet Math. Dokl. **17** (1976), 977–981.
107. I. V. Ostrovskiĭ and A. M. Ulanovskiĭ, *Classes of complex-valued Borel measures uniquely determined by their restrictions*, Zap. Nauchn. Sem. Leningrad. Otdel. Mat. Inst. Steklov. (LOMI) **170** (1989), 233–253; English transl. in J. Soviet Math. **63** (1993).

108. V. P. Palamodov, *Generalized functions and harmonic analysis*, Itogi Nauki i Tekhniki: Sovremennye Problemy Mat.: Fundamental'nye Napravleniya, vol. 72, VINITI, Moscow, 1991, pp. 5–134; English transl. in Encyclopaedia of Math. Sci., vol. 72, Springer-Verlag, Berlin, 1995.

109. R. E. A. C. Paley and N. Wiener, *Fourier transforms in the complex domain*, Amer. Math. Soc., New York, 1934.

110. A. Plancherel and G. Pólya, *Fonctions entières et intégrales de Fourier multiples* II, Comment. Math. Helv. **10** (1936), 110–163.

111. G. Pólya and G. Szegő, *Problems and theorems in analysis.* I, Springer-Verlag, Berlin, 1972.

112. I. I. Privalov, *Randeigenschaften analytischer Funktionen*, VEB Deutscher Verlag Wiss., Berlin, 1956.

113. A. Yu. Rashkovskiĭ, *Majorants of harmonic measures and uniform boundedness of a family of subharmonic functions*, Analytic Methods in the Probability Theory and Operator Theory, "Naukova Dumka", Kiev, 1990, pp. 115–126. (Russian)

114. ———, *On the radial projection of a harmonic measure*, Operator Theory, Subharmonic Functions, "Naukova Dumka", Kiev, 1991, pp. 95–102. (Russian)

115. R. M. Redheffer, *Completeness of sets of complex exponentials*, Adv. in Math. **24** (1977), 1–62.

116. L. I. Ronkin, *Introduction in the theory of entire functions of several variables*, Transl. Math. Monographs, vol. 44, Amer. Math. Soc., Providence, RI, 1974.

117. ———, *Functions of completely regular growth*, Kluwer, Dordrecht, 1992.

118. W. Rudin, *Functional analysis*, McGraw-Hill, New York, 1973.

119. L. Schwartz, *Théorie générale de fonctions moyenne-périodiques*, Ann. of Math. (2) **48** (1947), 857–929.

120. K. Seip, *On the connection between exponential bases and certain related sequences in* $L^2(-\pi,\pi)$, J. Funct. Anal. **130** (1995), 131–160.

121. A. A. Shkalikov, *A system of functions*, Mat. Zametki **18** (1975), no. 6, 855–860; English transl. in Math. Notes **18** (1975), 1097–1100.

122. A. Ph. Timan, *Theory of approximation of functions of a real variable*, MacMillan, New York, 1963.

123. M. Tsuji, *Potential theory in modern function theory*, reprint, Chelsea, New York, 1975.

124. R. S. Yulmukhametov, *Approximation of subharmonic functions*, Analysis Math. **11** (1985), no. 3, 257–283.

Author Index

Subject Index